Marine Biotechnology

A Gateway to Blue Economy

Editors

Anjana K. Vala
Department of Life Sciences
Maharaja Krishnakumarsinhji Bhavnagar University
Bhavnagar, Gujarat
India

Dushyant R. Dudhagara
Department of Life Sciences
Bhakta Kavi Narsinh Mehta University
Junagadh, Gujarat
India

CRC Press
Taylor & Francis Group
Boca Raton London New York

CRC Press is an imprint of the
Taylor & Francis Group, an **informa** business
A SCIENCE PUBLISHERS BOOK

Cover credit:

- Top image taken from Chapter 9: Created by authors using the website, BioRender.com. Reproduced by kind courtesy of the authors.
- Bottom Images belong to the authors of Chapter 6 and reproduced by their permission.

First edition published 2026
by CRC Press
2385 NW Executive Center Drive, Suite 320, Boca Raton FL 33431

and by CRC Press
4 Park Square, Milton Park, Abingdon, Oxon, OX14 4RN

CRC Press is an imprint of Taylor & Francis Group, LLC

Library of Congress Cataloging-in-Publication Data (applied for)

ISBN: 978-1-032-67363-9 (hbk)
ISBN: 978-1-032-69809-0 (pbk)
ISBN: 978-1-032-69810-6 (ebk)

DOI: 10.1201/9781032698106

Typeset in Times New Roman
by Prime Publishing Services

Preface

Covering 70% of the Earth's surface, the oceans carry out multitasking. The list of vital roles played by oceans includes a lot more besides supporting biodiversity, fisheries and livelihood.

The marine environment is home to several members from all domains of life. Marine biotechnology (Blue biotechnology) harnesses these bioresources to unlock immense opportunities to benefit mankind and their sustainable exploitation leads to economic progress.

Among the UN Sustainable Development Goals (SDG), SDG 14 (Life Below Water) is dedicated to oceans and the blue economy has a big role to play in achieving this Goal. The term 'Blue Economy' has many definitions and various components ranging from conventional ocean industries to emerging activities. Besides, reduction in greenhouse gas emission, reduction in pressure on land-based food systems and positive climate effects are also important features of a sustainable blue economy. Inclusive efforts would help overcoming challenges that limit the development of the blue economy.

Diverse marine biotechnological aspects including bioactive compound production, marine-derived pharmaceuticals, probiotics, bioremediation potential, sustainable agriculture and bioenergy are covered in this book. We are happy to anticipate the usefulness of this compilation to scientific communities from academia, research and industry and of course to students as well.

Thanks are due to the contributors, the eminent scientists, for sharing their knowledge in the form of a book chapter. Their significant contribution made us accomplish the task.

We are thankful to Dr. Raju Primlani, Senior Editor, Science Publishers and his team for continued support.

We are appreciative to our respective university authorities (Maharaja Krishnakumarsinhji Bhavnagar University and Bhakta Kavi Narsinh Mehta University) for encouragement and support. We thank our faculty and research colleagues, research scholars, students, family and friends.

Last but not the least, we are grateful to Prof. Bharti Dave, Former Head, Department of Life Sciences, MK Bhavnagar University and our teacher. Anjana K.

Vala especially expresses gratitude to her teacher Prof. H. C. Dube, Former Head, Department of Life Sciences, MK Bhavnagar University (then Bhavnagar University) who has always been inspirational for taking up scientific accomplishments.

Thank you all.

Anjana K. Vala
Dushyant R. Dudhagara

Contents

Chapter 1

Harnessing Seaweed Bioactive Compounds for Health

Sheyma Inoubli,† *Kai L. Baltrusch,*† *María Dolores Torres, Herminia Domínguez* and *Noelia Flórez-Fernández**

Introduction

Today, seaweed phyla Rhodophyta, Ochrophyta, and Chlorophyta red, brown, and green seaweeds, respectively, attract attention due to their potential for manufacturing products that have beneficial health outcomes. Numerous seaweed goods possess unique characteristics that have pharmacological and biomedical functionalities. The substantial components which promote health impacts among all seaweed classes are sulfated polysaccharides, polyunsaturated fatty acids, vitamins and pigments. Several research works have found interesting activities in these bioactive compounds, for example, antioxidant, antitumoral, anti-inflammatory, antiviral and anticoagulant among others (Peñalver et al. 2020). Figure 1 summarizes the bioactive compounds, health benefits and human applications associated to algae.

According to the Food and Agriculture Organization (FAO), there has been a significant increase in global aquatic algae production, primarily driven by seaweeds, which accounted for 99.84% of this increase, with 97% attributed to cultivation (FAO 2021). This is indicative for the growing interest and potential in seaweed-based industries.

Economic viability remains a critical consideration in seaweed farming, particularly concerning the extraction of valuable compounds. Orozco-González et al. (2022) highlight that extracting single molecules or compounds from seaweed

CINBIO, Universidade de Vigo, Department of Chemical Engineering, Facultade de Ciencias, Campus Ourense, As Lagoas, Ourense, 32004, Spain.
† both authors Sheyma Inoubli and Kai L. Baltrusch have equal contribution.
* Corresponding author: noelia.florez@uvigo.gal

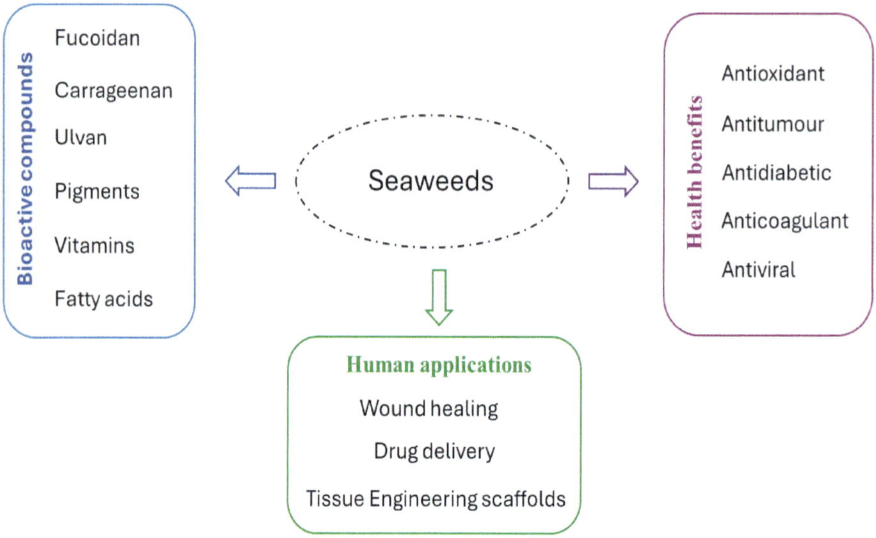

Figure 1. The potential bioactive compounds, health benefits and human applications of seaweeds.

can be economically challenging unless they possess exceptionally high market value. However, the development of more efficient extraction and processing methods, along with multisite cascade processes that yield a variety of valuable products from seaweed, may help reduce production costs of seaweed biostimulants and enable co-extraction with other bioactive ingredients, chemicals, and biofuels, and enhance economic viability (El-Shenody et al. 2023, Flórez-Fernández et al. 2021, Torres et al. 2019).

Moreover, cultivated seaweeds offer several advantages beyond economic considerations. They have minimal environmental impact, rapid growth rates, and contribute to carbon sequestration benefits (Kraan 2013). Additionally, valorizing invasive species and seaweed biomass blooms not only provides economic opportunities for affected communities but also contributes to the production of valuable compounds, further emphasizing the multifaceted benefits of seaweed use (Torres et al. 2019).

The current interest of society in having a healthy life using natural sources, promotes interest on marine sources like algae. In this context, the main objective of this chapter is to present an overview of the main bioactive compounds in seaweeds and their potential biomedical applications, such as wound healing, drug delivery, tissue engineering scaffolds, among others (Lomartire et al. 2023).

Bioactive compounds from seaweeds

Fucoidans, carrageenans, and ulvans

Sulfated polysaccharides (SPs) such as fucoidan, carrageenan, and ulvan produced from brown, red, and green seaweeds respectively, are the most appealing moieties for scientific communities due to their valuable bioactivities and lack of toxicity (Kwon et al. 2020, Wang et al. 2020).

Fucoidan, commonly known as sulfated fucans, is one of the chief components of brown seaweed cell walls. Predominantly obtained from *Sargassum* sp., *Ecklonia cava, Undaria pinnatifida, Cladosiphon okamuranus*, and *Ascophyllum nodosum* (Bilal and Iqbal 2019). Fucoidans are mostly obtained with water, acidic or alkaline aqueous extraction, and ethanol precipitation (Zayed and Ulber 2019). Moreover, column chromatography techniques like size exclusion and ion exchange can be used for extraction (Zhang et al. 2020). Fucoidan is mostly composed of L-fucose with sulfate ester groups and small amounts of other monosaccharides (such as mannose, arabinose, xylose, glucose, and galactose). Fucoidan often has two different kinds of chains. Fucose with α (1→3) links makes up the type I chain, whereas fucose with alternating α (1→3) and α (1→4) links makes up the type II chain (Wang et al. 2023). Molecular weights range from 100 to 1600 kDa. The potency of its pharmacological properties is due in large part to its molecular weight, and sulfate concentration (Zhang et al. 2019), and the lower molecular weight exhibits better efficiency (Zuo et al. 2015). Fucoidan has a high viscosity, and both its concentration and degree of sulfation can affect its behavior. Additionally, it gels in high concentrations or in the presence of cations with valence of 2+ (Rioux et al. 2007).

Carrageenan is a sulfated galactan hydrocolloid mainly obtained from Rhodophyta taxa, principally from species belonging to the Gigartinales. The overall structure of galactans is mainly characterized by the alternate of D-galactose and 3, 6-anhydrogalactose backboned via α-1,3 and β-1,4 glycosidic linkages (McKim et al. 2016). Carrageenans are categorized into different types, the most relevant commercially being λ (lambda), κ (kappa), and ι (iota) carrageenan. ι-carrageenan is often extracted from *Eucheuma denticulatum* ("spinosum"), whereas κ-carrageenan is derived from *Kappaphycus alvarezii* ("cottonii"). (λ)-carrageenan is derived from species in the *Gigartina* and *Chondrus* sp. ("Irish moss") (Cotas et al. 2020). Their molecular weights are within the range of 30 and 5000 kDa with an average range between 200 and 800 kDa (McKim et al. 2016). Native carrageenan moieties are usually complex hybrid structures or blends of various carrageenan types (Tuvikene 2021). The hot extraction approach provides higher amounts of carrageenan than the cold extraction approach. Yields vary from 28.7% to 44.0% based on the species. Conversely, the sulfate content is lower at high temperatures (Premarathna et al. 2023). The industrial functions of carrageenan are mainly based on gelling, stabilizing, and viscosity-building agents. Hence, this polymer is extensively used as a food, in personal care and in biomedical products (Younes et al. 2018).

Ulvan is a cell wall sulphated polysaccharide derived from green seaweeds of the Chlorophyceae taxa, it accounts for 9–36% DW of *Ulva* sp. (Kidgell et al. 2019). Its main chain frequently consists of α- and β-(1,4)-linked monosaccharides consisting of rhamnose, glucuronic acid, xylose, and iduronic acid, interlaced with repeating disaccharide units (Lahaye and Kaeffer 1997, Paradosi et al. 1999). Other monosaccharides (such as glucose, galactose, arabinose, and mannose) are referenced also in ulvan structural data. The principal repeating disaccharide units that comprise ulvan are aldobiouronic acids known as ulvanobiuronic acid

(A_{3s}, B_{3s}) and ulvanobioses. A_{3s} contain glucuronic acid attached to rhamnose 3-sulfate, whereas the similar B_{3s} also contain rhamnose 3-sulfate, but have iduronic acid instead of glucuronic acid. Ulvanobioses are made up of rhamnose 3-sulfate attached to xylose (Lahaye et al. 1998). The biological capacities of ulvans are intimately tied to their chemical structure in terms of sugar components and arrangement, glycosidic linkage, degree of branching, molecular weights, and the presence of functional groups (sulfate esters) (Kidgell et al. 2019). Ulvan is noteworthy since it contains the rare sulfated sugar rhamnose-3-sulfate (Reisky et al. 2019). Its extraction process was widely investigated through the use of physicochemical techniques such as ultrasound and microwave treatment (Gómez et al. 2020), chemical techniques such as sulfuric acid and hydrochloric acid treatment (Wahlstrom et al. 2020), or biological techniques including the use of enzymes (Kidgell et al. 2019). Ulvan was investigated for hydrogel preparation for biomedical applications, it compromises several bioactive properties with growing functionalities (Chai et al. 2017).

Pigments

Pigments contribute to the antioxidant defensive mechanism of seaweeds, acting as photoprotective agents and reducing the reactive oxygen species (ROS) impact generated by UV-radiation and other abiotic stimuli (Álvarez-Gómez et al. 2019). Chlorophylls (Chl), carotenoids, and phycobiliproteins (PBPs) are the three main categories of pigments that grant seaweeds their distinct colors (Kalasariya et al. 2021). These secondary metabolites can be classified into two categories based on their chemical structure: closed tetrapyrroles (Chl a, b, c) and open tetrapyrroles (PBPs and carotenoids) (Stengel et al. 2011). Chlorophyta phylum comprises Chl a, b, and the carotenoids β-carotenoids, lutein, neoxanthin, violaxanthin, and zeaxanthin. Rhodophyta phylum consists of Chl a, carotenoids: α and β-carotenoids, antheraxanthin, cryptoxanthin, lutein, violaxanthin, and zeaxanthin, the phycobiliproteins phycoerythrins, phycocyanin, and allophycocyanin. Ochrophyta phylum entails Chl a, c, the carotenoids β-carotenoids, fucoxanthin, and violaxanthin (Janarthanan et al. 2018, Aryee et al. 2018, Jesumani et al. 2019). Pigments have gained increasing economic interest as pharmacological and medical agents (Manivasagan et al. 2018, Pérez et al. 2016).

Vitamins

Some vital vitamins are not produced by the human body; therefore they must be obtained from other sources. They are recommended for skin protection, premature aging, and control of sebaceous gland secretions (Jacob et al. 2012). Seaweeds have been recognized to contain vitamins with multiple health benefits such as vitamins B, C, E, and A (β-carotene), which aid in reducing a number of health conditions such as high blood pressure, cardiovascular disease, and cancer risk (Ganesan et al. 2019, Škrovánková 2011). Retinol, a type of vitamin A, provides antioxidant and anti-wrinkle properties (Searle et al. 2020). Chlorophyta and Rhodophyta phyla include higher amounts of vitamin B (B1, B2, B3 or niacin, B6, B9, B12, or folic acid), which improves epidermal characteristics (Campiche et al. 2020). *C. rubrum* and *P. leucosticta* (Rhodophyta) possess high concentrations of vitamin C with versatile biological and health functionalities (Manela-Azulay and Bagatin 2009), red seaweeds are also an important source of vitamin A (Ganesan et al. 2019). *Ulva lactuca*, *Caulerpa lentillifera*, *Euchema cottonii*, *Sargassum polycstum*, and *Gracilaria* sp. contain considerable amounts of vitamin C (Noriega-Fernández et al. 2021). Ochrophyta seaweeds exhibit higher levels (99%) of vitamin E (α-tocopherols) than Chlorophyta and Rhodophyta (Holdt and Kraan 2011).

Fatty acids

Among primary metabolites, lipids are found in low amounts in seaweeds (from 0.60%–4.15% DW). Further, green seaweeds exhibit the lowest yield among their counterparts (Shannon and Abu-Ghannam 2019). Almost exclusively, algae are the only species that can produce long-chain polyunsaturated fatty acids (PUFAs) (14–24 carbons). They can do this due to the existence of certain enzymes involved in desaturation mechanisms, which have been demonstrated to have substantial health advantages (Kendel et al. 2015). In fact, seaweeds possess high amounts of PUFAs comprising ω-3 and ω-6 PUFAs, particularly linoleic acid (18:2 ω-6), α-linolenic acid (18:3 ω-3), octadecatetraenoic (18:4 ω-3), arachidonic (20:4 ω-6), eicosapentaenoic (20:5 ω-3) acids and stearidonic acid (18:4 ω-3), with a significant role in the proper functioning of human health by lowering cardiovascular risk and boosting immunological response neurodegenerative diseases and brain function (Gammone et al. 2019, Ibáñez et al. 2012, Gonçalves et al. 2017). Several techniques were used for the extraction of PUFAs, including enzyme-assisted extraction, supercritical fluid extraction, and ultrasound-assisted extraction (Michalak and Chojnacka 2014). Furthermore, it has been proven that sterols mainly cholesterols and phytosterols, which are preferred as a subclass of steroids with a sterane core, have a wide range of biofunctionalities as pharmaceuticals and therapeutic agents (Ibáñez et al. 2012, Alves et al. 2018, Santos et al. 2015).

All these bioactive compounds have a potential interest in application terms. However, their limitations are associated to several environmental factors directly related to the growth of seaweeds: such as depth, light intensity, water temperature, location, nutrients, pH, salinity (Endo et al. 2017). Some of these factors were evaluated for *Ulva pertusa*, the authors found that the factors studied (salinity, irradiance, temperature and pH) have an effect on the growth of the seaweed (Jin et al. 2016).

Figure 2 shows structural models of some of the bioactive compounds found in seaweeds.

a)fucoidan from Fucus vesiculosus

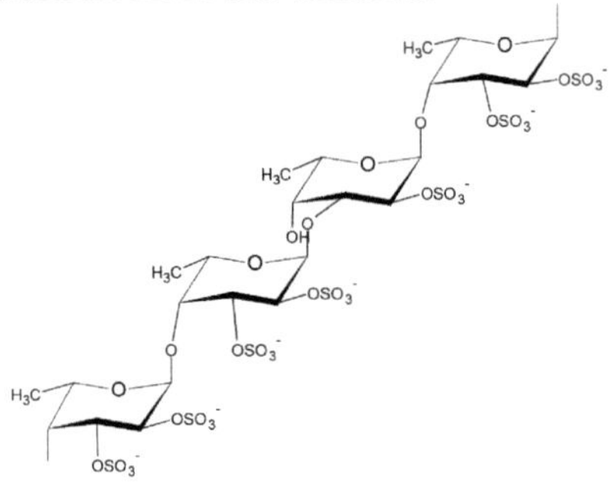

b) fucoidan from *Ecklonia* sp.

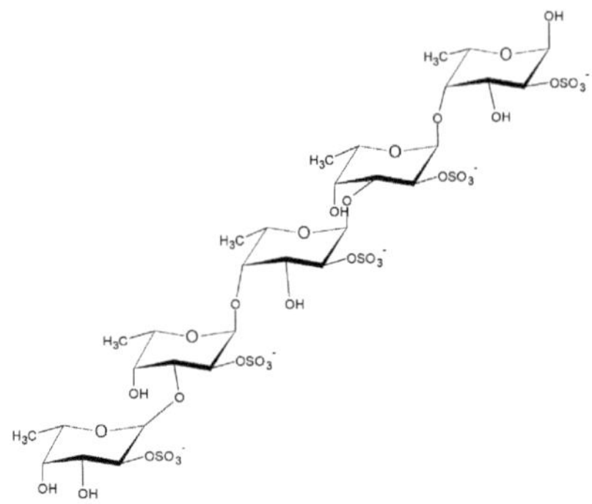

c) kappa – carrageenan

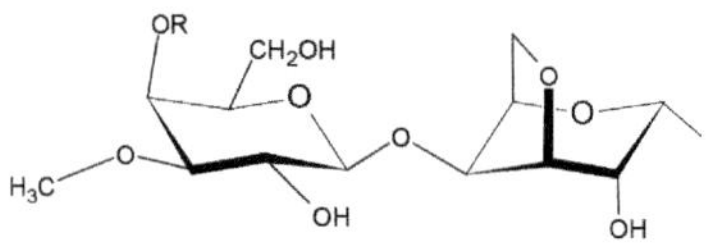

R: SO_3^-

d) iota – carrageenan

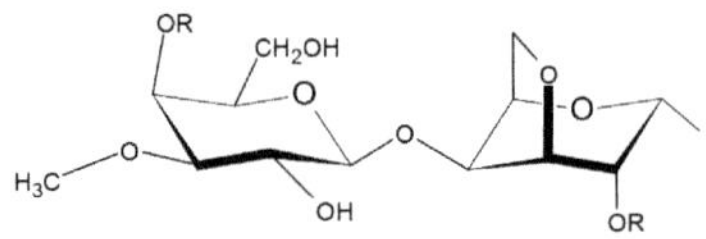

R: SO_3^-

e) ulvan type B

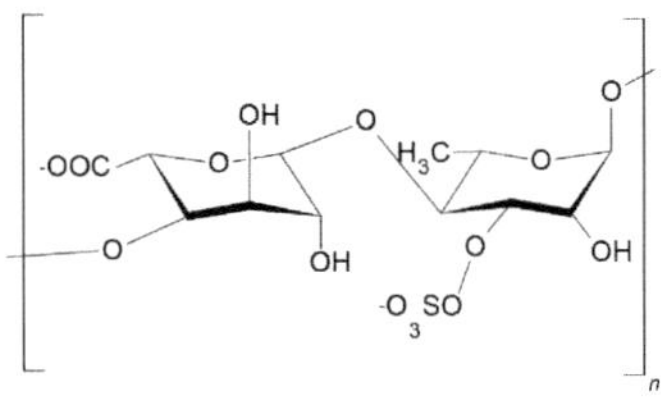

f) ulvanobioses type U3S

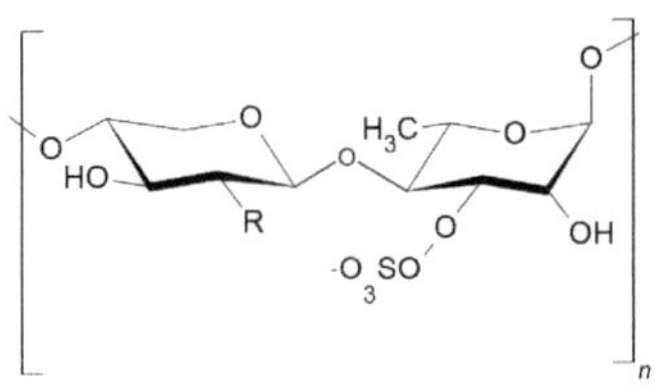

g) fucoxanthin

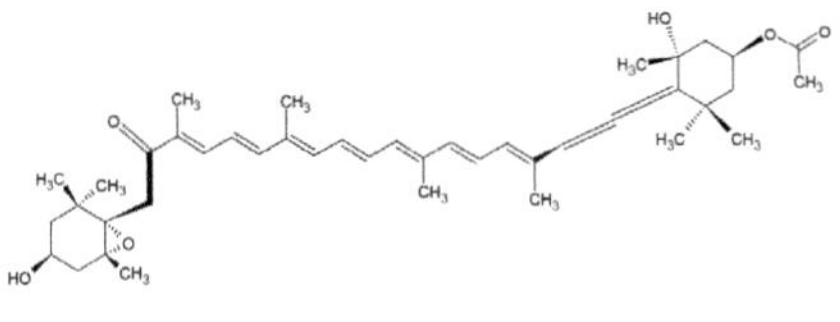

h) astaxanthin

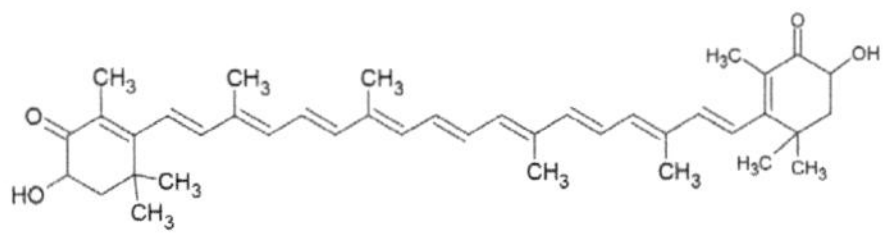

Figure 2 contd. ...

...Figure 2 contd.

i) Dieckol

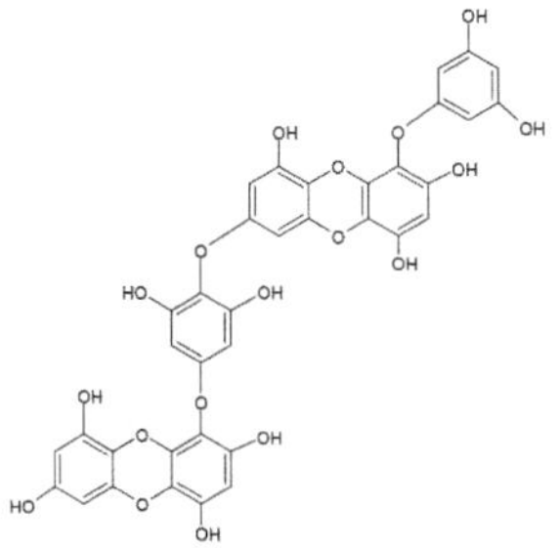

j) phloroglucinol

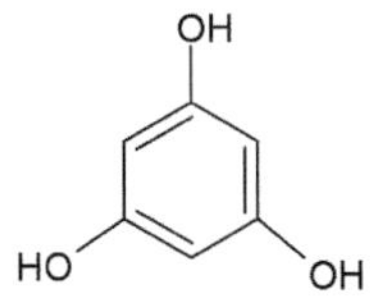

k) Vitamin E – alfa tocopherol

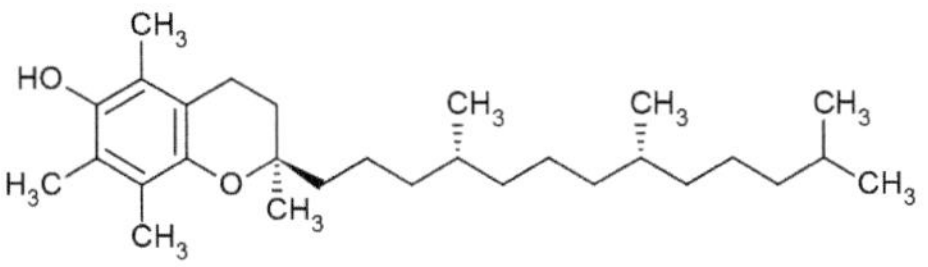

l) Vitamin C – ascorbic acid

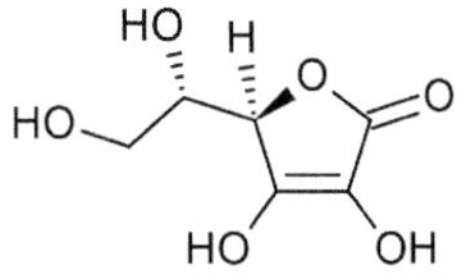

m) fatty acids

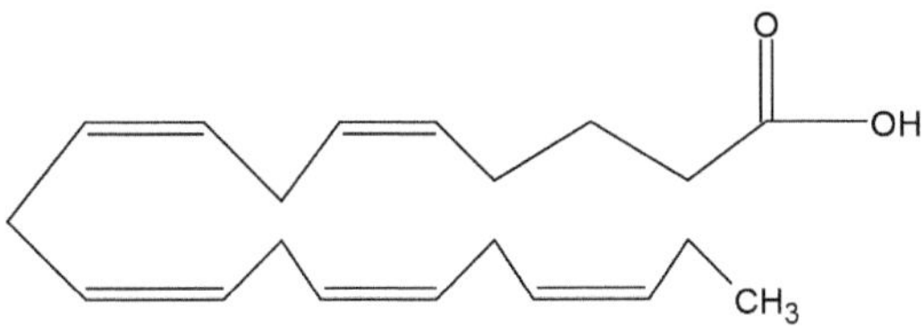

Figure 2. Structural model of fucoidan, carrageenan, ulvan, pigments, vitamins and fatty acids (Ale et al. 2011, Álvarez-Viñas et al. 2021, Pérez et al. 2016, Coulombier et al. 2021, Yamagata 2021, Flórez et al. 2017, Flórez-Fernández et al. 2023).

Health benefits

Antioxidant

Antioxidants shield the metabolism from the harmful impact of oxidative damage caused by ROS, which is the root cause of many health complications (cancer, chronic inflammation, and neurological and cardiovascular diseases). Catalase (CAT), glutathione-peroxidase (GPx), and superoxide dismutase (SOD) are the main enzymes that prevent ROS-induced damage in mammalian cells (Yuan et al. 2006). Antioxidant properties of seaweeds are attributed to a variety of substances, including SPs, carotenoids, phlorotannins and tocopherols (Arive et al. 2017). Low molecular fucoidan weight (LMF) isolated from *Sargassum autumnale* using enzyme-assisted-hydrolysis exhibits a vigorous antioxidant property evidenced by its counteraction against H_2O_2-induced oxidative stress in a zebrafish model, lowering ROS generation and inhibiting lipid peroxidation in it (Geun Lee et al. 2022). In a

UVB-irradiated skin-damaged mouse paradigm, LMF treatment inhibited the advancement of oxidative stress and enhanced innate antioxidant activity, as evidenced by an increase in (GPx) and a decrease in myeloperoxidase, malondialdehyde (MDA) produced by the peroxidation of PUFAs, and superoxide ($O_2^{\cdot -}$) (Kim et al. 2018). κ-carrageenan oligosaccharides and their derivatives possess both *in vitro* and *in vivo* antioxidant effects. Rat cells exposed to these biomolecules demonstrated a strong defense against oxidative damage caused by H_2O_2 and UVA (Yuan et al. 2006). Ulvan isolated from several *Ulva* sp. reduces the generation of oxidation products and increases antioxidant enzyme activity *via* activating their transcription (Abd-Ellatef et al. 2017, Berri et al. 2017, Forman et al. 2014). Ulvan with sulfate amounts (19.5% w/w) showed lower hydroxyl radical scavenging activity (45%, at 2.0 mg/mL) than those with high sulfate amounts (32.8% w/w) which demonstrated approximately 90% scavenging capability (Qi et al. 2005). Ulvan from *U. pertusa* induces the decline of (MDA) and the rise of CAT and SOD of treated hyperlipidemic Kunming rats (Li et al. 2018). Several studies revealed that the antioxidant effects of SPs are linked with their LMW, chemical composition, sulfate amount, and degree of polymerization (Hou et al. 2012, Jin et al. 2018, Sun et al. 2015). Mainly brown seaweeds possessed the highest content of phlorotannins (up to 15% DW) as compared to their counterparts. Notably, *Ascophyllum nodosum* and *Fucus* sp. represent prospective sources of antioxidant components (Jacobsen et al. 2019). *Fucus*, *Porphyra*, *Ascophyllum*, and *Ulva* possess high contents of tocopherols with relevant antioxidant properties (Holdt and Kraan 2011). Additionally, all carotenoids from the three-seaweed phylum demonstrated an antioxidant defense mechanism and mitigated chronic disease risks (Kurniawan et al. 2023, Vieira et al. 2018).

Antitumor potentialities

Uncontrolled cell proliferation, invasion, and metastasis culminate in the formation of a tumor mass (Hussain et al. 2016, Senthilkumar et al. 2013). Seaweed components have shown to be an effective alternative for cancer treatment and prevention. Fucoidans from *Sargassum polycystum*, *Sargassum duplicatum*, *Sargassum feldmannii* (Usoltseva et al. 2019), *Sargassum cinereum* (Narayani et al. 2019), *Sargassum hemiphyllum* (Yan et al. 2015), *Undaria pinnatifida* (Synytsya et al. 2010), demonstrated anticancer effect by hindering cell proliferation and promoting apoptosis towards a wide variety of cancer stem cells such as human colon cancer cells (DLD-1, HCT-116, and HT-29) (Usoltseva et al. 2019), human hepatocellular carcinoma (HCC) (Yan et al. 2015), human prostate (PC-3), human cervical (HeLa), human carcinomic alveolar basal epithelial (A549), and human hepatocellular liver carcinoma (HepG2) (Synytsya et al. 2010). K-carrageenan from *Hypnea musciformis* (Souza et al. 2018) and especially T hybrid carrageenan (λ/ξ) from *Gigartina pistillata* (Cotas et al. 2020) presented anti-proliferative effect on human breast cancer (MCF-7), human neuroblastoma (SH-SY5Y) (Souza et al. 2018) and colorectal cancer (CSC) stem cells (Cotas et al. 2020). The effectiveness of ulvan as an anticancer is mediated by the induction of apoptosis, the suppression of cell proliferation, and the stimulation of the innate immune system (Kidgell et al. 2019). The antitumor effect of Ulvan extracted from various *Ulva* sp. was demonstrated on gastric carcinoma (AGS)

(Tabarsa et al. 2012), hepatocellular carcinoma (HepG2) (Wang et al. 2014), cervical cancer (HeLa) (Thanh et al. 2016), gastric cancer (MKN45) (Shao et al. 2013), colon carcinoma (HT-29) and colon carcinoma (HCT-116) (Ahmed et al. 2014) and human cancer stem cells. Ulvan biobased nanoparticles possess relevant applications as drug delivery systems for antitumor drugs (Li et al. 2018). Pigments also seem to be potential anticancer agents. Carotenoids and chlorophylls a derived from *Halimeda opuntia* green seaweed display anticancer potentialities against HT-29 human colorectal adenocarcinoma (Nazarudin et al. 2022). Fucoxanthin, astaxanthin, and violaxanthin impede the apoptosis pathway (PI3K/Akt) (Sathasivam and Ki 2018). Furthermore, several reports have illustrated the effect of ω-3 PUFAs (Rajapakse and Kim 2011), ω-6 PUFAs (González-Fernández et al. 2019), and phytosterols (Jiang et al. 2019), especially β-sitosterol (Arul et al. 2012) and fucosterol (Ramos et al. 2019) as potent anticancer ingredients able to prevent many tumors. Secondary metabolites including polyphenols such as bromophenols (Liu et al. 2011) and phlorotannins (Cotas et al. 2020), terpenes (Gutierrez-Rodriguez et al. 2018) and vitamins C (Yun et al. 2015) and D (Song et al. 2016) also have potent anticancer effects.

Antidiabetic

Diabetes mellitus (DM) is a health issue that is threatening populations worldwide. Seaweeds display a wide spectrum of various bioactive crude extracts and isolated components with pivotal antidiabetic effects that might be used as functional drugs to mitigate the hazardous impact of synthetic drugs. Seaweed bioactive moieties act as enzymes hydrolyzing carbohydrate inhibitors. They promote the uptake of glucose by muscle cells, which enhances insulin resistance (IR) muscle (Bermano et al. 2020) and further control the expression of mRNA and pro- and anti-inflammatory cytokines associated with less weight in rats (Agarwal et al. 2023). Fucoidan, from *Undaria pinnatifida* demonstrated hypoglycemia effect by stimulating glucose absorption and increasing insulin sensitivity in rats (Sim et al. 2019). Water extracts of *Undaria pinnatifida*, *Codium fragile*, and *Gracilaria verrucosa* exert α-glucosidase inhibition activity and control anti-inflammatory cytokines in LPS-stimulated C2C12 myotubes by regulating and stimulating the AMP-activated protein kinase (AMPK) pathway, which hinders glucose uptake and promotes its utilization (Kim et al. 2021). *Ecklonia cava* extract decreases postprandial blood sugar levels in diabetic rats through the activation of AMPK and PI3K-Akt (phosphatidylinositol 3-kinase: PI3K and Protein Kinase B: Akt) pathways (Kang et al. 2010). Fucoidan from brown seaweeds decreases blood glucose levels, impedes β-cells damage, and enhances insulin production via enzymatic inhibition of α-glucosidase and α-amylase, control of AMP cycle, PI3K/AKT, JNK (Jun N-terminal kinase)/AKT and sirtuin-1 mechanisms (Wen et al. 2021). Fucoxanthin was demonstrated to also have the potential to decrease the expression of proinflammatory cytokines (TNF-α, IL-6, and MCP-1), related to IR and DM diseases (Hosokawa et al. 2010). Fucoxanthin from *Undaria pinnatifida* exhibits an α-glucosidase inhibitory effect (Zaharudin et al. 2019). Carotenoids from *Caulerpa racemosa* inhibit α-glucosidase and α-amylase enzymes and promote anti-inflammatory effect (Kurniawan et al. 2023). Several reports have illustrated that polyphenols extracted from brown, red and green seaweeds displayed antidiabetic effects through α-amylase and α-glucosidase inhibitory activities (Hwang et al. 2015, Park et al. 2018).

Anticoagulant activity

Probably, the coagulation mechanism is stimulated through intrinsic and extrinsic pathways implicating the activation and amplification of serine proteases and glycoproteins factors that catalyze the activation of a common pathway concurrently with the production of thrombin and fibrin which contribute to blood clot formation (Rang et al. 2016). Because of their strong hydrophobicity and biocompatibility, SPs are being widely studied as anticoagulant and antithrombotic agents to mitigate the adverse effects of commercial drugs (Song et al. 2018). Fucoidan inhibits intrinsic and common pathway factors and delays coagulation time (Jin et al. 2013). Fucoidan reveals anticoagulant and antithrombotic functionalities comparable to heparin—a commercial anticoagulant medication (Liao et al. 2017, Lin et al. 2022). Carrageenans also exhibit anticoagulant and antithrombotic characteristics, however, λ-carrageenan possesses greater potentialities than κ-carrageenan (Shanmugam et al. 2000). Sulfate clusters of this polymer resemble heparin structure (Liang et al. 2014, Silva et al. 2010). By mimicking the structure of heparin, Song et al. (2018) designed carrageenan biobased gel beads as hemoperfusion adsorbents. Ulvan exerts an anticoagulant effect by acting on intrinsic and common pathways (Adrien et al. 2017, Synytsya et al. 2015). The oversulfated ulvans with low molecular weights displayed the highest anticoagulant activity (Cui et al. 2018), however, its bioactivity remained relatively lower than heparin, ulvan is still a pertinent anticoagulant agent (Long et al. 2016). It has been reported that phenolics exert further anticoagulant effects that might be used for medical applications (Saidani et al. 2021).

Antiviral activity

Among seaweeds, bioactive substances SPs, are the most tested moieties for their antiviral effect. Fucoidans isolated from *Laminaria japonica* exert both *in vitro* and *in vivo* antiviral bioactivities against a variety of viruses, including influenza virus type I, parainfluenza virus, and adenovirus (Sun et al. 2018), by interfering with viral life cycle suppression (gene expression, replication, and invasion) (Lu et al. 2021). Molecular docking revealed that fucoidan from *Fucus evanescens* had an antiviral effect against ortho-hantavirus by lowering viral protein synthesis (Krylova et al. 2021). Fucoidan derivatives showed antiviral potential against COVID-19 by inhibiting heparin's association with COVID-19 spike protein mutants (Koike et al. 2021). This polymer has shown to influence immune system function which may lower COVID-19 infection (Barbosa et al. 2021). Furthermore, it modulates the gut flora and raises the generation of short-chain fatty acids, which reduces COVID-19 infection (Włodarczyk et al. 2022). It also has *in vitro* antiviral properties against herpes viruses (HSV-1 and 2), enteroviruses (ECHO-1), and human immunodeficiency virus (HIV-1) by decreasing viral adsorption and multiplication (Krylova et al. 2020). Carrageenan and its derivatives had potent antiviral activity reviewed by Álvarez-Viñas et al. (2021), especially against a wide range of enveloped viruses (HSV-1 and 2, varicela zoster virus (VZV), dengue virus type 2 (DENV-2), cytomegalovirus (CMV), HIV, Influenza virus (IAV), SARS-CoV2...) as well as some non-enveloped viruses (papillomavirus type 16 (HPV-16), human rhinovirus (HRV), and enterovirus 71 (EV-71)). Carrageenan could affect viral life cycle mechanisms regarding the inhibition of viral adsorption and penetration by binding to enveloped viral proteins (Shao et al. 2015) or host receptors (Grassauer et al. 2008), uncoating (Luo et al. 2015), gene expression and replication (Chiu et al. 2012, Wang et al. 2011). Ulvans from numerous *Ulva* sp. had both *in vitro* and *in vivo* antiviral activity against a wide spectrum of enveloped viruses like HSV (Lopes et al. 2017), avian influenza virus (AIV) (Song et al. 2016), influenza (H1N1) (Jiao et al. 2012), Virulent Newcastle disease (NDV) (Aguilar-Briseño et al. 2015), measles virus (MeV) (Morán-Santibañez et al. 2016), dengue virus (DENV), Japanese encephalitis virus (JEV), West Nile virus (WNV), and yellow fever virus (YFV) (Chiu et al. 2012). However, research into the antiviral effect of ulvan is still in its early stages, more investigations are required to understand its antiviral efficacy mechanisms in relation to its structural features (Kidgell et al. 2019). Hepatitis B virus (HBV) is susceptible to the antiviral effects of lutein (Pang et al. 2010). Most of carotenoids including α,β-carotene, zeaxanthin, astaxanthin, and lutein reduce oxidative stress and inflammation which boosts the immune system and attenuates the acuteness of COVID-19, particularly when coexisting with chronic diseases (Talukdar et al. 2020, Zafar et al. 2021).

In Table 1, a resume of relevant patents associated to the main polysaccharides of seaweeds (fucoidan, carrageenan and ulvan) due to their biological properties is included in this chapter.

Human applications

Seaweeds offer a source of a new generation of biomaterials for human health technologies. This section is dedicated to explore how seaweed-derived products are practically applied in wound healing, drug delivery systems and tissue engineering.

Wound healing

A wound is the disruption of the usual integrity, structure, and functions of the skin as a result of injury or illness. Wound healing consists of mainly three sequential phases: inflammation, proliferation and remodeling (Kondo and Ishida 2010).

New biomaterial based wound dressings have appeared as a convenient technology for wound treatment. Conventional dressings like gauze and bandages may protect but fail to maintain a moist environment, lacking in biocompatibility and risking secondary injury during removal. The ideal dressing should act as a microbial barrier, maintain moisture for healing, facilitate wound clearing, minimize adhesion and possess good elasticity, permeability and biocompatibility (Shen et al.

Table 1. Representative patents of the last five years related to the use of main Polysaccharides (Fucoidan, Ulvan and Carrageenan) from seaweeds.

Number of patent	Title	Reference
KR1020220109741	Composition For Prevention and Treatment of Coronavirus Containing Fucoidan Derived from *Agarum Clathratum* Extract As Active Component	Hwan Kwak et al. 2022
KR1020230007688	Adjuvant Composition Comprising Fucoidan from *Ecklonia cava* As Active Ingredient	Jun et al. 2023
CN111973749	Pharmaceutical Composition for Anti-tumor Immunotherapy	Hongfang et al. 2023
US20220023184	Fucoidan Hair Growth Agent	Quan et al. 2024
CN110420216	Application Of Ulvan in Inhibiting Aggregation of Beta-Amyloid Protein	Fufeng et al. 2019
CN106573013	Viscosupplement Composition Comprising Ulvan for Treating Arthritis	Sousa R.P. et al. 2020.
CN110627918	Ulvan With Anti-Tumor Activity and Production Method of Ulvan with Anti-Tumor Activity	Zhao et al. 2019
IN201841019841	Use Of Ulvan as Activator of Shrimp Immune System Against White Spot Syndrome Virus	Sivagnana Velmurugan and Arul 2023
CN111686125	Application Of Carrageenan to Resisting of Novel Coronavirus	Song et al. 2020
CN115926217	Carrageenan Antibacterial Film Containing Benzyl Isothiocyanate-Beta-Cyclodextrin Inclusion Compound as well as Preparation Method and Application of Carrageenan Antibacterial Film	Zhang et al. 2023
CN116808067	Application of Lambda-Carrageenan in Resisting African Swine Fever Virus Infection	Zhao et al. 2023
NL2032188	Application of L-Carrageenan Tetrasaccharide in Relieving Insulin Resistance and Preparing Related Product	Yingying et al. 2024

2021). In this sense, seaweed polysaccharides (SWP) present interesting properties as wound healing biomaterials, due to high biocompatibility, gelling features, storage capacity and biodegradability. The polysaccharide structure often presents carboxylic, hydroxyl, and sulphate groups, which can form non-covalent bonds with growth factors (GFs), bond bioactive compounds for drug delivery and support bioadhesion to the skin. To date, these SWP have been successfully incorporated into wound dressing technologies, including films, multilayer films, foams, non-woven nanofiber patches, injectable hydrogels and smart hydrogels. Often, these biomaterials are combined with copolymers or bioactive compounds to fine-tune the intended effect (Andryukov et al. 2020, Shen et al. 2021).

Alginate is widely used and a well-studied hydrocolloid for this purpose, while other SWP as ulvan, fucoidan, agar or carrageenan are gaining interest for their development (Szekalska et al. 2016). The versatile broad-spectrum antiviral, antibiotic, and antifungal properties of carrageenans in combination with other functional ingredients makes their use in wound-dressing technologies possible. Their structural resemblance to glycosaminoglycans promotes wound healing. Moreover, carrageenans possess tunable rheological properties, making them suitable for 3D bioprinting of dermal tissue (Liu et al. 2023). Ulvan has shown to be a promising material for advanced wound dressings. Incorporated into hydrogels, it enhances cell proliferation, demonstrates antimicrobial efficacy, and expedites wound closure. *In vivo* studies, notably in a type II diabetes mouse model, reveal accelerated wound healing (Shah et al. 2023). The similarity between ulvans and mammalian glycosaminoglycans suggests potential in wound healing and tissue regeneration, albeit facing challenges like variability and scalability (Chen et al. 2019). The structural similarity of fucoidan to heparin, originate antithrombotic behaviour and favour tissue regeneration. Hydrogel formulations which include fucoidan as active ingredient accelerate wound healing, enhancing granulation and angiogenesis (Moholkar et al. 2021).

Drug delivery

Drug delivery systems are chemical platforms for drug transportation to targeted sites. Ease of handling, water solubility, versatility, biocompatibility, non-immunogenicity, non-mutagenicity, bioadhesive properties and feasible chemical modifications are relevant characteristics for their development. Natural sources are often favored over synthetic materials for the design of drug delivery systems. The careful selection of carriers for drugs and the type of technology used is crucial, as it ensures that the active component effectively reaches its intended target, thereby impacting its overall success and resulting in a significant decrease in the drug dosage (Putro et al. 2023). SWP, have shown to be adequate biomaterials with huge potential (Lomartire and Gonçalves 2023). Different kinds of drug delivery system based on native/modified SWP, often mixed with adjuvants, drugs and copolymers, have been developed up to date, such as hydrogels, nanogels, microparticles, nanoparticles, porous scaffolds, graphene oxide systems, magnetic systems, etc. (Putro et al. 2023, Ulbrich et al. 2016).

In the context of drug delivery technology, alginate has been extensively studied as a versatile cancer nano therapy carrier, with effective delivery of chemotherapeutic agents, ensuring potent therapeutic outcomes in various combinatorial approaches. Delivery systems composed by this biomaterial show increased stability in acidic physiological fluids (Carrion et al. 2021, Lakkakula et al. 2021, Severino et al. 2019). Fucoidan is gaining a significant role in nanodrug delivery systems as well as a complementary therapeutic agent (George and Shrivastav 2023, Ramos-de-la-Peña et al. 2022). Carrageenans have proven effective as an eco-friendly alternative to gelatin for drug capsule materials and show potential for targeted stimuli-responsive drug delivery systems (Liu et al. 2023). Laminarin, although less studied, has been used for the preparation of nanogels for biomedical applications (Can and Sahiner

2021). Ulvan displays selective cytotoxicity against cancer cells, emphasizing its potential in drug delivery. Studies reveal successful utilization in managing glioblastoma, demonstrating biocompatibility, anti-inflammatory activity, and pH-responsive drug release. Ulvan's structural richness contributes to its cell-specific cytotoxicity, making it a promising candidate for targeted anticancer therapy (Shah et al. 2023).

Tissue engineering scaffolds

Scaffolds for tissue engineering must offer optimal cellular activities within suitable structures. These 3D culture systems, constructed from natural, synthetic polymers, or their composites form macroporous structures that emulate the native extracellular matrix and replicate natural physiological conditions, facilitating tissue engineering (Szekalska et al. 2016). Thus, their aim is to attain suitable architectural, biological, and mechanical characteristics, allowing cellular behavior as it is within the native environment (Chandika et al. 2015).

Farshidfar et al. (2023) extensively reviewed the use of alginate in tissue engineering, pointing out that alginates are promising biomaterials for bioinks because of their biocompatibility and fast ionic gelation. Nevertheless, these authors also mention that ionically crosslinked hydrogels of this biomaterial lack adequate mechanical features and stability in the long-term. Alginate-based hydrogels incorporating collagen or agarose exhibit improved potential as scaffolds for tissue engineering and regenerative medicine. Ulvan is a promising candidate in tissue engineering as well. Hybrid ulvan/gelatin scaffolds enhance bone tissue engineering, promoting cell adhesion and osteogenic differentiation. Ulvan nanofiber mats accelerate fibroblast growth with *in vivo* biocompatibility. Ulvan-loaded 3D-printed scaffolds enhance cell viability, growth, and extracellular matrix accumulation for skin regeneration. Moreover, ulvan scaffolds support enzyme-mediated apatite mineral formation without osteogenic growth factors, indicating potential applications in bone remodeling (Shah et al. 2023). Ulvan and ulvan/chitosan polyelectrolyte nanofibrous membranes have shown promising results as substrate material for the cultivation of osteoblasts (Toskas et al. 2012). Carrageenan-based bioinks exhibit excellent biocompatibility, mechanical strength, and rheological properties, although only kappa-carrageenans have been evaluated for this purpose. The addition of carrageenans to hydrogels improves their structural strength, printability, and biocompatibility. Nevertheless, unmodified carrageenans have bad printing quality and have to be complemented, for instance, with nanosilicates which increase printing resolution and shape fidelity (Liu et al. 2023). An overview of human applications is given in Table 2.

Table 2. Possible applications of seaweed Polysaccharides in health systems.

Polysaccharide	Systems	Description	Polysaccharide role	Application	References
Alginate	Hydrogels	Macrogel beads composed of chitosan and alginate, loaded with doxorubicin.	Structural/drug carrier biomaterial	Drug delivery in cancer therapy	(Wu et al. 2020)
		Nanogel particles formed by the combination of glycyrrhizin and alginate	Structural/drug carrier biomaterial	Drug delivery in cancer therapy	(Wang et al. 2019)
	Micro/ nanoparticles	Alginate microparticles encapsulated within mesoporous silica nanoparticles	Structural/drug carrier biomaterial	Drug delivery in cancer therapy	(Liao et al. 2014)
		Bismuth sulfide nanoparticles enclosed within alginate microspheres.	Structural/drug carrier biomaterial	Drug delivery in cancer therapy	(Zou et al. 2019)
		Graphene oxide functionalized with chitosan and sodium alginate.	Structural/drug carrier biomaterial	Drug delivery in cancer therapy	(Lei et al. 2016)
	Magnetic	Hydrogel composed of alginate, gelatin, and Fe_3O_4 magnetic nanoparticles.	Structural/drug carrier biomaterialm	Drug delivery in cancer therapy	(Jahanban-Esfahlan et al. 2020)
		Magnetic nanogel formulated with alginate	Structural/drug carrier biomaterial	Drug delivery in cancer therapy	(Peng et al. 2019)
Ulvan	Hydrogels	Curcumin-loaded acetylated ulvan nanogel	Structural/drug carrier biomaterial	Drug delivery system	(Bang et al. 2019)
	Micro/nano particles	Ulvan/chitosan/D-mannose loaded graphene oxide nanoparticle	Structural/drug carrier biomaterial	Targeted anticancer drug delivery	(Kesavan et al. 2021)
	Porous scaffolds	Ulvan/chitosan nanofibrous membranes	Structural/mechanical strength	Cultivation of osteoblasts	(Toskas et al. 2012)
Carrageenan	Hydrogels	Gelatin-carrageenan hydrogel bioink blend	To mantain adequate mechanical strength with large pore sizes	Adipose tissue regeneration	(Tytgat et al. 2019)
	Micro/ nanoparticles	ZnO encapsulated within carrageenan	Structural/drug carrier biomaterial	Topical. Antibacterial.	(Vijayakumar et al. 2020)
	Porous scaffolds	iota-carrageenan/chitosan/gelatin scaffold	Support for the attachment and cell proliferation	bone tissue engineering	(Li et al. 2015)
	Tablets	Controlled-release tablets	Controlled release beahviour	not specified	(Picker 1999)
	Magnetic	nanocomposite κ-carrageenan/chitosan hydrogels with magnetic montmorillonite	Structural/drug carrier biomaterial	Drug delivery in cancer therapy	(Jafari et al. 2021)

Table 2 contd. ...

...Table 2 contd.

Polysaccharide	Systems	Description	Polysaccharide role	Application	References
Fucoidan	Hydrogels	Hyaluronic acid/fucoidan/gelatin/genipin injectable hydrogel	Anti-inflamatory, Interaction with growth factors and sustained release	Cartilage regeneration and repair	(Lu et al. 2019)
		Glycol chitosan/fucoidan injectable nanogels loaded with KAFAK peptide	Anti-inflamatory	Osteoarthritis treatment	(Li et al. 2021)
	Micro/ nanoparticles	Nanoencapsulation of red gingseng with chitosan and fucoidan	Anti-coagulant	Antithrombotic treatment	(Kim et al. 2016)
		Chitosan/tripolyphosphate/fucoidan nanoparticles	Drug carrier	Drug delivery for stem cell mobilization in tissue engineering	(Huang and Liu 2012)
	Magnetic	Fucoidan coated Fe_3O_4 nanoparticles for magnetic hyperthermia cancer treatment	Enhanced magnetic hyperthermia	Cancer therapy	(Gonçalves et al. 2021)
Agar	Hydrogels	Injectable gellan gum/agar hydrogel	Mechanical strength	Cartilage tissue regeneration	(Baek et al. 2019)
	Porous scaffolds	Chitosan–gelatin–agar-based porous scaffold	Mechanical strength	Tissue engineering	(Ahmad Ruzaidi et al. 2021)
	Magnetic	Agar-conjugated Fe_3O_4 magnetic nanoparticles	Structural/drug carrier biomaterial	Drug delivery system	(Hsieh et al. 2010)
	Film	Immobilized silver nanoparticles in agar film	Scaffold/structural	Antibacterial	(Davidović et al. 2019)
Laminarin	Hydrogels	Methacrylated laminarin micropatterned hydrogel patch	Structural/drug carrier biomaterial	Dermal drug release system	(Amaral et al. 2023)
	Micro/ nanoparticles	Dialdheyde modification of laminarin silver nanoparticles	Structural/drug carrier biomaterial	Wound healing	(Sharma et al. 2022)
	Porous scaffolds	Laminarin-based scaffold gel with graphene foam	Scaffold/structural	Tissue engineering	(Feng et al. 2019)

Future trends

Polysaccharide modification, processing, and material design are crucial for the successful development of novel biomaterials. Strategies for modifying seaweed-derived polysaccharides or new combinations with other biopolymers will likely play a pivotal role in shaping the next generation of biomaterials for medical applications. Nanotechnology continues to revolutionize drug delivery systems. Alginate-based nanoparticles, known for their biocompatibility and safety, hold promise for diverse biomedical applications, including wound healing, drug delivery, and tissue engineering. These nano delivery systems may address challenges such as bioavailability, stability, and controlled release of bioactives (Severino et al. 2019). Moreover, alginate-based drug delivery systems show great promise in targeted chemotherapy. These systems, when functionalized with targeting moieties, exhibit stimulus-responsive behavior, releasing drugs only at tumor sites. This approach minimizes side effects, increases cellular uptake, and enhances chemotherapeutic efficiency (Lakkakula et al. 2021). The carrageenan industry's rapid growth suggests a burgeoning interest in its biomedical applications. Carrageenans' biofunctionality and rheological characteristics make them excellent candidates for intelligent drug delivery systems and tissue engineering. Ongoing challenges, such as improving cell binding and optimizing stimuli-responsive mechanisms, highlight avenues for future research (Liu et al. 2023).

Future dressings may incorporate smart hydrogels, asymmetric/multi-layer films, and nanofiber mats, expanding their applications in wound therapy with enhanced antibacterial, antioxidant, and pro-regenerative properties. Underexplored polysaccharides like laminarin or ulvan exhibit wound healing potential. Although being currently underutilized, ongoing research suggests they could be developed into novel wound dressings with unique therapeutic attributes (Shen et al. 2021). In case of ulvan, despite its potential for hybrid biomaterials in biomedicine, the lack of systematic studies on structure-activity relationships necessitates thorough investigations into biocompatibility before targeted therapeutic applications can be realized (Tziveleka et al. 2019).

Acknowledgements

The authors thank the Consellería de Cultura, Educación e Universidade da Xunta de Galicia (GRC-ED431C 2022/08). M.D.T. acknowledges the Ministry of Science, Innovation and Universities of Spain for her postdoctoral grant (RYC2018-024454-I) and thanks the Consellería de Cultura, Educación e Universidade da Xunta de Galicia (ED431F 2020/01). N.F.F. thanks the Xunta de Galicia for her postdoctoral fellowship (ED481D-2022/018).

References

Abd-Ellatef, G.E.F., Ahmed, O.M., Abdel-Reheim, E.S. and Abdel-Hamid, A.H.Z. (2017). *Ulva lactuca* polysaccharides prevent Wistar rat breast carcinogenesis through the augmentation of apoptosis, enhancement of antioxidant defense system, and suppression of inflammation. Breast Cancer: Targets and Therapy 9: 67–83.

Adrien, A., Dufour, D., Baudouin, S., Maugard, T. and Bridiau, N. (2017). Evaluation of the anticoagulant potential of polysaccharide-rich fractions extracted from macroalgae. Nat. Prod. Res. 31: 2126–2136.

Agarwal, S., Singh, V. and Chauhan, K. (2023). Antidiabetic potential of seaweed and their bioactive compounds: a review of developments in last decade. Crit. Rev. Food Sci. Nutr. 63: 5739–5770.

Aguilar-Briseño, J.A., Cruz-Suarez, L.E., Sassi, J.F., Ricque-Marie, D., Zapata-Benavides, P., Mendoza-Gamboa, E. et al. (2015). Sulfated polysaccharides from *Ulva clathrata* and *Cladosiphon okamuranus* seaweeds both inhibit viral attachment/entry and cell-cell fusion, in NDV infection. Mar. Drugs. 13: 697–712.

Ahmad Ruzaidi, D.A., Mahat, M.M., Mohamed Sofian, Z., Nor Hashim, N.A., Osman, H., Nawawi, M.A. et al. (2021). Synthesis and characterization of porous, electro-conductive chitosan–gelatin–agar-based pedot: Pss scaffolds for potential use in tissue engineering. Polymers 13(17): 2901.

Ahmed, O.M. and Ahmed, R.R. (2014). Anti-proliferative and apoptotic efficacies of ulvan polysaccharides against different types of carcinoma cells *in vitro* and *in vivo*. J. Cancer. Sci. Ther. 6: 202–208.

Ale, M.T., Mikkelsen, J.D. and Meyer, A.S. (2011). Important determinants for fucoidan bioactivity: a critical review of structure-function relations and extraction methods for fucose-containing sulfated polysaccharides from brown seaweeds. Mar. Drugs 9: 2106–2130.

Álvarez-Gómez, F., Korbee, N. and Figueroa, F.L. (2019). Effects of UV radiation on photosynthesis, antioxidant capacity and the accumulation of bioactive compounds in *Gracilariopsis longissima*, *Hydropuntia cornea* and *Halopithys incurva* (Rhodophyta). J. Phycol. 1273: 1258–1273.

Álvarez-Viñas, M., Souto, S., Flórez-Fernández, N., Dolores Torres, M., Bandín, I., Domínguez, H. et al. (2021). Antiviral activity of carrageenans and processing implications. Mar. Drugs 19: 437.

Alves, C., Silva, J., Pinteus, S., Gaspar, H., Alpoim, M.C., Botana, L.M. et al. (2018). From marine origin to therapeutics: The antitumor potential of marine algae-derived compounds. Front. Pharmacol. 9: 777.

Amaral, K.R., Silva, A.S., Santos, L.F., Castanheira, E.J., Mendes, M.C., Costa, D.C.S. et al. (2023). Biomimetic adhesive micropatterned hydrogel patches for drug release. Adv. Healthc. Mater. 12(28): 2301513.

Andryukov, B.G., Besednova, N.N., Kuznetsova, T.A., Zaporozhets, T.S., Ermakova, S.P., Zvyagintseva, T.N. et al. (2020). Sulfated polysaccharides from marine algae as a basis of modern biotechnologies for creating wound dressings: Current achievements and future prospects. Biomedicines 8(9): 301.

Arive, P.L.C., Inquimboy, I.H. and No, N.L.I. (2017). *In vitro* antioxidant activity of selected seaweeds in the Philippines. Int. J. Theor. Appl. Sci. 9: 212–216.

Arul, A.B., Al Numair, K., Al Saif, M. and Savarimuthu, I. (2012). Effect of dietary β-sitosterol on fecal bacterial and colonic biotransformation enzymes in 1, 2-dimethylhydrazine–induced colon carcinogenesis. Turkish. J. Med. Sci. 42: 1307–13.

Aryee, A.N., Agyei, D. and Akanbi, T.O. (2018). Recovery and utilization of seaweed pigments in food processing. Curr. Op. Food Sci. 19: 113–9.

Baek, J.S., Carlomagno, C., Muthukumar, T., Kim, D., Park, J.H. et al. (2019). Evaluation of cartilage regeneration in gellan gum/agar blended hydrogel with improved injectability. Macromol. Res. 27(6): 558–564.

Bang, T.H., Van, T.T.T., Hung, L.X., Ly, B.M., Nhut, N.D. et al. (2019). Nanogels of acetylated ulvan enhance the solubility of hydrophobic drug curcumin. Bull. Mater. Sci. 42: 1.

Barbosa, J.R. and de Carvalho Junior, R.N. (2021). Polysaccharides obtained from natural edible sources and their role in modulating the immune system: Biologically active potential that can be exploited against COVID-19. Trends Food Sci. Technol. 108: 223–235.

Bermano, G., Stoyanova, T., Hennequart, F. and Wainwright, C.L. (2020). Seaweed-derived bioactives as potential energy regulators in obesity and type 2 diabetes. Adv. Pharmacol. 87: 205–256.

Berri, M., Olivier, M., Holbert, S., Dupont, J., Demais, H., Le Goff, M. et al. (2017). Ulvan from *Ulva armoricana* (Chlorophyta) activates the PI3K/Akt signalling pathway via TLR4 to induce intestinal cytokine production. Algal Res. 28: 39–47.

Bilal, M. and Iqbal, H.M. (2019). Marine seaweed polysaccharides-based engineered cues for the modern biomedical sector. Mar. Drugs 18(1): 7.

Campiche, R., Curpen, S.J., Lutchmanen-Kolanthan, V., Gougeon, S., Cherel, M., Laurent, G. et al. (2020). Pigmentation effects of blue light irradiation on skin and how to protect against them. Int. J. Cosmet. Sci. 42: 399–406.

Can, M. and Sahiner, N. (2021). A facile one-pot synthesis of microgels and nanogels of laminarin for biomedical applications. J. Colloid Interface Sci. 588: 40–49.

Carrion, C.C., Nasrollahzadeh, M., Sajjadi, M., Jaleh, B., Soufi, G.J. and Iravani, S. (2021). Lignin, lipid, protein, hyaluronic acid, starch, cellulose, gum, pectin, alginate and chitosan-based nanomaterials for cancer nanotherapy: Challenges and opportunities. Int. J. Biol. Macromol. 178: 193–228.

Chai, Q., Jiao, Y. and Yu, X. (2017). Hydrogels for biomedical applications: their characteristics and the mechanisms behind them. Gels 3(1): 6.

Chandika, P., Ko, S.C. and Jung, W.K. (2015). Marine-derived biological macromolecule-based biomaterials for wound healing and skin tissue regeneration. Int. J. Biol. Macromol. 77: 24–35.

Chen, X., Yue, Z., Winberg, P.C., Dinoro, J.N., Hayes, P. and Beirne, S. (2019). Development of rhamnose-rich hydrogels based on sulfated xylorhamno-uronic acid toward wound healing applications. Biomater. Sci. 7(8): 3497–3509.

Chiu, Y.H., Chan, Y.L., Tsai, L.W., Li, T.L. and Wu, C.J. (2012). Prevention of human enterovirus 71 infection by kappa carrageenan. Antivir. Res. 95: 128–134.

Chiu, Y.H., Chan, Y.L., Li, T.L. and Wu, C.J. (2012). Inhibition of Japanese encephalitis virus infection by the sulfated polysaccharide extracts from *Ulva lactuca*. Mar. Biotechnol. 14: 468–478.

Cotas, J., Leandro, A., Monteiro, P., Pacheco, D., Figueirinha, A., Gonçalves, A.M. et al. (2020). Seaweed phenolics: from extraction to applications. Mar. Drugs 18: 384.

Cotas, J., Marques, V., Afonso, M.B., Rodrigues, C.M.P. and Pereira, L. (2020). Antitumour potential of *Gigartina pistillata* carrageenans against colorectal cancer stem cell-enriched tumourspheres. Mar. Drugs 18: 50.

Coulombier, N., Jauffrais, T. and Lebouvier, N. (2021). Antioxidant compounds from microalgae: a review. Mar. Drugs 19: 549.

Cui, J., Li, Y., Wang, S., Chi, Y., Hwang, H. and Wang, P. (2018). Directional preparation of anticoagulant-active sulfated polysaccharides from *Enteromorpha prolifera* using artificial neural networks. Sci. Rep. 8: 3062.

Davidović, S., Lazić, V., Miljković, M., Gordić, M., Sekulić, M., Marinović-Cincović, M. et al. (2019). Antibacterial ability of immobilized silver nanoparticles in agar-agar films co-doped with magnesium ions. Carbohydr. Polym. 224: 115187.

El-Shenody, R.A., Elshobary, M.E., Ragab, G.A., Huo, S. and Essa, D. (2023). Towards biorefinery: Exploring the potential of seaweed-derived biodiesel and its residual biomass in improving the traits of *Eruca vesicaria* (L.) cav. S. Afr. J. Bot. 155: 361–371.

Endo, H., Suehiro, K., Gao, X. and Agatsuma, Y. (2017). Interactive effects of elevated summer temperature, nutrient availability, and irradiance on growth and chemical compositions of juvenile kelp, *Eisenia bicyclis*. Phycological Research 65: 118–126.

FAO. (2021). Global Seaweed and Microalgae Production, 1950–2019. World Aquaculture Performance Indicators (WAPI) Factsheet.

Farshidfar, N., Iravani, S. and Varma, R.S. (2023). Alginate-based biomaterials in tissue engineering and regenerative medicine. Mar. Drugs 21(3): 189.

Feng, L., Hao, Y., Zhu, M., Zhai, Y., Yang, L. et al. (2019). Incorporation of laminarin-based hydrogel with graphene foam to enhance the toughness of scaffold and regulate the stem cell behavior. ACS Biomater. Sci. Eng. 5(10): 5295–5304.

Flórez, N., González-Muñoz, M.J., Ribeiro, D., Fernandes, E., Domínguez, H. and Freitas, M. (2017). Algae polysaccharides' chemical characterization and their role in the inflammatory process. Curr. Med. Chem. 24: 149 – 175.

Flórez-Fernández, N., Illera, M., Sánchez, M., Lodeiro, P., Torres, M.D., López-Mosquera, M.E. et al. (2021). Integrated valorization of *Sargassum muticum* in biorefineries. Chem. Eng. J. 404: 125635.

Flórez-Fernández, N., Rodríguez-Coello, A., Latire, T., Bourgougnon, N., Torres, M.D., Buján, M. et al. (2023). Anti-inflammatory potential of ulvan. Int. J. Biol. Macromol. 253: 126936.

Forman, H.J., Davies, K.J. and Ursini, F. (2014). How do nutritional antioxidants really work: nucleophilic tone and para-hormesis versus free radical scavenging *in vivo*. Free Radic. Biol. Med. 66: 24–35.

Fufeng, L., Wenping, Z. and Fuping, L. (2019). Application of ulvan in inhibiting aggregation of beta-amyloid protein. CN110420216.

Gammone, M.A., Riccioni, G., Parrinello, G. and D'Orazio, N. (2019). Omega-3 polyunsaturated fatty acids: Benefits and endpoints in sport. Nutrients 11: 46.

Ganesan, A.R., Tiwari, U. and Rajauria, G. (2019). Seaweed nutraceuticals and their therapeutic role in disease prevention. Food Sci. Hum. Wellness 8: 252–263.

George, A. and Shrivastav, P.S. (2023). Fucoidan, a brown seaweed polysaccharide in nanodrug delivery. Drug Deliv. and Transl. Res. 13: 2427–2446.

Geun Lee, H., Jayawardena, T.U., Liyanage, N.M., Song, K.-M., Choi, Y.-S., Jeon, Y.-J. et al. (2022). Antioxidant potential of low molecular weight fucoidans from *Sargassum* autumnal against H_2O_2-induced oxidative stress *in vitro* and in zebrafish models based on molecular weight changes. Food Chem. 384: 132591.

Gomez, L.P., Alvarez, C., Zhao, M., Tiwari, U., Curtin, J., Garcia-Vaquero, M. et al. (2020). Innovative processing strategies and technologies to obtain hydrocolloids from macroalgae for food applications. Carbohydr. Polym. 248: 116784.

Gonçalves, A.M.M., Marques, J.C. and Gonçalves, F. (2017). Fatty acids' profiles of aquatic organisms: Revealing the impacts of environmental and anthropogenic stressors. *In*: Catala, A. (ed.). Fatty Acids. IntechOpen, London, UK.

Gonçalves, J., Nunes, C., Ferreira, L., Cruz, M.M., Oliveira, H. et al. (2021). Coating of magnetite nanoparticles with fucoidan to enhance magnetic hyperthermia efficiency. Nanomaterials 11(11): 1–20.

González-Fernández, M.J., Fabrikov, D., Ramos-Bueno, R.P., Guil-Guerrero, J.L. and Ortea, I. (2019). Swath differential abundance proteomics and cellular assays show *in vitro* anticancer activity of arachidonic acid-and docosahexaenoic acid-based monoacylglycerols in Ht-29 colorectal cancer cells. Nutrients 11: 2984.

Grassauer, A., Weinmuellner, R., Meier, C., Pretsch, A., Prieschl-Grassauer, E. and Unger, H. (2008). Iota-Carrageenan is a potent inhibitor of rhinovirus infection. Virol. J. 5: 5–7.

Gutierrez-Rodriguez, A., Juarez-Portilla, G.C., Olivares-Banuelos, T. and Zepeda, R.C. (2018). Anticancer activity of seaweeds. Drug. Discov. Today 23: 434–47.

Holdt, S. and Kraan, S. (2011). Bioactive compounds in seaweed: functional food applications and legislation. J. Appl. Phycol. 23: 543–597.

Hongfang, D., Guangli, Y., Bo, L., Qizhuang, M., Xia, Z., Guangsheng, L. et al. (2023). Pharmaceutical composition for anti-tumour immunotherapy. CN111973749.

Hosokawa, M., Miyashita, T., Nishikawa, S., Emi, S., Tsukui, T., Beppu, F. et al. (2010). Fucoxanthin regulates adipocytokine mRNA expression in white adipose tissue of diabetic/obese KK-Ay mice. Arch. Biochem. Biophys. 504: 17–25.

Hou, Y., Wang, J., Jin, W., Zhang, H. and Zhang, Q. (2012). Degradation of *Laminaria japonica* fucoidan by hydrogen peroxide and antioxidant activities of the degradation products of different molecular weights. Carbohydr. Polym. 87: 153–159.

Hsieh, S., Huang, B.Y., Hsieh, S.L., Wu, C.C., Wu, C.H. et al. (2010). Green fabrication of agar-conjugated Fe_3O_4 magnetic nanoparticles. Nanotechnology 21(44): 445601.

Huang, Y.C. and Liu, T.J. (2012). Mobilization of mesenchymal stem cells by stromal cell-derived factor-1 released from chitosan/tripolyphosphate/fucoidan nanoparticles. Acta Biomater. 8(3): 1048–1056.

Hussain, E., Wang, L.J., Jiang, B., Riaz, S., Butt, G.Y. and Shi, D.Y. (2016). A review of the components of brown seaweeds as potential candidates in cancer therapy. RSC Adv. 6: 12592.

Hwang, P.A., Hung, Y.L., Tsai, Y.K., Chien, S.Y. and Kong, Z.L. (2015). The brown seaweed *Sargassum hemiphyllum* exhibits alpha-amylase and alpha-glucosidase inhibitory activity and enhances insulin release *in vitro*. Cytotechnology 67: 653–660.

Hwan Kwak, J., Su Yoon, H. and Jeon, Y. (2022). Composition for prevention and treatment of coronavirus containing fucoidan derived from *Agarum clathratum* extract as active component. KR1020220109741.

Ibañez, E., Herrero, M., Mendiola, J.A. and Castro-Puyana, M. (2012). Extraction and characterization of bioactive compounds with health benefits from marine resources: Macro and microalgae, cyanobacteria, and invertebrates. pp. 55–98. *In*: Hayes, M. (ed.). Marine Bioactive Compounds. Springer, Boston, MA.

Jacob, L., Baker, C. and Farris, P. (2012). Vitamin-based cosmeceuticals. Cosmet. Dermatol. 25: 405.

Jacobsen, C., Sørensen, A.d.M., Holdt, S.L., Akoh, C.C. and Hermund, D.B. (2019). Source, extraction, characterization, and applications of novel antioxidants from seaweed. Annu. Rev. Food Sci. Technol. 10: 541–568.

Jafari, H., Atlasi, Z., Mahdavinia, G.R., Hadifar, S. and Sabzi, M. (2021). Magnetic κ-carrageenan/chitosan/montmorillonite nanocomposite hydrogels with controlled sunitinib release. Mater. Sci. Eng. C 124: 112042.

Jahanban-Esfahlan, R., Derakhshankhah, H., Haghshenas, B., Massoumi, B., Abbasian, M. et al. (2020). A bio-inspired magnetic natural hydrogel containing gelatin and alginate as a drug delivery system for cancer chemotherapy. Int. J. Biol. Macromol. 156: 438–445.

Janarthanan, M. and Senthil Kumar, M. (2018). The properties of bioactive substances obtained from seaweeds and their applications in textile industries. J. Ind. Text. 48: 361–401.

Jesumani, V., Du, H., Aslam, M., Pei, P. and Huang, N. (2019). Potential use of seaweed bioactive compounds in skincare – a review. Mar. Drugs 17(12): 688.

Jiang, L., Zhao, X., Xu, J., Li, C., Yu, Y., Wang, W. et al. (2019). The protective effect of dietary phytosterols on cancer risk: a systematic meta-analysis. J. Oncol. 7479518.

Jiao, G., Yu, G., Wang, W., Zhao, X., Zhang, J. and Ewart, S.H. (2012). Properties of polysaccharides in several seaweeds from Atlantic Canada and their potential anti-influenza viral activities. J. Ocean Univ. 11: 205–212.

Jin, W., Ren, L., Liu, B., Zhang, Q. and Zhong, W. (2018). Structural features of sulfated glucuronomannan oligosaccharides and their antioxidant activity. Mar. Drugs 16: 291.

Jin, W., Zhang, Q., Wang, J. and Zhang, W. (2013). A comparative study of the anticoagulant activities of eleven fucoidans. Carbohydr. Polym. 91: 1–6.

Jin, Q., Wu, H.-G., Zhang, X.-X. and Ke, Z.-L. (2016). Influence of non-nutrient environmental factors on *Ulva pertusa*'s inhibitory effects on *Heterosigma akashiwo* growth. J. Mar. Biol. Assoc. U. K., 96: 1041–1052.

Jun, J., Juyoung, H. and Gyoung, A.E. (2023). Adjuvant composition comprising fucoidan from *Ecklonia cava* as active ingredient. KR1020230007688.

Kalasariya, H.S., Yadav, V.K., Yadav, K.K., Tirth, V., Algahtani, A., Islam, S. et al. (2021). Seaweed-based molecules and their potential biological activities: an eco-sustainable cosmetics. Molecules 26: 1–22.

Kang, C., Jin, Y.B., Lee, H., Cha, M., Sohn, E.-t., Moon, J. et al. (2010). Brown alga *Ecklonia cava* attenuates type 1 diabetes by activating AMPK and Akt signaling pathways. Food Chem. Toxicol. 48: 509–516.

Kendel, M., Wielgosz-Collin, G., Bertrand, S., Roussakis, C., Bourgougnon, N. and Bedoux, G. (2015). Lipid composition, fatty acids and sterols in the seaweeds *Ulva armoricana*, and *Solieria chordalis* from Brittany (France): An analysis from nutritional, chemotaxonomic, and antiproliferative activity perspectives. Mar. Drugs 13: 5606–5628.

Kesavan, S., Meena, K.S., Sharmili, S.A., Govindarajan, M., Alharbi, N.S. et al. (2021). Ulvan loaded graphene oxide nanoparticle fabricated with chitosan and D-mannose for targeted anticancer drug delivery. J. Drug Deliv. Sci. Technol. 65(March): 102760.

Kidgell, J.T., Magnusson, M., de Nys, R. and Glasson, C.R.K. (2019). Ulvan: a systematic review of extraction, composition and function. Algal. Res. 39: 101422.

Kim, E., Cui, J., Kang, I., Zhang, G. and Lee, Y. (2021). Potential antidiabetic effects of seaweed extracts by upregulating glucose utilization and alleviating inflammation in C2C12 myotubes. Int. J. Environ. Res. Public Health. 18: 1367.

Kim, E.S., Lee, J.S. and Lee, H.G. (2016). Nanoencapsulation of red ginseng extracts using chitosan with polyglutamic acid or fucoidan for improving antithrombotic activities. J. Agric. Food Chem. 64(23): 4765–4771.

Kim, Y.-I., Oh, W.-S., Song, P.H., Yun, S., Kwon, Y.-S., Lee, Y.J. et al. (2018). Anti-photoaging effects of low molecular-weight fucoidan on ultraviolet B-irradiated mice. Mar. Drugs 16: 286.

Koike, T., Sugimoto, A., Kosono, S., Komaba, S., Kanno, Y., Kitamura, T. et al. (2021). Synthesis of low-molecular weight fucoidan derivatives and their binding abilities to SARS-CoV-2 spike proteins. RSC Med. Chem. 12: 2016–2021.

Kondo, T. and Ishida, Y. (2010). Molecular pathology of wound healing. Forensic. Sci. Int. 203: 93–98.

Kraan, S. (2013). Mass-cultivation of carbohydrate rich macroalgae, a possible solution for sustainable biofuel production. Mitig. Adapt. Strateg. Glob. Change 18: 27–46.

Krylova, N.V., Silchenko, A.S., Pott, A.B., Ermakova, S.P., Iunikhina, O.V., Rasin, A.B. et al. (2021). *In vitro* anti-orthohantavirus activity of the high-and low-molecular-weight fractions of fucoidan from the brown alga *Fucus evanescens*. Mar. Drugs 19: 577.

Krylova, N.V., Ermakova, S.P., Lavrov, V.F., Leneva, I.A., Kompanets, G.G., Iunikhina, O.V. et al. (2020). The comparative analysis of antiviral activity of native and modified fucoidans from brown algae *Fucus evanescens in vitro* and *in vivo*. Mar. Drugs 18: 224.

Kurniawan, R., Nurkolis, F., Taslim, N.A., Subali, D., Surya, R., Gunawan, W.B. et al. (2023). Carotenoids composition of green algae *Caulerpa racemosa* and their antidiabetic, anti-obesity, antioxidant, and anti-inflammatory properties. Mol. 28: 3267.

Kwon, P.S., Oh, H., Kwon, S.J., Jin, W., Zhang, F., Fraser, K. et al. (2020). Sulfated polysaccharides effectively inhibit SARS-CoV-2 *in vitro*. Cell Discovery 6: 50.

Lahaye, M. and Kaeffer, B. (1997). Seaweed dietary fibres: structure, physico-chemical and biological properties relevant to intestinal physiology. Sci. Aliment. 17: 563–584.

Lahaye, M., Inizan, F. and Vigoureux, J. (1998). NMR analysis of the chemical structure of ulvan and of ulvan-boron complex formation. Carbohydr. Polym. 36: 239–249.

Lakkakula, J.R., Gujarathi, P., Pansare, P. and Tripathi, S. (2021). A comprehensive review on alginate-based delivery systems for the delivery of chemotherapeutic agent: Doxorubicin. Carbohydr. Polym. 259: 117696.

Lei, H., Xie, M., Zhao, Y., Zhang, F., Xu, Y. and Xie, J. (2016). Chitosan/sodium alginate modificated graphene oxide-based nanocomposite as a carrier for drug delivery. Ceram. Int. 42(15): 17798–17805.

Li, J., Jiang, F., Chi, Z., Han, D., Yu, L. and Liu, C. (2018). Development of *Enteromorpha prolifera* polysaccharide-based nanoparticles for delivery of curcumin to cancer cells. Int. J. Biol. Macromol. 112: 413–421.

Li, J., Yang, B., Qian, Y., Wang, Q., Han, R. et al. (2015). Iota-carrageenan/chitosan/gelatin scaffold for the osteogenic differentiation of adipose-derived MSCs *in vitro*. J. Biomed. Mater. Res. B Appl. Biomater. 103(7): 1498–1510.

Li, T., Yang, J., Weng, C., Liu, P., Huang, Y. et al. (2021). Intra-articular injection of anti-inflammatory peptide-loaded glycol chitosan/fucoidan nanogels to inhibit inflammation and attenuate osteoarthritis progression. Int. J. Biol. Macromol. 170: 469–478.

Li, W., Jiang, N., Li, B., Wan, M., Chang, X., Liu, H. et al. (2018). Antioxidant activity of purified ulvan in hyperlipidemic mice. Int. J. Biol. Macromol. 113: 971–975.

Liang, W., Mao, X., Peng, X. and Tang, S. (2014). Effects of sulfate group in red seaweed polysaccharides 674 on anticoagulant activity and cytotoxicity. Carbohydr. Polym. 101: 776–785.

Liao, M., Wang, W., Chen, S. and Liao, S. (2017). Composition analysis and antithrombotic activity of fucoidan from *Sargassum henslowianum* C. Agardh. Food Sci. 38: 189–194.

Liao, Y.-T., Wu, K.C.-W. and Yu, J. (2014). Synthesis of mesoporous silica nanoparticle-encapsulated alginate microparticles for sustained release and targeting therapy. J. Biomed. Mater. Res. Part B Appl. Biomater. 102(2): 293–302.

Lin, P., Chen, S., Liao, M. and Wang, W. (2022). Physicochemical characterization of fucoidans from *Sargassum henslowianum* C. Agardh and their antithrombotic activity *in vitro*. Mar. Drugs 20: 300.

Liu, F., Duan, G. and Yang, H. (2023). Recent advances in exploiting carrageenans as a versatile functional material for promising biomedical applications. Int. J. Biol. Macromol. 235: 123787.

Liu, M., Hansen, P.E. and Lin, X. (2011). Bromophenols in marine algae and their bioactivities. Mar. Drugs 9: 1273–1292.

Lomartire, S. and Gonçalves, A.M.M. (2023). Algal phycocolloids: bioactivities and pharmaceutical applications. Mar. Drugs 21(7): 1–30.

Long, A.T., Kenne, E., Jung, R., Fuchs, T.A. and Renné, T. (2016). Contact system revisited: an interface between inflammation, coagulation, and innate immunity. J. Thromb. Haemost. 14: 427–437.

Lopes, N., Ray, S., Espada, S.F., Bomfim, W.A., Ray, B., Faccin-Galhardi, L.C. et al. (2017). Green seaweed *Enteromorpha compressa* (Chlorophyta, Ulvaceae) derived sulfated polysaccharides inhibit herpes simplex virus. Int. J. Biol. Macromol. 102: 605–612.

Lu, H.T., Chang, W.T., Tsai, M.L., Chen, C.H., Chen, W.Y. et al. (2019). Development of injectable fucoidan and biological macromolecules hybrid hydrogels for intra-articular delivery of platelet-rich plasma. Mar. Drugs 17(4).

Lu, W., Yang, Z., Chen, J., Wang, D. and Zhang, Y. (2021). Recent advances in antiviral activities and potential mechanisms of sulfated polysaccharides. Carbohydr. Polym. 272: 118526.

Luo, Z., Tian, D., Zhou, M., Xiao, W., Zhang, Y., Li, M. et al. (2015). lambda;-carrageenan P32 is a potent inhibitor of rabies virus infection. PLoS One 10: e0140586.

Manela-Azulay, M. and Bagatin, E. (2009). Cosmeceuticals vitamins. Clin. Dermatol. 27: 469–474.

Manivasagan, P., Bharathiraja, S., Santha Moorthy, M., Mondal, S., Seo, H., Dae Lee, K. et al. (2018). Marine natural pigments as potential sources for therapeutic applications. Crit. Rev. Biotechnol. 38(5): 745–61.

McKim, J.M., Baas, H., Rice, G.P., Willoughby, J.A., Weiner, M.L. and Blakemore, W. (2016). Effects of carrageenan on cell permeability, cytotoxicity, and cytokine gene expression in human intestinal and hepatic cell lines. Food. Chem. Toxicol. 96: 1–10.

Michalak, I. and Chojnacka, K. (2014). Algal extracts: Technology and advances. Eng. Life Sci. 14: 581–591.

Moholkar, D.N., Sadalage, P.S., Peixoto, D., Paiva-Santos, A.C. and Pawar, K.D. (2021). Recent advances in biopolymer-based formulations for wound healing applications. Eur. Polym. J. 160.

Morán-Santibañez, K., Cruz-Suárez, L.E., Ricque-Marie, D., Robledo, D., Freile-Pelegrín, Y., Peña-Hernández, M.A. et al. (2016). Synergistic effects of sulfated polysaccharides from Mexican seaweeds against measles virus. Biomed. Res. Int. 2016: 8502123.

Narayani, S.S., Saravanan, S., Ravindran, J., Ramasamy, M.S. and Chitra, J. (2019). *In vitro* anticancer activity of fucoidan extracted from *Sargassum cinereum* against Caco-2 cells. Int. J. Biol. Macromol. 138: 618–628.

Nazarudin, M., Yasin, I., Mazli, N., Saadi, A., Azizee, M., Nooraini, M. et al. (2022). Preliminary screening of antioxidant and cytotoxic potential of green seaweed, *halimeda opuntia* (linnaeus) *lamouroux*. Saudi. J. Biol. Sci. 29: 2698–2705.

Noriega-Fernández, E., Sone, I., Astráin-Redín, L., Prabhu, L., Sivertsvik, M., Álvarez, I. et al. (2021). Innovative ultrasound-assisted approaches towards reduction of heavy metals and iodine in macroalgal biomass. Foods 10: 649.

Orozco-González, J.G., Amador-Castro, F., Gordillo-Sierra, A.R., García-Cayuela, T., Alper, H.S. and Carrillo-Nieves, D. (2022). Opportunities surrounding the use of *Sargassum* biomass as precursor of biogas, bioethanol, and biodiesel production. Front. Mar. Sci. 1–11.

Pang, R., Tao, J.Y., Zhang, S.L., Zhao, L., Yue, X., Wang, Y.F. et al. (2010). *In vitro* antiviral activity of lutein against hepatitis B virus. Phytotherapy. Res. 24: 1627–30.

Paradossi, G., Cavalieri, G.F., Pizzoferrato, L. and Liquori, A.M. (1999). A physicochemical study on the polysaccharide ulvan from hot water extraction of the macroalga *Ulva*, Int. J. Biol. Macromol. 25: 309–315.

Park, S.R., Kim, J.H., Jang, H.D., Yang, S.Y. and Kim, Y.H. (2018). Inhibitory activity of minor phlorotannins from *Ecklonia cava* on alpha-glucosidase. Food Chem. 57: 128.

Peñalver, R., Lorenzo, J., Ros, G. Amarowicz, R., Pateiro, M. and Nieto, G. (2020). Seaweeds as a functional ingredient for a healthy diet. Mar. Drugs 18: 1–27.

Peng, N., Ding, X., Wang, Z., Cheng, Y., Gong, Z. et al. (2019). Novel dual responsive alginate-based magnetic nanogels for onco-theranostics. Carbohydr. Polym. 204: 32–41.

Pérez, M.J., Falqué, E. and Domínguez, H. (2016). Antimicrobial action of compounds from marine seaweed. Mar. Drugs 14(3): 1–38.

Picker, K.M. (1999). Matrix tablets of carrageenans. II. Release behavior and effect of added cations. Drug Dev. Ind. Pharm. 25(3): 339–346.

Premarathna, A.D., Ahmed, T.A.E., Kulshreshtha, G., Humayun, S., Darko, C.N.S., Rjabovs, V. et al. (2023). Polysaccharides from red seaweeds: Effect of extraction methods on physicochemical characteristics and antioxidant activities. Food. Hydrocoll. 147: 109307.

Putro, J.N., Soetaredjo, F.E., Lunardi, V.B., Irawaty, W., Yuliana, M., Santoso, S.P. et al. (2023). Polysaccharides gums in drug delivery systems: A review. Int. J. Biol. Macromol. 253(P4): 127020.

Qi, H., Zhang, Q., Zhao, T., Chen, R., Zhang, H., Niu, X. et al. (2005). Antioxidant activity of different sulfate content derivatives of polysaccharide extracted from *Ulva pertusa* (Chlorophyta) *in vitro*. Int. J. Biol. Macromol. 37: 195–199.

Quan, Y.-S., Yamashita, H., Kamiyama, F., Yamamoto, A. and Katsumi, H. (2024). Fucoidan hair growth agent. US20220023184.

Rajapakse, N. and Kim, S.-K. (2011). Nutritional and digestive health benefits of seaweed. Adv. Food. Nutr. Res. 64: 17–28.

Ramos, A.A., Almeida, T., Lima, B. and Rocha, E. (2019). Cytotoxic activity of the seaweed compound fucosterol, alone and in combination with 5-fluorouracil, in colon cells using 2d and 3d culturing. J. Toxicol. Environ. Health. A. 82: 537–49.

Ramos-de-la-Peña, A.M., Contreras-Esquivel, J.C., Aguilar, O. and González-Valdez, J. (2022). Structural and bioactive roles of fucoidan in nanogel delivery systems. A review. Carbohydr. Polymer Technol. Appl. 4: 100235.

Rang, H.P., Ritter, J.M., Flower, R.J. and Henderson, G. (2016). Rang & Dale's Pharmacology. Elsevier. Churchill Livingstone.

Reisky, L., Pr´echoux, A., Zühlke, M.K., Baumgen, Ṁ., Robb, C.S., Gerlach, N. et al. (2019). A marine bacterial enzymatic cascade degrades the algal polysaccharide ulvan. Nat. Chem. Biol. 15: 803–812.

Rioux, L.E., Turgeon, S.L. and Beaulieu, M. (2007). Rheological characterization of polysaccharides extracted from brown seaweeds. J. Sci. Food. Agric. 87: 1630–1638.

Saidani, K., Ziani, N., Touati, N., Merzouk, H., Bejou, F. (2021). Anticoagulant activity of sulfated polysaccharides and polyphenols extracted from marine algae. Curr. Bioact. Compd. 17. 3: 246–255.

Santos, S.A.O., Vilela, C., Freire, C.S.R., Abreu, M.H., Rocha, S.M. and Silvestre, A.J.D. (2015). Chlorophyta and Rhodophyta macroalgae: A source of health-promoting phytochemicals. Food Chem. 183: 122–128.

Sathasivam, R. and Ki, J-S.A. (2018). Review of the biological activities of microalgal carotenoids and their potential use in healthcare and cosmetic industries. Mar Drugs 16: 26.

Searle, T., Al-Niaimi, F. and Ali, F.R. (2020). The top 10 cosmeceuticals for facial hyperpigmentation. Dermatol. Ther. 33: 14095.

Senthilkumar, K., Manivasagan, P., Venkatesan, J. and Kim, S.K. (2013). Brown seaweed fucoidan: Biological activity and apoptosis, growth signaling mechanism in cancer. Int. J. Biol. Macromol. 60: 366–374.

Severino, P., da Silva, C.F., Andrade, L.N., de Lima Oliveira, D., Campos, J. and Souto, E.B. (2019). Alginate nanoparticles for drug delivery and targeting. Curr. Pharm. Des. 25(11): 1312–1334.

Shah, S., Famta, P., Shahrukh, S., Jain, N., Vambhurkar, G., Srinivasarao, D.A. et al. (2023). Multifaceted applications of ulvan polysaccharides: Insights on biopharmaceutical avenues. Int. J. Biol. Macromol. 234(February): 123669.

Shanmugam, M. and Mody, k.H. (2000). Heparinoid-active sulfated polysaccharides from marine algae as potential blood anticoagulant agents. Curr. Sci. 79: 1672–1683.

Shannon, E. and Abu-Ghannam, N. (2019). Seaweeds as nutraceuticals for health and nutrition. Phycologia. 58: 563–577.

Shao, P., Chen, X. and Sun, P. (2013). *In vitro* antioxidant and antitumor activities of different sulfated polysaccharides isolated from three algae. Int. J. Biol. Macromol. 62: 155–161.

Shao, Q., Guo, Q., Xu, W.P., Li, Z. and Zhao, T.T. (2015). Specific inhibitory effect of κ-carrageenan polysaccharide on swine pandemic 2009 H1N1 influenza virus. PLoS One 10: e0126577.

Sharma, G., Alle, M., Ki Son, H. and Kim, J.-C. (2022). Dialdehyde modification of laminarin for facile synthesis of ultrafine silver nanoparticles with excellent antibacterial and wound healing properties. Int. J. Biol. Macromol. 222: 1364–1375.

Shen, S., Chen, X., Shen, Z. and Chen, H. (2021). Marine polysaccharides for wound dressings application: An overview. Pharmaceutics 13(10).

Silva, F.R.F., Dore, C.M.P.G., Marques, C.T., Nascimento, M.S., Benevides, N.M.B., Rocha, H.A.O. et al. (2010). Anticoagulant activity, paw edema and pleurisy induced 677 carrageenan: Action of major types of commercial carrageenans. Carbohydr. Polym. 79: 26–33.

Sim, S.Y., Shin, Y.E. and Kim, H.K. (2019). Fucoidan from *Undaria pinnatifida* has anti-diabetic effects by stimulation of glucose uptake and reduction of basal lipolysis in 3T3-L1 adipocytes. Nutr. Res. 65: 54–62.

Sivagnana Velmurugan, M. and Arul, V. (2023). Use of ulvan as activator of shrimp immune system against white spot syndrome virus. IN201841019841.

Škrovánková, S. (2011). Seaweed vitamins as nutraceuticals. *In*: Advances in Food and Nutrition Research 64: 357–369.

Song, L., Chen, X., Liu, X., Zhang, F., Hu, L., Yue, Y. et al. (2016). Characterization and comparison of the structural features, immune-modulatory and anti-avian influenza virus activities conferred by three algal sulfated polysaccharides. Mar. Drugs 14: 4.

Song, M., Nishihara, R., Wang, M., Chan, A.T., Qian, Z.R., Inamura, K. et al. (2016). Plasma 25-Hydroxyvitamin D and colorectal cancer risk according to tumor immunity status. Gut. 65: 296–304.

Song, S., Wen, C., Dong, X., Zhu, B., Ai, C., Li, D. et al. (2020). Application of carrageenan to resisting of novel coronavirus. CN111686125.

Song, X., Wang, K., Tang, C.Q., Yang, W.W., Zhao, W. and Changsheng Zhao, C. (2018). Design of carrageenan-based heparin-mimetic gel beads as self-anticoagulant hemoperfusion adsorbents. Biomacromolecules. Biomac 19(6): 1966–1978.

Sousa, R.P., Romero, A.G., Freire, A.C., Correia, C., Morais, A.J.S., Goncalves C.M. M. et al. (2020). Viscosupplement composition comprising ulvan for treating arthritis. CN106573013.

Souza, R.B., Frota, A.F., Silva, J., Alves, C., Neugebauer, A.Z., Pinteus, S. et al. (2018). *In vitro* activities of kappa-carrageenan isolated from red marine alga *Hypnea musciformis*: Antimicrobial, anticancer and neuroprotective potential. Int. J. Biol. Macromol. 112: 1248–1256.

Stengel, D.B., Connan, S. and Popper, Z.A. (2011). Algal chemodiversity and bioactivity: sources of natural variability and implications for commercial application. Biotechnol. Adv. 29: 483–501.

Sun, T., Zhang, X., Miao, Y., Zhou, Y., Shi, J., Yan, M. et al. (2018). Studies on antiviral and immuno-regulation activity of low molecular weight fucoidan from *Laminaria japonica*. J. Ocean Univ. 17: 705–711.

Sun, Y., Yang, B., Wu, Y., Liu, Y., Gu, X., Zhang, H. et al. (2015). Structural characterization and antioxidant activities of κ-carrageenan oligosaccharides degraded by different methods. Food Chem. 178: 311–318.

Synytsya, A., Kim, W.J., Kim, S.M., Pohl, R., Synytsya, A., Kvasniˇcka, F. et al. (2010). Structure and antitumour activity of fucoidan isolated from sporophyll of Korean brown seaweed *Undaria pinnatifida*. Carbohydr. Polym. 81: 41–48.

Synytsya, A., Choi, D.J., Pohl, R., Na, Y.S., Capek, P., Lattová, E. et al. (2015). Structural features and anti-coagulant activity of the sulfated polysaccharide SPS-CF from a green alga *Capsosiphon fulvescens*. Mar. Biotechnol. 17: 718–735.

Szekalska, M., Puciłowska, A., Szymańska, E., Ciosek, P. and Winnicka, K. (2016). Alginate: Current use and future perspectives in pharmaceutical and biomedical applications. Int. J. Polym. Sci. 2016.

Tabarsa, M., Han, J.H., Kim, C.Y. and You, S.G. (2012). Molecular characteristics and immunomodulatory activities of water-soluble sulfated polysaccharides from *Ulva pertusa*. J. Med. Food. 15: 135–144.

Talukdar, J., Bhadra, B., Dattaroy, T., Nagle, V. and Dasgupta, S. (2020). Potential of natural astaxanthin in alleviating the risk of cytokine storm in COVID-19. Biomed Pharmacotherapy 132: 1–24.

Thanh, T.T.T., Quach, T.M.T., Nguyen, T.N., Luong, D.V., Bui, M.L. and Tran, T.T.V. (2016). Structure and cytotoxic activity of ulvan extracted from green seaweed *Ulva lactuca*. Int. J. Biol. Macromol. 93: 695–702.

Torres, M.D., Flórez-Fernández, N. and Domínguez, H. (2019). Integral utilization of red seaweed for bioactive production. Mar. Drugs 17: 314.

Toskas, G., Heinemann, S., Heinemann, C., Cherif, C., Hund, R.D., Roussis, V. et al. (2012). Ulvan and ulvan/chitosan polyelectrolyte nanofibrous membranes as a potential substrate material for the cultivation of osteoblasts. Carbohydr. Polym. 89(3): 997–1002.

Tuvikene, R. (2021). Carrageenans. Handbook of Hydrocolloids. Publisher: Woodhead Publishing.

Tytgat, L., Van Damme, L., Ortega Arevalo, M., del P., Declercq, H., Thienpont, H. et al. (2019). Extrusion-based 3D printing of photo-crosslinkable gelatin and κ-carrageenan hydrogel blends for adipose tissue regeneration. Int. J. Biol. Macromol. 140: 929–938.

Tziveleka, L.A., Ioannou, E. and Roussis, V. (2019). Ulvan, a bioactive marine sulphated polysaccharide as a key constituent of hybrid biomaterials: A review. Carbohydr. Polym. 218: 355–370.

Ulbrich, K., Holá, K., Šubr, V., Bakandritsos, A., Tuček, J. and Zbořil, R. (2016). Targeted drug delivery with polymers and magnetic nanoparticles: covalent and noncovalent approaches, release control, and clinical studies. Chem. Rev. 116(9): 5338–5431.

Usoltseva, R.V., Anastyuk, S.D., Surits, V.V., Shevchenko, N.M., Thinh, P.D., Zadorozhny, P.A. et al. (2019). Comparison of structure and *in vitro* anticancer activity of native and modified fucoidans from *Sargassum feldmannii* and *S. duplicatum*. Int. J. Biol. Macromol. 124: 220–228.

Viera, I., Pérez-Gálvez, A. and Roca, M. (2018). Bioaccessibility of marine carotenoids. Mar. Drugs 16(10): 397.

Vijayakumar, S., Saravanakumar, K., Malaikozhundan, B., Divya, M., Vaseeharan, B. et al. (2020). Biopolymer K-carrageenan wrapped ZnO nanoparticles as drug delivery vehicles for anti MRSA therapy. Int. J. Biol. Macromol. 144: 9–18.

Wahlstrom, N., Steinhagen, S., Toth, G., Pavia, H. and Edlund, U. (2020). Ulvan dialdehyde gelatin hydrogels for removal of heavy metals and methylene blue from aqueous solution. Carbohydr. Polym. 249: 116841.

Wang, L., Oh, J.Y., Jayawardena, T.U., Jeon, Y.J. and Ryu, B.M. (2020). Anti-inflammatory and anti-melanogenesis activities of sulfated polysaccharides isolated from *Hizikia fusiforme*: short communication. Int. J. Biol. Macromol. 142: 545–550.

Wang, M., Veeraperumal, S., Zhong, S. and Cheong, K.L. (2023). Fucoidan derived functional oligosaccharides: Recent developments, preparation, and potential applications. Foods 12(4): 878.

Wang, Q.S., Gao, L.N., Zhu, X.N., Zhang, Y., Zhang, C.N. et al. (2019). Co-delivery of glycyrrhizin and doxorubicin by alginate nanogel particles attenuates the activation of macrophage and enhances the therapeutic efficacy for hepatocellular carcinoma. Theranostics 9(21): 6239–6255.

Wang, W., Zhang, P., Hao, C., Zhang, X.E., Cui, Z.Q. and Guan, H.S. (2011). *In vitro* inhibitory effect of carrageenan oligosaccharide on influenza A H1N1 virus. Antivir. Res. 92: 237–246.

Wang, X.X., Chen, Y., Wang, J.J., Liu, Z.X. and Zhao, S.G. (2014). Antitumor activity of a sulfated polysaccharide from *Enteromorpha intestinalis* targeted against hepatoma through mitochondrial pathway. Tumor Biol. 35: 1641–1647.

Wen, Y., Gao, L., Zhou, H., Ai, C., Huang, X., Wang, M. et al. (2021). Opportunities and challenges of algal fucoidan for diabetes management. Trends. Food Sci. Technol. 111: 628–641.

Włodarczyk, J., Czerwinski, B. and Fichna, J. (2022). Short-chain fatty acids–microbiota crosstalk in the coronavirus disease (COVID-19). Pharmacol. Rep. 74: 1198–1207.

Wu, T., Yu, S., Lin, D., Wu, Z., Xu, J., Zhang, J. et al. (2020). Preparation, characterization, and release behavior of doxorubicin hydrochloride from dual cross-linked chitosan/alginate hydrogel beads. ACS Appl. Bio Mater. 3(5): 3057–3065.

Yan, M.D., Yao, C.J., Chow, J.M., Chang, C.L., Hwang, P.A., Chuang, S.E. et al. (2015). Fucoidan elevates microRNA-29b to regulate DNMT3B-MTSS1 axis and inhibit EMT in human hepatocellular carcinoma cells. Mar. Drugs 13: 6099–6.

Yamataga, K. (2021). Prevention of cardiovascular disease through modulation of endothelial cell function by dietary seaweed intake. Phytomedicine Plus 2: 100026.

Yingying, T., Jingfeng, W., Changhu, X., Yaoguang, C., Hao, Y. and Yanqi, L. (2024). Application of L-carrageenan tetrasaccharide in relieving insulin resistance and preparing related product. NL2032188.

Younes, M., Aggett, P., Aguilar, F., Crebelli, R., Filipič, M., Frutos, M.J. et al. (2018). Re-evaluation of carrageenan (E 407) and processed *Eucheuma* seaweed (E 407a) as food additives. EFSA. J. 16.

Yuan, H., Song, J., Zhangm, W., Li, X., Li, N. and Gao, X. (2006). Antioxidant activity and cytoprotective effect of κ-carrageenan oligosaccharides and their different derivatives. Bioorg. Med. Chem. Lett. 16: 1329–1334.

Yun, J., Mullarky, E., Lu, C., Bosch, K.N., Kavalier, A., Rivera, K. et al. (2015). Vitamin C selectively kills kras and braf mutant colorectal cancer cells by targeting gapdh. Science 350: 1391–6.

Zafar, J., Aqeel, A., Shah, F.I., Ehsan, N. and Gohar, U.F., Moga, M.A. et al. (2021). Biochemical and immunological implications of lutein and zeaxanthin. Int. J. Mol Sci. 22: 10910.

Zaharudin, N., Staerk, D. and Dragsted, L.O. (2019). Inhibition of alpha-glucosidase activity by selected edible seaweeds and fucoxanthin. Food Chem. 270: 48–486.

Zayed, A. and Ulber, R. (2019). Fucoidan production: Approval key challenges and opportunities. Carbohydr. Polym. 211: 289–297.

Zhang, G., Huang, Y., Hou, H. and Bi, J. (2023). Carrageenan antibacterial film containing benzyl isothiocyanate-beta-cyclodextrin inclusion compound as well as preparation method and application of carrageenan antibacterial film. CN115926217.

Zhang, R., Zhang, X., Tang, Y. and Mao, J. (2020). Composition, isolation, purification and biological activities of *Sargassum fusiforme* polysaccharides: A review. Carbohydr. Polym. 228: 115381.

Zhao, C., Wu, D., Lin, G. and Liu, D. (2019). Ulvan with anti-tumor activity and production method of ulvan with anti-tumor activity. CN110627918.

Zhao, S., Bi, T., Han, K., Sun, L., Liu, H., Tao, D. et al. (2023). Application of lambda-carrageenan in resisting african swine fever virus infection. CN116808067.

Zou, Q., Hou, F., Wang, H., Liao, Y., Wang, Q. and Yang, Y. (2019). Microfluidic one-step preparation of alginate microspheres encapsulated with *in situ*-formed bismuth sulfide nanoparticles and their photothermal effect. Eur. Polym. J. 115: 282–289.

Zuo, T., Li, X., Chang, Y., Duan, G., Yu, L., Zheng, R. et al. (2015). Dietary fucoidan of *Acaudina molpadioides* and its enzymatically degraded fragments could prevent intestinal mucositis induced by chemotherapy in mice. Food. Funct. 6: 415–422.

Chapter 2

Fish Collagen Peptides

Preparation, Characteristics and Potential Applications

Hari Eko Irianto[1,2] and *Giyatmi Gyatmi*[2,*]

Introduction

Collagen peptides can be manufactured through the breakdown of collagen or gelatin using chemical or biochemical processes (Chen et al. 2020). Therefore, collagen peptides are sometimes referred to as hydrolyzed collagen or hydrolyzed gelatin. The molecular weight of gelatin has been reduced from more than 50,000 Da to less than 5,000 Da for collagen peptides, making absorption and digestion easier in the human body (Geltech 2019).

Collagen peptides have unique properties and functions. These functions can determine their uses (Chen et al. 2020, Biocatalysts 2023). Collagen peptide or collagen hydrolysate is made up of 20 amino acids. Glycine, proline, and hydroxyproline account for more than 50% of total amino acids, which is quite unusual and considerably greater as compared to its content in other proteins (Geltech 2019). Fish has emerged as a promising source of raw materials for collagen peptide production.

The recent surge in scientific interest surrounding fish collagen peptides can be attributed to their potential to utilize an underutilized resource—fish waste generated by the fishing and aquaculture industries. This sustainable approach minimizes waste and maximizes resource utilization (Rajabimashhadi et al. 2023). Additionally, fish collagen offers distinct advantages over traditional sources like bovine or

[1] Research Center for Marine and Land Bioindustry, National Research and Innovation Agency, Jakarta, Indonesia.
[2] Faculty of Food Technology and Health, Sahid University, Jakarta, Indonesia.
* Corresponding author: giyatmi@hotmail.com

porcine collagen. It avoids ethical concerns related to animal husbandry practices and religious restrictions on consuming bovine or porcine products for Hindus and Muslims, respectively (Gaikwad and Kim 2024). Furthermore, fish collagen eliminates the risk of diseases like Creutzfeld-Jacob disease (CJD) and mad cow disease (bovine spongiform encephalopathy - BSE) (Gauza-Włodarczyk et al. 2017).

The fish collagen peptides market size was USD 750 million in 2022 and is predicted from 2023 to 2032 to achieve a 4.5% CAGR. The growing use of fish collagen peptides in the pharmaceutical sector will contribute to its worldwide industrial growth. The market value of fish collagen peptides from the skin source segment is forecasted to exceed USD 875 million by 2032. This form of collagen may be derived from a number of fish byproducts, including skins, which are consumed regularly in many regions of the world but are discarded in vast quantities in others (Global Market Insight 2022). The current underutilization of fish waste coincides with a growing market for collagen peptides, driven by rising awareness of their diverse health benefits. This convergence suggests significant potential for developing collagen peptide production from fish waste.

Aquatic source of collagen peptides

Aquatic animal production is predicted to increase by 202 million tons by 2030 (FAO 2022). Thus, it is estimated that the availability of fish and waste that can be utilized a s raw material for making collagen peptides will increase. These raw materials can come from vertebrate and invertebrate fish.

Vertebrate fish

Fish processing discards, including skin, scales, and bones, represent a valuable resource for collagen peptide production. These waste materials can be processed individually or as a mixture.

Fish skins from emperor fish (*Lethrinuslentjan*) (Firdayanti et al. 2023), silver carp (*Hypophthalmichthys molitrix*) (Huang et al. 2023), milkfish (*Chanoschanos*) (Baehaki et al. 2016), sturgeon fish (*Acipenser baerii*) (Hou and Chen 2023a), Atlantic cod (*Gadus morhua*) (Coscueta et al. 2021), *Gadiformes*, *Pleuronectidae* (Yamada et al. 2013), nile tilapia (*Oreochromis niloticus*) (Hu et al. 2017), mackerel scad (*Decapterusmacarellus*) (Herawati et al. 2023), unicorn leatherjacket (*Aluterus monoceros*) (Kumar et al. 2019b), and chum salmon (*Oncorhynchus keta*) (Zhang et al. 2011) have been utilized to produce collagen peptides. Fish skin is a rich source of collagen and gelatin. These can be hydrolyzed to produce bioactive peptides, which are short amino acid chains ranging from 2 to 20 units in length (Abuine et al. 2019).

The pre-treatment for tilapia skins is to remove the scales before cutting the fish into small pieces (Hu et al. 2017). The silver carp skin is cleaned, dried, and chopped into 1 × 1 cm pieces (Huang et al. 2023). The remaining flesh on the skin is removed manually (Firdayanti et al. 2023). For fish skin that is preserved in salt, pretreatment can be accomplished by washing it with tap water three times to remove remaining fat and fish flesh (Coscueta et al. 2021).

Fish scales contain type I collagen and calcium hydroxyapatite ($Ca_5(PO_4)_3OH$). Type I collagen fish scales are known to be covered by 16–59% calcium salts with a crystal thickness of about 1.5–4 nm. The collagen layer on fish scales is rigid and has low flexibility due to mineralization (Almada 2018). Pre-treatments are applied to the fish scales by cleaning, drying for two hours at 100°C, and then grinding it into powder (Huang et al. 2015).

Yellowfin tuna bones processed into collagen peptides are freed from the remaining flesh, then rinsed several times with cold water, cut into small pieces, and crushed using a hammer (Natsir et al. 2021).

Invertebrate fish

Invertebrates that have been explored for the production of collagen peptides are sea cucumbers and jellyfish. The major edible portion of sea cucumber is the body wall, which is made up of epithelial tissue and dermal connective tissue. The extracellular matrix connecting the dermal connective tissue of sea cucumbers consists mostly of collagen, proteoglycans, and other glycoproteins, of which more than 70% of the total protein based on dry weight is contributed by collagen (Wang and Wang 2022). The use of sea cucumber (*Acaudinamolpadioides*) for making collagen peptides requires pretreatment by immersing the body wall in 10 times the volume of 0.2 M EDTA (pH 8.0) for forty-eight hours to eliminate heavy metals (Jin et al. 2019, Zhao et al. 2022).

Preparation of fish collagen peptides

Basically, the production of collagen peptides involves two stages, i.e., collagen extraction and collagen hydrolysis into collagen peptides. Collagen peptides are produced by chemically or biochemically hydrolyzing collagen or gelatin. This process, also known as gelatin hydrolysis, breaks down the larger, complex collagen or gelatin molecules into smaller, more bioavailable peptides. Proteolytic enzymes are commonly used to achieve this breakdown (Chen et al. 2020). The degree of hydrolysis achieved by the proteolytic enzymes will impact the size and hence, the functionality of the peptides generated (Biocatalysts 2023).

Extraction of collagen

There are two methods that can be used to extract collagen from aquatic organisms, particularly from the waste products of the fish processing industry, i.e., chemical and enzymatic extraction.

Chemical method

Chemical techniques have traditionally been used to extract collagen from fish skin, bones, and scales. Those collagen sources that have been reduced in size are washed thoroughly, then the non-collagenous proteins are removed by soaking them in a NaOH solution (Kumar et al. 2021). Removal of non-collagenous protein from milkfish skin is extracted by combining it with 0.1 mol/L NaOH in a solid solution (fish skin) and alkaline solution (NaOH) in a 1:10 (w/v) ratio, followed by six hours of continuous stirring with an overhead stirrer (Baehaki et al. 2016). Yellowfin tuna bones are soaked in 0.1 N NaOH for six hours with a 1:10 (w/v) sample-solvent ratio, with the solvent refreshed every two hours to eliminate non-collagenous protein (Natsir et al. 2021).

The previously processed fish skin is soaked in 1.5% acetic acid for twenty-four hours to extract collagen, with a 1:2 (w/v) solid-solvent ratio. After rinsing with cold water to a neutral pH, the skin is extracted for three hours at 50°C using distilled water with a 2:1 (w/v) solid-solvent ratio and collagen is obtained with a 0.54 mg/ml protein concentration (Baehaki et al. 2016). Non-collagenous protein-free yellowfin tuna bones are neutralized with cold water. The bone samples are steeped in 1.5% acetic acid for twenty-four hours with a 1:2 (w/v) sample-solvent ratio and then neutralized with water. The final step is that the samples are extracted with water for three hours at 45°C with a 2:1 (w/v) sample-solvent ratio (Natsir et al. 2021). Extraction improvement is conducted to get a higher yield of collagen by employing sonication after acetic acid extraction (Hou and Chen 2023a).

Enzymatic method

The enzymatic method is introduced for the extraction of sea cucumber (Acaudinamolpadioides) collagen by mixing heavy metal-free sea cucumbers with 0.1 M NaOH ten times to remove non-collagen proteins. After being centrifuged and rinsed with water until the pH is neutral, the sediment is then treated with 15% (v/v) butyl alcohol ten times and soaked for forty-eight hours. The sediment is rinsed with deionized water before being soaked for forty-eight hours in ten times the volume of 0.5 M acetic acid. A final concentration of 2.5 M, NaCl is added to the filtrate collected using two layers of cotton fabric. After centrifugation, the precipitate is dissolved in 0.5 M acetic acid solution. The solution is dialyzed with deionized water for twenty-four hours, and the dialysate is lyophilized to generate acid-soluble collagen (Jin et al. 2019).

Hydrolyzation of collagen

One technique for producing collagen peptides is the hydrolysis or degradation of collagen. This process can begin with triple-helix collagen denaturation, which fractures the triple-helix structure into three open chains, making peptide linkages prone to breakage. Furthermore, a hydrolysis procedure is used to break peptide bonds, resulting in collagen peptides made of 2–20 amino acids (Abuine et al. 2019, Zhang et al. 2013). Collagen hydrolysis techniques include acid hydrolysis (usually dilute H_2SO_4, HCl, or H_3PO_4), alkaline hydrolysis (KOH, NaOH, or $Ba(OH)_2$), enzymatic hydrolysis, or microbial degradation enzymatic hydrolysis (Koliada and Plavan 2015). *In silico* hydrolysis of collagen produces thirteen active peptides which are non-allergenic, non-toxic, as well as salty and bitter, but some have undetectable taste. The molecular interactions between α-amylase, α-glucosidase, dipeptidyl peptidase IV, and Glucose-6-phosphate dehydrogenase with active peptides involve van der Waals forces, hydrogen bonds, hydrophobic, pi-sulfur, electrostatic and unfavorable interactions. The active peptide competitively inhibits all four enzymes (Berliana 2021).

One method of collagen hydrolysis that is currently being developed is enzymatic hydrolysis (Abuine et al. 2019, Hong et al. 2019). Various enzymes are explored for this purpose. The use of 1% collagenase is for the hydrolysis of the unicorn leather jacket collagen and the enzyme is then inactivated with 4 mL of 6 N HCl (Kumar et al. 2019b). Collagenase is also used to hydrolyze mackerel skin to produce collagen peptides (Herawati et al. 2023).

Alkaline protease is employed to hydrolyze silver carp skin collagen and inactivate it by boiling (Huang et al. 2023). Alkaline protease is used to extract collagen peptides from sea cucumbers and their activity is deactivated by ten minutes of boiling in water (Li et al. 2019). Alcalase is another type of enzyme used to hydrolyze tilapia fish scale collagen and is heated at 90°C to deactivate the enzyme (Izzati et al. 2017).

Collagenase from *Bacillus* sp. 6-2 is investigated to hydrolyze yellowfin tuna bone collagen which is carried out with a substrate concentration of 10%, enzyme concentration of 10%, hydrolysis time of four hours and boiling to stop enzyme activity (Natsir et al. 2021). Another use of bacterial enzyme for producing peptide hydrolysate from skin collagen of snakehead murrel and milkfish is the use of collagenase which is produced using *Bacillus licheniformis* F11.4 (Sinaga et al. 2020, Baehaki et al. 2016). A series of microbial enzymes, with proteolytic activity to break down extracted collagen into lower molecular weight peptides while maintaining neutral, non-bitter-tasting peptides have been developed (Biocatalysts 2023).

Plant-based cysteine proteases have also been studied to hydrolyze collagen. *Carica papaya* protease is used as a lypaine-dominated cysteine protease. The ratio of protease to cysteine mass to collagen mass is adjusted to the desired hydrolysis conditions (Bonnet et al. 2016). Enzymes from papaya sap are another vegetable enzyme that are used to make collagen hydrolysate from snakehead fish scales by incubating at 50°C for three hours and inactivating it at 90–95°C for ten minutes (Almada 2018). A mixture of papain and flavourzyme is used to hydrolyze dried fish scale powder at 50°C for two hours to produce collagen peptides (Huang et al. 2015). Pepsin is used in a two-stage hydrolysis followed by flavourzyme to synthesize high yield pepsin soluble collagen peptides from 500 Da molecular weight sturgeon skin collagen, resulting in low molecular weight pepsin soluble collagen peptides (Hou and Chen 2023a). Protease complex (3000 U/g protein) which includes 65% papain, 28% alkaline proteinase, and 7% trypsin is used to hydrolyze chum salmon (*O. keta*) skin collagen at a temperature of 40°C and a pH value of 8 for three hours to obtain collagen peptides (Pei et al. 2010, Wang et al. 2015).

Apart from the enzymatic process, non-enzymatic hydrolysis is also developed to produce collagen hydrolysate from collagen through subcritical water hydrolysis with collagen to liquid ratio (w/v) of 1:200. Hydrolysis is carried out in a reactor, which is heated to 120–300°C at a pressure of 50 bar/or slightly beyond saturated steam pressure, 80 bar for 280°C, and 100 bar for 300°C (Ahmed and Chun 2018).

Deodorize

Methylional, dimethyl tetrasulfide and dimethyl trisulfide are off-odor chemicals found in marine collagen peptides and gelatin. Unknown compounds with sulfurous, cool/refreshing and medicinal odors impact the off-scents of marine collagen peptide and gelatin collagen peptide. As the production process progresses, the volatile components of marine collagen peptide and gelatin collagen peptide grow steadily. Off-odor components are not eliminated by activated carbon deodorization (Cho et al. 2015).

A deodorization technique for abyssal fishbone and skin collagen peptides involves adjusting the solution to weakly acidic pH, urease deamination, activated carbon and resin decolorization, and finally, spray drying to obtain odorless and colorless collagen peptides (Wuyin and Minjie 2012). According to a current study, *Lactobacillus plantarum* fermentation deodorized off-odors most efficiently (Cho et al. 2015).

Characteristics of collagen peptides

Nutrition is provided by collagen hydrolysates or peptides, which include amino acids and small peptides (Chen et al. 2020). Fish collagen peptide is a type I collagen-based health component derived from fish. Fish collagen peptide is distinguished by its unique amino acid makeup, which includes a high concentration of glycine and proline (Alfa-chemistry 2023). Silver carp skin collagen peptides that are rich in Gly (Glycine) and amino acids and have a stable-helical structure with anti-photoaging activity potential have been found as GPPGPPGTPGPQ, SGLPGPIGPPGPR, and GLPGPIGPPGP (Huang et al. 2023). Small peptides have shown to exhibit biological actions based on their amino acid content and sequence. The hydrophobic amino acids found in fish skin peptides contribute to their antioxidant and angiotensin-converting enzyme inhibitory action (Abuine et al. 2019).

Collagen peptides extracted from various types of fish waste have a low molecular weight. The molecule weight distributions of collagen peptides from four fish scales are < 10 kDa, with an average of 1.3 kDa (Huang et al. 2015). Collagen peptides from tilapia skin have a molecular weight of < 5 kDa, with random coils as the primary molecular shape (Hu et al. 2017). The molecular weight distribution of the silver carp skin collagen peptides is mostly 665.93 and 980.95 Da, indicating good antioxidant activity (Huang et al. 2023). The collagen peptide profile in snakehead fish scale collagen hydrolysate has a molecular weight of < 10 kDa (Almada 2018). The yields of four peptide fractions, namely > 30 kDa, 10–30 kDa, 3–10 kDa, and < 1 kDa produced from the collagen hydrolysis of common carp fish heads, skins, and skeletons employing alcalase enzyme are 21.4, 26.6, 18, and 31.7%, respectively (González-Serrano et al. 2022).

Salt-cured cod skin hydrolysate is found to be biocompatible by being non-cytotoxic to any of the cell lines examined, so it may be a viable alternative to mammalian type I collagen in many applications (Coscueta et al. 2021). The smallest size of fish collagen peptides is most successfully transported across

Caco-2 monolayers, according to studies on the intestinal permeability of fish collagen peptides utilizing Caco-2 monolayers (Shimizu et al. 2010). The pH of the peptide fractions influences their solubility. The peptide fraction with > 30 kDa has the highest emulsifying activity index, foam stability, and foam capacity, but the lowest emulsion stability. The solubility of the peptide fractions is affected by pH. The peptide fraction with > 30 kDa has the greatest foam capacity, foam stability, and emulsifying activity index, but the lowest emulsion stability (González-Serrano et al. 2022). Collagen peptides from *Chanos* have the highest Fe(II)-binding activity, followed by *Lates calcarifer* and *Mugil cephalus*, while *Oreochromis* spp. has the lowest (Hu et al. 2017).

The temperature of the hydrolysis affects the properties of the collagen peptide generated. For example, hydrolysis of unicorn leatherjacket skin with collagenase at 5°C produces collagen peptides with better anti-diabetic, anti-cancer, and wound healing performance *in vitro* than collagen peptides prepared at 25°C and 50°C (Kumar et al. 2019b).

Quality and safety

Protein decomposition through enzymatic hydrolysis releases flavor compounds from the protein that can affect the taste of the hydrolysate. So, for protein hydrolysates that are used as food ingredients, the hydrolysis process must not cause an unpleasant taste which then requires further processing or disguise in the final food product. Taste changes vary depending on the specific hydrolysis process and collagen starting material (Biocatalysts 2023). Most collagen supplements have been hydrolyzed, which means the manufacturer has broken down the collagen in the supplement into peptides that are simpler for the body to absorb and use (Fletcher 2019). The quality of the collagen peptides produced must be in accordance with the standard requirements applied by each type of product that will use it as an ingredient. For example, the use of collagen in cosmetic products in Indonesia can refer to Indonesian National Standard no. SNI 8076-2014 concerning Crude Collagen from Fish Scales – Quality and Processing Requirements.

Gelatin and collagen peptides are generally considered safe. Gelatin was granted GRAS certification by the FDA in 1975. The World Health Organization and the European Commission on Health and Consumer Protection have declared collagen peptides safe based on research findings worldwide (Czech 2016). Although collagen peptide is safe, the risk of overconsumption, toxic components, and pathogenic agents must be recognized. Collagen peptides are absorbed by the intestinal mucosa after digestion, cross the liver, and reach the peripheral circulation, where 25 different hydroxyproline-containing peptides have been found (Wang et al. 2023). Following intake of collagen peptides, oligopeptides containing hydroxyproline (Hyp) are found in peripheral blood, which are known to have various physiological functions. The effects of collagen-peptide supplementation on immunological response, however, remain unknown. Collagen-peptide consumption lowers allergic responses by changing the balance of CD4+ T cells toward Th1 and Treg cells and therefore looks to be a viable medication for allergy and inflammatory illness prevention (Nishikimi et al. 2018).

A study found that long-term toxicity of collagen peptides from Chum Salmon skin fed to Sprague-Dawley rats in their diet did not pose any substantial health hazards (Liang et al. 2012). In addition, low molecular weight collagen peptide nanoemulsions demonstrated safety when administered to diabetic mice for 15 d (Hou and Chen 2023b). Collagen supplement side effects are generally modest and might include diarrhea, stomach heaviness, and rash. Adults who use collagen supplements do not pose any substantial health risks (Fletcher 2019).

Therapeutic benefit of fish collagen peptides

Many studies have been carried out to explore the therapeutic potential of fish collagen peptides *in-vitro*, *in-vivo* using animals and clinical trials against various types of diseases or health disorders.

Antiaging and skin care

Collagen peptides demonstrate antiaging activities. Peptides containing hydroxyproline accumulate in the skin, enhancing its condition by reducing oxidative stress, regulating cell activity, decreasing inflammatory responses, and preserving extracellular matrix homeostasis (Wang et al. 2023). The type I fish collagen hydrolysates improve skin biomechanics, reduce wrinkles, provide excellent subject satisfaction, and have no associated side effects (Duteil et al. 2016). Collagen hydrolysate stimulated rat skin metabolism by decreasing matrix metalloproteinase 2 activity and boosting types I and IV collagen expression (Zague et al. 2011).

A significant reduction in periorbital wrinkles is observed after twelve weeks of collagen peptide consumption which is also demonstrated by an increase in facial skin moisture and elasticity (Koizumi et al. 2017). Collagen peptides can enhance the physiological features of the skin including wrinkles, elasticity, hydration, melanin index, and erythema index, without significant side effects (Seong et al. 2023). Clinical evidence indicates that collagen peptides with an average molecular weight of 2000–5000 Da can increase skin moisture as well as prevent and reduce dermal collagen tissue fragmentation, thereby counteracting one of the signs of skin aging that is most likely associated with increased collagen and glycosaminoglycan synthesis in the respective skin layers (Asserin et al. 2015).

A combination of fish collagen peptides, bonito elastin peptides, rose water, and fish stomach powder shows synergistic effectiveness in stimulating collagen production through fibroblast stimulation (Douglass 2022). Fish collagen peptide drink could have powerful antiaging effects by inhibiting reactive oxygen species, facilitating the synthesis of extracellular matrix proteins, increasing mitochondrial activity, and improving gene expression in terms of correct protein folding, base excision repair, and DNA mismatch repair (Lin et al. 2020).

Bone regeneration

Fish collagen peptides have the potential to be used as biomaterials for bone repair and regeneration (Yamada et al. 2013). Low concentrations of fish collagen peptides increase osteoblast proliferation, differentiation, and the expression of genes associated with collagen-modifying enzymes, as well as accelerating matrix mineralization. Collagen administration shows higher levels of lysine hydroxylation, higher levels of hydroxylysine aldehyde derivative cross-links, and quicker cross-link maturation and matrix mineralization (Yamada et al. 2013, Yamada et al. 2021). Fish collagen peptides can stimulate bone-forming cells, and better calcium absorption, as well as anti-inflammatory and antioxidant properties that make it possible for dietary interventions in osteoporosis prevention and bone tissue engineering (Kewat et al. 2023, Yamada et al. 2013).

Wound healing

Administration of marine collagen peptides can accelerate wound healing. Animal studies in rabbits (Hu et al. 2017) and mice (Zhang et al. 2011, Wang et al. 2015, Herawati et al. 2023) support this claim. Collagen peptides extracted from Nile tilapia skin are found to improve wound healing in rabbits with severe partial-thickness scald wounds and that finding shows promise in burn therapy (Hu et al. 2017).

Oral administration of fish collagen peptides significantly accelerates the wound healing process in mice treated with fish collagen peptide (dose 1.3 g/kgBW) (Herawati et al. 2023). Oral administration of marine collagen peptides increases wound contraction, collagen deposition, and angiogenesis wound healing of mice after a cesarean section (Wang et al. 2015). Ingestion of marine collagen peptides promotes angiogenesis with thicker and more ordered collagen fiber deposition resulting in quicker closure of wound and more regeneration of tissue at the wound site (Zhang et al. 2011). Collagen peptides generated from the jellyfish-*Rhopilema esculentum* can expedite wound healing (Felician et al. 2019).

Acute kidney injury

Collagen peptides from *Acaudina molpadioides* significantly reduce urea nitrogen and serum creatinine levels in mice with acute kidney injury. Additionally, the levels of superoxide dismutase and catalase, two antioxidant enzymes, significantly increase in these mice when given collagen peptides. Moreover, collagen peptides significantly upregulate heme oxygenase-1 (HO-1), Nrf2, p-PI3K, and p-AKT in the kidneys. Collagen peptides have shown to considerably reduce the levels of nuclear factor tumor necrosis factor α (TNF-α), NF-kappa-B (NF-κB), and the interleukin 1β (IL-1β), and inflammatory cytokines interleukin 6 (IL-6). Collagen peptide has shown to protect against acute kidney damage by reducing inflammation and oxidative stress via the PI3K/AKT/NF-κB and PI3K/AKT/Nrf2 pathways (Zhao et al. 2022).

Anti-diabetes

The prospect of fish collagen peptides for diabetes treatment has demonstrated that active peptides extracted from tilapia skin collagen show the potential to be candidates for type 2 diabetes treatment (Berliana 2021). The anti-diabetic benefits of marine collagen peptides are proven by changing the amounts of molecules implicated in diabetes pathogenesis. Patients receiving marine collagen peptides treatment have a significant decrease in fasting blood glucose, creatinine levels, HbA1c, diastolic blood pressure, and mean arterial pressure, but an increase in the Insulin Secretion Index and Insulin Sensitivity Index (Zhu et al. 2010).

The study in mice found that a four week therapy with marine collagen peptide decreases blood glucose levels while also lowering inflammatory exudation and endothelial thinning in carotid-artery vascular endothelial cells (Zhu et al. 2017). The anti-diabetic efficacy of fish collagen peptide is more effective as a nanoemulsion than as an extract (Hou and Chen 2023b).

Muscle recovery and improvement

Collagen peptide supplementation for 9 d can help speed up muscle healing and lessen muscular discomfort after severe physical activity. This is assumed to be due to increased collagen production in the connective tissue surrounding the muscle, as well as regulation of the inflammatory response to exercise, which can speed up the initial remodeling process. Collagen peptides have a mild effect on countermovement jump recovery and muscle soreness but do not affect inflammation or bone collagen formation (Cliford et al. 2019).

Fish collagen peptides can enhance muscular function and increase muscle mass in older people suffering from sarcopenia symptoms. A twenty-four week therapy period indicates that a 6-month or longer treatment period considerably improves sarcopenia symptoms in the elderly. Thus, in senior persons, regular supplementation with fish collagen peptides can relieve sarcopenia symptoms, increase muscle mass, and enhance upper and lower-body muscular performance (Tyree et al. 2021).

Cardiovascular disease

To reduce the risk of cardiovascular disease, blood cholesterol must be lowered or kept at an acceptable level. Peptide hydrolysate from snakehead murrel skin collagen exhibits potent anti-cholesterol and anti-oxidant properties, so it can reduce cholesterol levels and prevent the formation of radicals in the blood (Sinaga et al. 2020). Serum triglycerides, low-density lipoprotein, free fatty acids, total cholesterol, nitric oxide, cytochrome P450, and prostacyclin levels are considerably reduced in patients taking marine collagen peptides, whereas high-density lipoprotein, adiponectin, and bradykinin levels are dramatically increased (Zhu et al. 2010).

Cardiovascular disease can also be a complication of type 2 diabetes mellitus. Studies show that marine collagen peptides protect against cardiovascular endothelial cell injury and reduce cardiovascular dysfunction in type 2 diabetes mellitus mouse models by lowering systemic and local coupling factor 6 and microparticle levels, decreasing cell apoptosis, and increasing the expression and activity of peroxisome proliferator-activated receptor. These findings provide credence to the idea that a modest oral marine collagen peptide dosage (4.5 g/kg body weight/day) might be a viable therapy option to avoid early cardiovascular issues associated with type 2 diabetes (Zhu et al. 2017).

Memory deficit

By lowering oxidative damage in the brain as well as raising postsynaptic density protein 95 expression and brain-derived neurotrophic factor levels, marine collagen peptides can improve learning and memory in aged mice. The increase in life expectancy in the twenty-first century has been accompanied by an increase in the prevalence of age-related disorders, i.e., depression, Alzheimer's disease, and other dementias. Therefore, the neuroprotective effects of marine collagen peptides are useful in the development of functional foods and drugs to improve memory in the treatment of memory loss caused by age-related neurodegenerative diseases, including cognitive deficiencies (Pei et al. 2010).

Anticancer

The anticancer activity of fish collagen peptides has been explored. The cytotoxicity effects exhibited by collagen peptides obtained from the skin of unicorn leatherjacket range between 17–50% on COLO320 cells (Kumar et al. 2019b). At 1 mg/mL concentration, collagen peptides derived from milkfish (*Chanoschanos*) inhibited HeLa and HCT-166 cells by 70% and 18%, respectively (Baehaki et al. 2016). *In vitro* testing of the antiproliferative effects of fish protein hydrolysates showed that hydrolysates from three blue whiting, three cod, three plaice, and one salmon had substantial growth inhibition against two human breast cancer cell lines (Picot et al. 2006).

Potential applications

By considering the various therapeutic benefits of fish collagen peptides, it is appropriate to develop products related to maintaining human health and healing various diseases. This fact can enhance the physiological functions of organisms and make collagen peptides useful as ingredients in food, pharmaceuticals, or cosmetics as well as other useful functions including antifreeze, metal chelation and forming edible film (Xu et al. 2023, Chen et al. 2020). Therefore, there are opportunities for the development of various types of products using fish collagen peptides, including nutraceuticals, food supplements, cosmetics, and pharmaceutical products.

Nutraceuticals and food supplements

Peptides obtained through protein hydrolysis hold promise as nutraceuticals and functional food ingredients due to their potential health benefits (Zague 2008). Nutraceuticals include products derived from natural sources or enriched with bioactive compounds that boost wellness benefits and provide additional basic nutrition. They are grouped under several heads, including snacks, digestive supplements, and herbal products (Rehan and Asghar 2023). Supplementation of marine collagen peptides in food and beverage products is an effective way for consumers to be benefitted by them. For example, biscuits, which are a type of food that is popular with consumers of all ages, have been supplemented with collagen peptides which have succeeded in increasing the protein content and antioxidant potential. The protein content of biscuits added with 10% collagen peptides increased by 78% (from 7.09 to 12.61%) (Kumar et al. 2019a).

A collagen elastin peptide complex juice drink has been developed commercially with ingredients consisting of fish collagen peptide powder (12%), and fruit juices as well as fruit powders, seed powders, where the drink is intended to improve aging skin (Lin et al. 2020). After consuming nutraceutical drinks made from collagen peptides every day, there is an improvement not only in skin quality, especially the appearance and texture of the skin, but also in skin healing (Anderson 2020).

Food supplements of fish collagen peptides can be developed in the form of powder, pills, capsules, tablets, or liquids.

Cosmetics

The findings of the studies significantly justify the usage of collagen peptides in cosmetic manufacture. Because of its excellent penetration into the skin and ease of distribution in the human body, collagen with a low molecular weight is frequently employed in the cosmetic industry (Izzati et al. 2017). Collagen hydrolysate is identified as a safe cosmetic component for topical formulations with good moisturizing properties in the stratum corneum layer of the skin (Aguirre-Cruz et al. 2020). These findings imply that topical collagen tripeptide administration may ameliorate clinical aging phenotypes by inhibiting glycation and oxidative stress, resulting in a delay in cellular aging (Lee et al. 2022). Within eight weeks, fish collagen peptides have demonstrated enhanced skin hydration by 12%. In twelve weeks, increased skin collagen density has shown enhanced dermal echogenicity (Asserin et al. 2015). Collagen peptides in the 0.6–12 kDa range can be absorbed on keratinized structures such as skin and hair via topical administration, forming a protective layer that prevents damage and moisture loss, whereas larger molecules bind more tightly to these structures, with their solubility in water decreasing as size increases. Collagen peptides penetrate fibrous hair cells in hair (Schrieber and Gareis 2007).

Pharmaceutical products

Many different types of pharmaceuticals may be made from fish collagen peptides, depending on their possible therapeutic benefits, such as wound healing, diabetes treatment, and bone regeneration (Yamada et al. 2021). Clinical studies have also been conducted to reveal some potential therapeutic effects on humans. For example, the consumption of collagen hydrolysate containing high amounts of Pro-Hyp and Hyp-Gly has been investigated and found to aid the healing of decubitus ulcers as an adjunct to normal therapy. These findings suggest that the Pro-Hyp and Hyp-Gly absorbed into the blood after collagen hydrolysate consumption act on fibroblasts in the dermal layers of pressure ulcers and may also affect stem cells, inducing re-epithelialization and improved healing (Sugihara et al. 2018). Trials of administering marine collagen peptides have been conducted in Chinese patients with Type 2 Diabetes Mellitus and hypertension to improve glucose and lipid metabolism, renal function, insulin sensitivity, and hypertension control (Zhu et al. 2010, Wong et al. 2024). According to a clinical investigation, fish-derived collagen peptides have a lot of potential as a natural supplement for healthy skin with anti-aging skin characteristics (Koizumi et al. 2017, Seong et al. 2023).

Conclusion

Fish resources from both capture fisheries and aquaculture in the world are quite large, where waste produced from processing industrial activities is a promising source of raw material for manufacturing collagen peptides. Collagen peptide production technology has been developed using various extraction method approaches and some of them have even been applied for commercial production. The potential therapeutic benefits of collagen peptides have also been revealed from studies that have been carried out, including anti-aging and skin care, bone regeneration, wound healing, acute kidney injury, anti-inflammatory. antidiabetic, muscle recovery. and repair, cardiovascular disease, memory deficits, and anticancer.

Collagen peptides have undergone clinical trials for several therapeutic benefits so that they are ready to be applied as ingredients for various food supplements, nutraceuticals, cosmetics and pharmaceutical products. Several commercial products for food supplements, nutraceuticals, and cosmetics have been found on display in the market, especially products for anti-aging and skin care that appear to be the most widely produced by industries. Even though the study results state that collagen peptides are safe for use by humans and have even received GRAS certification from the FDA in 1975, investigation may still need to be carried out to determine the optimum dose for various purposes of application and observe the effects of routine long-term use.

Looking ahead, the future of collagen peptides appears bright. We can expect advancements in extraction technologies that improve efficiency and sustainability. Research will likely continue to uncover new applications and potential health benefits of these peptides. As consumer awareness of these benefits grows, the market for collagen peptide-based products is likely to expand significantly. The development of personalized dosage recommendations and long-term safety studies will further solidify the role of collagen peptides as a valuable tool for promoting health and well-being.

References

Abuine, R., Rathnayake, A,U. and Byun, H.G. (2019). Biological activity of peptides purified from fish skin hydrolysates. Fisheries and Aquatic Sciences 22: 10.

Aguirre-Cruz, G., León-López, A., Cruz-Gómez, V., Jiménez-Alvarado, R. and Aguirre-Álvarez, G. (2020). Collagen hydrolysates for skin protection: oral administration and topical formulation. Antioxidants 9: 181.

Ahmed, R. and Chun, B.S. (2018). Subcritical water hydrolysis for the production of bioactive peptides from tuna skin collagen. The Journal of Supercritical Fluids 41(November): 88–96.

Alfa-chemistry. (2023). Fish Collagen Peptide. https://www.alfa-chemistry.com/products/fish-collagen-peptide.html#DownloadTechnicalDataShee. Accessed on 25 November 2023.

Almada, W. (2018). ProfilPeptidaKolagendalamHidrolisatKolagenSisik Ikan Gabus (*Channa striata*) Hasil EkstraksiMenggunakanEnzim Papain. Sarjana Thesis, Universitas Brawijaya, Malang.

Anderson, K.L. (2020). Collagen peptide nutraceutical drink on the skin. J. Drugs Dermatol. 19(1Suppl1): s5–10.

Asserin, J., Lati, E., Shioya, T. and Prawitt, J. (2015). The effect of oral collagen peptide supplementation on skin moisture and the dermal collagen network: evidence from an *ex vivo* model and randomized, placebo-controlled clinical trials. Journal of Cosmetic Dermatology 14: 291–301.

Baehaki, B., Suhartono, M.T., Sukarno, Syah, D. and Setyahadi, S. (2016). Collagen peptides from fish skin with Angiotensin I-Converting Enzyme (ACE) inhibitor and cancer antiproliferative activity. Research Journal of Pharmaceutical, Biological and Chemical Sciences 7(1): 1996–2000.

Berliana, J.D. (2021). Kajian PotensiPeptidaAktifdariKolagenKulit Ikan Nila (*Oreochromis niloticus*) SebagaiKandidatAntidiabetes Tipe-2 BerdasarkanStudi *Molecular Docking*. Sarjana Thesis, Universitas Pendidikan Indonesia, Bandung.

Biocatalysts. (2023). Enzymatic Hydrolysis for the Production of Collagen Peptides. https://biocatalysts.com/media-resources/enzyme-hydrolysis-collagen-peptides. Accessed on 23 December 2023.

Bonnet, C.B., Auffret, Y., Robert, P.J. and Corbille, P. (2016). Fish Skin Collagen Peptide Composition and Its Use as a Medicinal Product. France Patent#FR3058142A1.

Chen, M., Li, Y. and Huang, G. (2020). Potential health functions of collagen bioactive peptides: a review. American Journal of Biochemistry and Biotechnology 16(4): 507.519.

Cho, H.N., Cho, D.W., Hurk, B.S., Bae, H.A. Kim, D.E. and Baek, H.H. (2015). Characterization of off-odor compounds of collagen peptides from tilapia scale using GC-MS-olfactometry. Food Sci. Biotechnol. 24: 403–410.

Cliford, T., Ventress, M., Allerton, D.M., Stansfeld, S., Tang, J.C.Y., Fraser, W.D. et al. (2019). The efects of collagen peptides on muscle damage, inflammation and bone turnover following exercise: a randomized, controlled trial. Amino Acids 51: 691–704.

Coscueta, E.R., Brassesco, M.E. and Pintado, M. (2021). Collagen-based bioactive bromelain hydrolysate from salt-cured cod skin. Appl. Sci. 11: 8538.

Czech, I. (2016). Collagen Peptides – Source, Properties and Benefits. Seagarden AS, www.seagarden.no, February 2016.

Douglass, S.J. (2022). The synergistic effects of fish collagen peptide, bonito elastin peptide, fish maw powder and rose water on the improvement of skin conditions: a mechanistic study. Am. J. Biomed. Sci. & Res. 17(1): 67–71.

Duteil, L., Queille-Roussel, C., Maubert, Y., Esdaile, J., Bruno-Bonnet, C. and Lacour, J.P. (2016). Specific natural bioactive type 1 collagen peptides oral intake revenue skin aging signs in mature woman. J. Aging Res. Clin. Practice 5(2): 84–92.

FAO. (2022). In Brief to The State of World Fisheries and Aquaculture 2022. Towards Blue Transformation. FAO, Rome.

Felician, F.F., Yu, R.H., Li, M.Z., Li, C.J., Chen, H.Q., Jiang, Y. et al. (2019). The wound healing potential of collagen peptides derived from the jellyfish *Rhopilema esculentum*. Chinese Journal of Traumatology 22: 12e20.

Firdayanti, W., Trilaksani, W. and Purwaningsih, S. (2023). Valorization of emperor fish (*Lethrinuslentjan*) skin for collagen isolation: effect of acetic acid ratio and extraction time. IOP Conf. Series: Earth and Environmental Sciencc 1137: 012047.

Fletcher, J. (2019). What to know about collagen supplements. Medical News Today - June 1, 2019.

Gaikwad, S. and Kim, M.J. (2024). Fish by-product collagen extraction using different methods and their application. Mar. Drugs 22: 60.

Gauza-Włodarczyk, M., Kubisz, L., Mielcarek, S. and Włodarczyk, D. (2017). Comparison of thermal properties of fish collagen and bovine collagen in the temperature range 298–670 K. Materials Science and Engineering C 80: 468–471.

Geltech. (2019). Collagen – Collagen Peptides For Healthy Skin, Bone, Joint and Hair. https://www.geltech.co.kr/img/pdf/Geltech+Collagen+leaflet+2019+revised.pdf. Accessed on 23 December 2023.

Global Market Insight. (2022). Fish Collagen Peptides Market Size - Forecasts Report 2023–2032. https://www.gminsights.com/industry-analysis/fish-collagen-peptides-market. Accessed on 25 November 2023.

González-Serrano, D.J., Hadidi, M., Varcheh, M., Jelyani, A.Z., Moreno, A. and Lorenzo, J.M. (2022). Bioactive peptide fractions from collagen hydrolysate of common carp fish by-product: antioxidant and functional properties. Antioxidants 11: 509.

Herawati, E., Setyawan, V.A. and Listyawati, S. (2023). PeptidaKolagen Ikan LayangBiru (*Decapterusmacarellus*) MempercepatPenyembuhan Luka Pada Mencit. Journal of Pharmaceutical Science and Clinical Research 02: 278–290.

Hong, H., Fan, H., Chalamaiah, M. and Wu, J. (2019). Preparation of low-molecular-weight, collagen hydrolysates (peptides): Current progress, challenges, and future perspectives. Food Chemistry 301: 125.

Hou, N.T. and Chen, B.H. (2023a). Extraction, purification and characterization of collagen peptide prepared from skin hydrolysate of sturgeon fish. Food Quality and Safety 7: 1–15.

Hou, N.T. and Chen, B.H. (2023b). Preparation of nanoemulsions with low-molecular-weight collagen peptides from sturgeon fish skin and evaluation of anti-diabetic and wound-healing effects in mice. Pharmaceutics 15: 2304.

Hu, Z., Yang, P., Zhou, C., Li, S. and Hong, P. (2017). Marine collagen peptides from the skin of Nile Tilapia (*Oreochromis niloticus*): Characterization and wound healing evaluation. Mar. Drugs 15: 102.

Huang, C.Y., Wu, C.H., Yang, J.L., Li, Y.H. and Kuo, J.M. (2015). Evaluation of iron-binding activity of collagen peptides prepared from the scales of four cultivated fishes in Taiwan. Journal of Food and Drug Analysis 23(4): 671–678.

Huang, J.J., Hai-lan Li, Xiong, G.Q., Cai, J., Liao, T. and Zu, X.Y. (2023). Extraction, identification and anti-photoaging activity evaluation of collagen peptides from silver carp (Hypophthalmichthys molitrix) skin. LWT – Food Science and Technology 173: 114384.

Izzati, H., Zainol and Hanim, H. (2017). Low molecular weight collagen from tilapia fish scales for potential cosmetic application. Der Pharma Chemica 9(7): 108–114.

Jin, H.X., Xu, H.P., Li, Y., Zhang, Q.W. and Xie, H. (2019). Preparation and evaluation of peptides with potential antioxidant activity by microwave assisted enzymatic hydrolysis of collagen from sea cucumber *Acaudinamolpadioides* obtained from Zhejiang province in China. Mar. Drugs 17: 169.

Kewat, A., Singh, A.P., Mahajan, S.K., Sharma, M., Dubey, S., Mishra, P. et al. (2023). Review on osteoporosis and its treatment using fish collagen peptides. The Pharma Innovation Journal 12(3): 1318–1322.

Koizumi, S., Inoue, N., Shimizu, M., Kwon, C.J., Kim, H.Y. and Park, K.S. (2017). Effects of dietary supplementation with fish scales-derived collagen peptides on skin parameters and condition: a randomized, placebo-controlled, double-blind study. Int. J. Pept. Res. Ther. 24: 397–402.

Koliada, M. and Plavan, V. (2015). Problems of efficient processing and use of collagen-containing materials. Pure Appl. Chem. 87(1): 43–49.

Kumar, A., Elavarasan, K., Hanjabam, M.D., Binsi, P.K., Mohan, C.O., Zynudheen, A.A. et al. (2019a). Marine collagen peptide as a fortificant for biscuit: Effects on biscuit attributes. LWT - Food Science and Technology 109: 450–4.

Kumar, L.V., Shakila, R.J. and Jeyasekaran, G. (2019b). *In vitro* anti-cancer, anti-diabetic, anti-inflammation and wound healing properties of collagen peptides derived from unicorn leatherjacket (*Aluterus Monoceros*) at different hydrolysis. Turk. J. Fish. & Aquat. Sci. 19(7): 551–560.

Lee, Y.I., Lee, S.G., Jung, I., Suk, J., Lee, M.H., Kim, D.U. et al. (2022). Effect of a topical collagen tripeptide on antiaging and inhibition of glycation of the skin: a pilot study. Int. J. Mol. Sci. 23: 1101.

Li, Y., Li, J., Lin, S.J., Zui-Su, Z.S. and Jin, H.X. (2019). Preparation of antioxidant peptide by microwave assisted hydrolysis of collagen and its protective effect against H_2O_2-induced damage of RAW264.7 cells. Mar. Drugs 17: 642.

Liang, J., Pei, X.R., Zhang, Z.F., Wang, N., Wang, J.B. and Li, Y. (2012). A chronic oral toxicity study of marine collagen peptides preparation from chum salmon (*Oncorhynchus keta*) skin using Sprague-Dawley rat. Mar. Drugs 10: 20–34.

Lin, P., Hua, N., Hsu, Y.C., Kan, K.W., Chen, J.H., Lin, Y.H. et al. (2020). Oral collagen drink for antiaging: antioxidation, facilitation of the increase of collagen synthesis, and improvement of protein folding and DNA repair in human skin fibroblasts. Oxidative Medicine and Cellular Longevity Article ID 8031795.

Natsir, H., Dali, S., Sartika, Leliani and Arif, A.R. (2021). Enzymatic hydrolysis of collagen from yellowfin tuna bones and its potential as antibacterial agent. Rasayan J. Chem. 14(1): 594–600.

Nishikimi, A., Koyama, Y., Ishihara, S., Kobayashi, S., Tometsuka, C., Kusubata, M. et al. (2018). Collagen-derived peptides modulate CD4þ T-cell differentiation and suppress allergic responses in mice. Immunity, Inflammation and Disease 6(2): 245–255.

Pei, X., Yang, R., Zhang, Z., Gao, L,, Wang, J., Xu, Y. et al. (2010). Marine collagen peptide isolated from Chum Salmon (*Oncorhynchus keta*) skin facilitates learning and memory in aged C57BL/6J mice. Food Chemistry 118: 333–340.

Picot, L., Berdenave, S., Didelot, S., Fruiter-Arnaudin, I., Sannier, F., Thorkelsson, G. et al. (2006). Antiproliferative activity of fish protein hydrolysates on human breast cancer cell lines. Proc Biochem. 41: 1217–1222.

Rajabimashhadi, Z., Gallo, N., Salvatore, L. and Lionetto, F. (2023). Collagen derived from fish industry waste: progresses and challenges. Polymers 15: 544.

Rehan, H. and Asghar, M. (2023). Understanding the concept of "nutraceutical". Glob. J. Pharmaceu. Sci. 11(1): 555805.

Schrieber, R. and Gareis, H. (2007). Gelatine Handbook: Theory and Industrial Practice. Wiley-VCH, Weinheim, Germany.

Seong, S.H., Lee, Y.I., Lee, J., Choi, S., Kim, J.A., Suk, J. et al. (2023). Low-molecular-weight collagen peptides supplement promotes a healthy skin: A randomized, double-blinded, placebo-controlled study. J. Cosmet. Dermatol. 00: 1–9.

Shimizu, K., Sato, M., Zhang, Y., Kouguchi, T., Takahata, Y., Morimatsu, F. et al. (2010). Molecular size of collagen peptide reverses the permeability of Caco-2 cells. Biosci. Biotechnol. Biochem. 74(5): 1123–1125.

Sinaga, W.S.L., Ismaya, W.T., Retroningrum, D.S., Tjandrawinat, R.R. and Suhartono, M.T. (2020). Peptides hydrolysate derived from collagen of snakehead murrel (*Channa striata*) skin demonstrate anti-cholesterol and antioxidant activities. Hayati J. Biosci. 27(2): 136–141.

Sugihara, F., Inoue, N. and Venkateswarathirukumara, S. (2018). Ingestion of bioactive collagen hydrolysates enhanced pressure ulcer healing in a randomized double-blind placebo-controlled clinical study. Scientific Reports 8: 11403.

Tyree, S.M., Galvin, G.D., Bonnet, C.B. and Dinan, T.G. (2021). A randomized, double-blind, placebocontrolled study trial to evaluate the potential effects of Naticol®, fish collagen peptides on symptoms of sarcopenia in the elderly. Juniper Online Journal of Case Studies 12(5): ID.555850.

Wang, J., Xu, M., Liang, R., Zhao, M., Zhang, Z. and Li, Y. (2015). Oral administration of marine collagen peptides prepared from chum salmon (*Oncorhynchus keta*) improves wound healing following cesarean section in rats. Food & Nutrition Research 59: 1, 26411.

Wang, Y. and Wang, J. (2022). The extraction, separation technology, and new product development of collagen peptides from sea cucumber. pp 293–313. *In*: Xue, C. (ed.). Advances in Sea Cucumber Processing Technology and Product Development. Springer Nature, Switzerland.

Wang, L., Zhang, Y., Zhu, Z., Zheng, F. and Gao, R. (2023). Food-derived collagen peptides: safety, metabolism, and anti-skin-aging effects. Current Opinion in Food Science 51, June 2023, 101012.

Wong, R.P.M., Zhou, Z.K. and Strappe, P.M. (2024). The anti-obesogenic and anti-diabetic properties of marine collagen peptides. Front. Food. Sci. Technol. 3: 1270392.

Wuyin, W. and Minjie, C. (2012). Decolorizing and deodorizing method of collagen peptide of abyssal fishes. China Patent#CN102718862B.

Xu, S., Zhao, Y., Song, W., Zhang, C., Wang, Q., Li, R. et al. (2023). Improving the sustainability of processing by-products: extraction and recent biological activities of collagen peptides. Foods 12: 1965.

Yamada, S., Nagaoka, H., Terajima, M., Tsuda, N., Hayashi, Y. and Yamauchi, M. (2013). Effects of fish collagen peptides on collagen post-translational modifications and mineralization in an osteoblastic cell culture system. Dent Mater J. 32(1): 88–95.

Yamada, S., Yamamoto, K., Nakazono, A., Matsuura, T. and Yoshimura, A. (2021). Functional roles of fish collagen peptides on bone regeneration. Dental Materials Journal 40(6): 1295–1302.

Zague, V. (2008). A new view concerning the effects of collagen hydrolysate intake on skin properties. Archives of Dermatological Research 300(9): 479–483.

Zague, V., Freitas, V., Costa Rosa, M., Alvares de Castro, G., Jaeger, R. and MachadoSantelli, G.M. (2011). Collagen hydrolysate intake increases skin collagen expression and suppresses matrix metalloproteinase 2 activity. Journal of Medical Food 14(6): 618–624.

Zhang, Y., Olsen, K., Grossi, A. and Otte, J. (2013). Effect of pretreatment on enzymatic hydrolysis of bovine collagen and formation of ACE-inhibitory peptides. Food Chemistry 141(3): 2343–2354.

Zhang, Z., Wang, J., Ding, Y., Dai, X. and Li, Y. (2011). Oral administration of marine collagen peptides from Chum Salmon skin enhances cutaneous wound healing and angiogenesis in rats. J. Sci. Food Agric. 91: 2173–2179.

Zhao, W., Li, J., Li, Y., Chen, Y. and Jin, H. (2022). Preventive effect of collagen peptides from *Acaudina molpadioides* on acute kidney injury through attenuation of oxidative stress and inflammation. Oxidative Medicine and Cellular Longevity Article ID 8186838.

Zhu, C., Zhang, W., Lius, J., Mu, B., Zhang, F., Lai, N. et al. (2017). Marine collagen peptides reduce endothelial cell injury in diabetic rats by inhibiting apoptosis and the expression of coupling factor 6 and microparticles. Molecular Medicine Report 16: 3947–3957.

Zhu, C.F., Li, G.Z., Peng, H.B., Zhang, F., Chen, Y. and Li, Y. (2010). Effect of marine collagen peptides on markers of metabolic nuclear receptors in type 2 diabetic patients with/without hypertension. Biomedical and Environmental Sciences 23: 113–120.

Chapter 3

Halophilic Microorganisms and their Biotechnological Potential in Agricultural Systems

Mayara Baptistucci Ogaki

Introduction

Halophiles include a range of organisms that grow optimally in the presence of salt, thriving under extreme conditions. Depending on their place of occurrence, they may be exposed to UV radiation, extreme temperature fluctuations, and/or alkaline pH levels (Thombre et al. 2017). Those microorganisms able to survive under a wide spectrum of extreme environmental factors are considered polyextremophiles (Seckback et al. 2013). All these extreme organisms are able to survive in such harsh environments and thrive in challenging growth conditions due to their unique physiological characteristics that are biotechnologically interesting to science.

Marine environments offer natural habitats where numerous halophilic and halotolerant organisms can be discovered and explored for various applications, making these locations intriguing bioprospecting niches. Research with microbial halophiles often emphasizes their utility in medicinal applications, as these microorganisms are well-known to produce valuable bioactive compounds with pharmaceutical-grade quality for therapeutic purposes, such as antimicrobial potential (Henciya et al. 2020, Santhaseelan et al. 2022), antioxidant and anticancer compounds (Hou et al. 2018, Hamidi et al. 2019). Some of these bioactive compounds encompass molecules such as biosurfactants, exopolysaccharides, pigments, enzymes and other extracellular metabolites which can be physiologically screened to discover new natural products with various therapeutic purposes (Corral et al. 2019).

Universidade Estadual do Norte do Paraná - UENP, Centro de Ciências Biológicas, Paraná, Brazil.
Email: mayaraogaki@gmail.com

Nevertheless, there is also a diverse range of applications in agroindustry for these organisms and their compounds that necessitate further exploration. Some of them are in industrial enzymes, bioremediation, plant growth promotion, biocontrol activity, protective biomolecules and food-processing (Sharma et al. 2016a). Therefore, this chapter aims to explore information about halophilic and halotolerant microorganisms, their bioprospection in marine environments, as well as their potential applications in the agriculture sector.

Classification and occurrence of microbial halophiles

Saline ecosystems select halophilic or halotolerant organisms based on the salt concentration in their environment, and differences exist among microorganisms in terms of halophilism and halotolerance. However, the concept of halotolerance is not easy to define in natural ecosystems. Because tolerance and salt requirements depend on temperature, pH, and growth substrate. Then, any artificial criteria do not necessarily have to adjust to all possible situations in nature (Ventosa and Arahal 2009).

The most extended classification of halophilism of Kushner and Kamekura (1988) is based on this NaCl requirement for growth of categorized halophiles as: slight (1–3%); moderate (3–15%) and extreme (15–30%). Therefore, in definition, halotolerant organisms resist ionic stress with optimal growth under NaCl concentrations, while all halophilic organisms require sodium ions for their metabolism and in contrast, non-halophilic microorganisms show optimum growth with NaCl concentration less than 1% (0.2 M) (Reang et al. 2022). Lastly, obligate halophiles require NaCl concentrations higher than 3%, equivalent to or exceeding seawater, with approximately 3.5% NaCl for their optimal growth (Rodriguez-Valera et al. 1981).

In terms of diversity, halophiles are found in all three domains of life: Archaea, Bacteria, and Eukarya (Ma et al. 2010). Archaea prevail in some of the most extreme hypersaline environments, with the ability to endure salt concentrations nearing saturation levels. Halophilism in Archaea is predominantly confined to members of the Haloarchaea class and "Nanohaloarchaeota" subphylum (Loukas et al. 2018). Halobacteria are the most salt-requiring and salt-resistant microorganisms within the domain. For example, the species *Halobacterium pharaonis* described by Soliman and Trüper (1982) from the alkaline brines of eutrophic desert lakes of Wadi Natrun, Egypt by culture-dependent methods.

The class includes 275 species allocated in 20 genera, 49.8% described from marine salt environments (thalassohaline), while 50.2% from inland salt environments (athalassohaline) (Cui and Dyall-Smith 2021), and these data are increasingly supplemented by advancements in diversity studies utilizing culture-independent methods (Vavourakis et al. 2016, Wu et al. 2022).

Furthermore, within Eubacteria, halophiles represent a highly diverse group, distributed across eight distinct phyla (Abaramak et al. 2020). Among them, Proteobacteria and Firmicutes are most common, with dominance of slight and moderate species, while Actinobacteria, Bacteroidetes, Cyanobacteria, Spirochaetes, Rhodothermaeota, and Thermotogae occur at lower frequencies (Loukas et al. 2018).

In contrast, unlike other divisions of microorganisms, halophilic fungi do not require sodium chloride for growth, and can be adapted to a wide range of salt concentrations, with the exception of a few obligate halophiles (Plemenitaš et al. 2014). For example, an obligate halophilic fungus, *Wallemia ichthyophaga*, a fungus belonging from the ancient basidiomycetous genus *Wallemia* (Wallemiales, Wallemiomycetes) grows only at salinities between 10% (w/v) NaCl to saturated NaCl solution (Zajc et al. 2014).

The most halophilic fungi have been classified primarily into Ascomycota (84%) and some species into Basidiomycota (9%) (Loukas et al. 2018), and their primary representatives are yeast-like fungi and different related melanized species of the genus *Cladosporium*, anamorphic *Aspergillus* and *Penicillium*, the teleomorphic *Eurotium* and *Emericella*, additionally, other filamentous species from *Wallemia*, *Alternaria*, and *Scopulariopsis* (Gunde-Cimerman et al. 2009). Because of their ability to thrive in adverse conditions, such as high salinity, many of these fungi have already been studied and documented for their capacity to produce biomolecules such as enzymes, pigments, biosurfactants, and osmoprotectants (Śliżewska et al. 2022).

Another group underreported is certain viruses known as halophages, which play an equally significant ecological role by impacting microbial community composition and population dynamics, restructuring prokaryotic and eukaryotic communities (Sabet et al. 2012, Truitt et al. 2021).

All these halophilic microorganisms can constitute the natural microbial communities of hypersaline ecosystems, which are widely distributed around the world (Oren 2008). In ecological terms, salt-adapted microbial communities thrive within diverse ecosystems marked by salinity levels exceeding that of seawater, which is typically around 3.5% NaCl (35 g/L, 0.6 M) (Edbeib et al. 2016). These niches encompass saltern pond brines, natural salt lakes, hypersaline soils, springs, salt lakes, sabkhas, as well as naturally occurring coastal saline habitats, marshes, marine abyssal sediments, and even endophytes (Ventosa and Arahal 2009).

According to Ventosa and Arahal (2009), hypersaline environments can be categorized into two primary types: thalassohaline and athalassohaline environments, depending on whether they have their origins in seawater or not, respectively. Among these settings, the most extensively examined natural habitats are the hypersaline lakes of oceanic (thalassohaline) or non-oceanic (athalassohaline) origin and solar salterns.

Thalassohaline environments arise from the evaporation of adjacent seawater or saltwater lakes, resulting in a salinity of 35%, approximately ten times greater than the average salinities of Earth's seawater. Some thalassohaline areas include solar salterns, salt mine drainage waters, playas and brine springs from underground salt deposits (Edbeib et al. 2016). Furthermore, thalassohaline environments also mirror the ionic composition of seawater and maintain a pH level that is nearly neutral to slightly alkaline (Ma et al. 2010).

On the other hand, athalassohaline environments have non-marine composition, and microorganisms of these areas must survive with extremely high salt levels in conjunction with alkaline pH and elevated temperatures (Bowers et al. 2009). The athalassohaline environment has been more extensively studied at Dead Sea. The total of dissolved salts from Dead Sea is on average 322.6 g/L, with dominance of divalent cations as Mg, Na, Ca and K with (concentrations of 41, 39, 17 and 7 g/L, respectively) (Nissenbaum 1975).

Other examples of athalassohaline environments include haloalkaline regions, with predominance of bicarbonate and sodium ions, hypersaline anoxic basins and ancient evaporite deposits composed of magnesium chloride (Montoya et al. 2013, Sorokin et al. 2011, Wielen et al. 2005) and some alkaline lakes (soda lakes) that are poor in magnesium or calcium ions and abundant in carbonates, resulting in a very alkaline pH (pH 9–11.5) (Ventosa et al. 2011).

Osmoadaptation of halophilic microorganisms

Halophilic microorganisms demonstrate a low nutritional demand and possess the capability to resist high salt concentrations, employing a survival mechanism known as osmoadaptation (Thombre et al. 2017). This adaptation involves achieving an internal osmotic pressure equal to or higher than that of the external environment to maintain turgor.

To thrive in high salt concentrations, halophilic and halotolerant microorganisms employ various strategies. Generally, they require the stability and functionality of all intracellular proteins in the presence of elevated KCl and other molecules concentrations (Ventosa et al. 1998) to regulate the balance of Na within the cytoplasm and counteract external osmotic pressure.

A distinct strategy is evident in the Halobacteriaceae (as well as certain bacteria like Salinibacter and anaerobic Halanaerobiales) (Weinisch et al. 2018), often termed "true halophiles". True halophiles have a halophilic cytoplasm, requiring the adaptation of all enzymes and biological processes to these conditions. They can accumulate KCl over 37% (5 M) as part of a "salt-in" strategy (Bremer and Krämer 2019). In general, halophiles can accumulate inorganic cations ($K+$ and in some cells $Na+$) as a key player in this osmotic balance and osmotic response (Roberts 2005).

On the other hand, there is also a "salt-out" strategy where salt ions are excluded from the cytoplasm, while simultaneously synthesizing or accumulating high concentrations of organic molecules compatible with their internal metabolic activities (i.e., "compatible solutes"). There are various types of compatible solutes utilized for osmotic balance including amino acids and derivatives, betaines, ectoines, polyols and derivatives, carbohydrates and derivatives and occasionally peptides suitably altered to remove charges (Galinski 1995). In general, bacteria and eukaryotes typically store neutral compatible solutes, while archaea tend to favor negatively charged solutes (Martin et al. 1999, Roberts 2004).

For instance, some osmolytes identified in bacteria and archaea include uncharged organic osmolytes such as sucrose, trehalose and alpha-glucosylglycerol. Other Zwitterionic organic osmolytes as ectoine, betaine, beta-glutamine and anionic organic osmolytes containing carboxylates as beta-glutamate, hydroxybutyrate or containing phosphate or sulfate moieties as sulfotrehalose and mannosyl-DIP have been detected (Roberts 2005). Glutamate and glutamine predominate at intermediate salinities, and proline and ectoine dominate at high salinities (Ma et al. 2010).

This salt-out strategy is typical of moderately halophilic and halotolerant bacteria and eukaryotes (Bremer and Krämer 2019). These microorganisms exhibit physiological diversity, encompassing aerobic and anaerobic, heterotrophic, phototrophic, and chemoautotrophic forms (Corral et al. 2019). The biosynthesis of compatible solutes ectoine and hydroxyectoine can be found in halophilic and halotolerant microorganisms induced by salt stress. These solutes provide protection against salt stress and against temperature stress in *Bacillus subtilis* and other salt-tolerant bacilli (Ma et al. 2010).

Additionally, the "low salt or organic solutes" strategy involves the biosynthesis and/or storage of uncharged or zwitterionic low molecular mass organic compounds within the cell metabolism at high cytoplasmic concentrations. This approach provides osmotic balance and minimally disrupts the functionality of non-salt-adapted enzymes and other macromolecular structures, preventing inactivation, inhibition, and denaturation (Weinisch et al. 2018). Nevertheless, organisms utilizing this strategy must still integrate salt-adapted proteins into the membrane exposed to the saline environment (Ma et al. 2010).

Bioprospection in marine environments

Compared to other categories of extremophilic microorganisms, like thermophiles and alkaliphiles, the potential applications of halophiles from all three domains in biotechnological processes have been relatively underutilized (Ma et al. 2010). Bioprospecting of halophilic microorganisms has focused on the applicability of their properties, such as the production of hydrolytic enzymes and biosurfactants (Fariq and Yasmin 2020), their adsorption of heavy metals and bioremediation (Diba et al. 2021), in addition to the promotion function of plant growth (Kapadia et al. 2022). Different sources have been explored to prospect halophilic microorganisms, including thalassohaline and athalassohaline environments (Cui and Dyall-Smith 2021).

Athalassohaline ecosystems, created by evaporation of fresh water (Orellana et al. 2018), are often found inland and therefore not directly connected to marine waters, such as the Dead Sea and the soda lakes (Naghoni et al. 2017). These ecosystems have ionic compositions, differing from seawater (Oren 2002), shaped by the dissolution of mineral salt deposits of continental origin (leaching process), which are dominated by potassium-, magnesium-, sodium- and carbonate ions (Youssef et al. 2012).

Physical-chemical aspects must be considered in microbial isolation from these environments. The analysis of the mineral composition and pH (whether alkaline or acidic) is crucial in both enrichment methods and the culture media used. Depending on the media composition, certain species may face limitations. Thus, there is a risk of underestimating local diversity, as certain species may not be represented in the study.

The Dead Sea represents a hypersaline desert lake with 34.8% salinity which is characterized by elevated Mg^{2+} concentrations, approximately 1.9 mol/L, and other minerals as 0.5 M Ca^{2+}, 1.5 M Na^+, 0.2 M K^+, 6.5 M Cl^- and 0.1 M Br^-, which significantly restrict microbial growth (Satbhai et al. 2015). The pH levels in these hypersaline surroundings can exhibit considerable variation; for instance, the Dead Sea maintains a pH around 6.0, whereas alkaline soda lakes in regions like Africa, China, and India can surpass 11 or even higher pHs (Oren 2002, Sorokin et al. 2015, Vavourakis et al. 2016).

Studies conducted with Dead sea samples involved methods of enrichment, filtration, and subsequent inoculation in specific culture media developed with varying compositions and concentrations of salts (Satbhai et al. 2015), or using commercial culture media, such as Czapek-dox, Marine Broth (MB), Luria Broth medium (LB) and Mannitol-Glutamate medium (MG), all supplemented with a specific salt composition (Nazareth et al. 2011, Al-Karablieh 2017), in order to better approximate the composition found naturally in this ecosystem.

Thalassohaline waters or brines have marine origins, with an ionic composition like seawater (predominantly NaCl). They are commonly found near seas and oceans, including industrial solar salterns and natural shallow basins that have become detached from the sea or ocean (Naghoni et al. 2017).

On the other hand, in marine environments, halophiles can be found in various habitats, including ponds, lakes, salt pans (Aljohny 2015), seawater (Radouane et al. 2023), and sediments (Dell'Anno and Danovaro 2005, Dang et al. 2009). Additionally, other halophilic and halotolerant microorganisms may act as endophytes or be associated with mangrove substrates, algae, seagrass, plants rhizosphere, as well as marine invertebrates (sponges, corals, crustaceans, ascidians, holothurians, and bivalves) and vertebrates (mainly fish) (Mora-Ruiz et al. 2016, Bibi et al. 2018, El-Bondkly et al. 2021, Abdulrahman et al. 2022).

To isolate microorganisms from these environments, it is also essential to consider the physical-chemical parameters such as the pH, temperature, and the composition of predominant salts. This ensures the simulation of optimal cultivation conditions. In these environments, it is common to use low-nutrient media with greatly reduced levels of complex organic nutrients (e.g., yeast extract, tryptone or peptone) supplemented by NaCl (Singh and Jha 2016, Fariq and Yasmin 2020) or incorporate a sea salt mixture into the culture medium or even use commercial media as Inorganic Salt Starch Agar (ISSA) and Marine Agar (MA) to mimic the salinity of samples obtained from the natural environment (El-Tarabily et al. 2020, 2021).

An example of a thalassohaline system is the deep-sea brine found at the bottom of the Red Sea. This depression reaches depths of approximately 1,500–2,800 m. Apart from its significantly high salinity, the environment is anaerobic, characterized by elevated temperature, acidity, and an unusually low average nutrient level (Renn et al. 2021). Due to the unique conditions, several polyextremophilic microbes in this system have been explored for their microbial enzymes and natural compounds, demonstrating potential industrial applications (Ziko et al. 2019).

Other halophilic microorganisms bioprospected from marine environments and their applications in agricultural systems are represented in Table 1.

Table 1. Halophilic microorganisms bioprospected from marine environments and their biotechnological applications in agricultural systems.

Halophilic microorganism	Application	Bioproducts	Source	Location	Reference
Alcanivorax, Bacillus, Cobetia, Halomonas, Methylarcula, Micrococcus, Myroides, Paracoccus, Planococcus, Pseudomonas, Psychrobacter, Sporosarcina, Sufflavibacter, Wangia	Biocontrol	Enzymes: Amylase, protease, lipase, Dnase	Deep-sea sediments	Southern Okinawa Trough (China)	Dang et al. 2009
Halomonas sulfidaeris, Marinobacter excellens	Biocontrol	Enzymes: Lipase	Deep-sea sediments of the South Atlantic	South Atlantic Ocean: the Walvis Ridge Sector and the South Equatorial MAR	Delabary et al. 2020
Bacillus sp. (*B. amyloliquefaciens, B. subtilis, B. velezensis, B. vallismortis*), *Ochrobactrum* sp. (*Ochrobactrum thiophenivorans*).	Biocontrol: against *Penicillium digitatum* and *Penicillium italicum* in citrus desease	Bioactive compounds: Screening for antifungal activity	Saline water	Dead Sea, the Agadir Sea, the Rabat Sea	Radouane et al. 2023
Actinobacteria spp., *Nocardiopsis dassonvillei, Pseudonocardia* sp., *Arthrobacter* sp., *Streptomyces* sp.	Biocontrol: against fungal pathogens, e.g., *Phytophthora capsica, Pythium ultimum, Magnaporthe grisea, Altenaria malli* and *Fusarium moniliforme*	Bioactive compounds: Thiabendazole, sulfamethoxypyridazine, N-nitrosodiethylamine, ibuprofen, laberalol, isoxsuprine, xibendazole, sulfamethoxypyridazine, dimetridazole, salbutero, moxidectin, atenolol, timolol, benzydamine, acebutolol, and bambuterol Enzymes: Protease, amylase	Rhizosphere and endophytic association with *Salsola imbricata* (mangrove)	Marine coastal, Jeddah Saudi Arabia	Bibi et al. 2018

Table 1 contd. ...

...Table 1 contd.

Halophilic microorganism	**Application**	**Bioproducts**	**Source**	**Location**	**Reference**
Peniciullium spp., *Acremonium* sp.	Biocontrol: Herbicidal and antifungal activities against phytopathogen *Cladosporium cladosporioides*	Bioactive compounds: Aromatic compounds derived from fungi that exhibit considerable broad-spectrum antimicrobial effects	Deep sea sediments	Antarctica	Ogaki et al. 2019
Deleya venusta	Bioremediation	Phenol degradation	Seawater (500 m depth)	Almeria Coast, Spain	Munoz et al. 2001
Halococcus sp.	Biorredmediation: EPS capacity as biosurfactant Industrial: emulsification Medical:enzymatic and non-enzymatic antioxidants, anticancer activity	Exopolyssacharides	Marine sediment (25 m depth)	Chennai, Tamil Nadu, India	Kannan et al. 2022
Bacillus cereus	Biorremediation	Exopolyssacharides	Superficial marine sediment	Uppanar estuary, Cuddalore, Southeast coast of India	Krishnamurthy et al. 2020
Actinoplanes deccanensis, Streptomyces euryhalinus	Plant growth promotion	Increase of endogenous polyamines and plant growth regulators (IAA, gibberellic acid, and cytokinins), but a concomitant reduction in abscisic acid in *S. bigelovii* tested tissues	*Salicornia bigelovii* rhizosphere (mangrove plant that grows in seawater of the intertidal zone)	United Arab Emirates	El-Tarabily et al. 2020

Pseudoalteromonas maricaloris	Plant growth promotion	Production of ACC deaminase	*Avicennia marina* rhizosphere (mangrove)	Abu Dhabi coast, United Arab Emirates	El-Tarabily et al. 2011
Agrobacierium tumefaciens, Klebsiella sp., *Ochrobactrum anthropi, Pseudomonas stutzeri, Pseudomonas* sp.	Plant growth promotion: Peanut (*Arachis hypogaea* L.)	Production of IAA, ACC deaminase, N-fixation and phosphate solubilization activities	Rhizosphere of *Arthrocnemum indicum* found in tropical salt marshes that are frequently inundated with seawater	Bhavnagar, India	Sharma et al. 2016
Azotobacter, A. chroococcum, A. virelandii, A. beijerinckii	Plant growth promotion	Nitrogen fixation and in vitro production of phytohormone (Indole Acetic Acid, IAA)	Rhizosphere of 16 mangrove plant species	Southeast coast of India	Ravikumar et al. 2004
Pantoea, Sphingosinicella, Xanthomonas, Ochrobactrum, Bacillus	Plant growth promotion of reforestation tree (*Acacia polyphylla*)	Nitrogen fixation, production of IAA and siderophore, phosphate solubilization	Endophytic association with *Rhizophora mangle, Laguncularia racemosa* and *Avicennia nitida* (mangrove)	Bertioga, São Paulo, Brazil	Castro et al. 2018
Streptomyces chartreusis, S. tritolerans, S. rochei	Plant growth promotion	Production of IAA, polyamines and ACC deaminase	*Salicornia bigelovii* rhizosphere (mangrove plant that grows in seawater of the intertidal zone)	West-north coast of the United Arab Emirates	Mathew et al. 2020

Perspectives in agriculture

There are also a variety of applications in agriculture for halophilic microorganisms and their compounds that still need to be thoroughly explored. Some of them are involved in plant growth promotion, biocontrol activity and protective biomolecules (Sharma et al. 2016a).

Salinity and hydric stress on plants

In the last decades, global temperatures have been influenced by both natural and artificial factors that have led to a significant and abrupt shift in climate. The impact of anthropogenic factors surpasses that of natural ones, resulting not only in substantial alterations to precipitation patterns, but also triggering a series of climate events. When considering agriculture, it is crucial to acknowledge the heightened resistance of pests and parasites, coupled with reduced plant productivity stemming from elevated temperatures that diminish the yield of essential crops due to increased evapotranspiration (Ulla and Khan 2021). These factors also contribute to soil salinization, prompting concerns about global water and food security.

Soil salinization represents a significant environmental constraint, contributing to soil degradation and harming global agricultural production. Furthermore, salinity also negatively affects soil properties and the ecological balance, resulting in decreased agricultural productivity, economic returns, and increased soil erosion (Shrivastava and Kumar 2015).

According to Bui (2013), the prolonged accumulation of salts in arid and semi-arid regions is the main cause of the development of salt-affected lands. In addition to affecting coastal agriculture land, soil salinity also raises sea levels, modifying salinity from 1 to 33% over twenty-five consecutive years (Rahman et al. 2018). In short, saline stress in the soil can reduce its agroecological potential, posing an ecological and socioeconomic threat to sustainable development.

In terms of definition, soil salinity is characterized by increase in the concentration of dissolved mineral salts, which negatively influences plant growth at physiological, biochemical, and molecular levels through osmotic and ionic stress (Ashraf 2004, Rojas-Tapias et al. 2012). These impacts span various plant stages, from seed germination to maturation, affecting the regulation of phytohormones, root/shoot development, water/nutrient uptake, and photosynthesis (Sharma et al. 2016a).

Salinity exerts its deleterious impact on plants through two mechanisms: osmotic stress and ion toxicity (Ullah and Khan 2021). Osmotic stress diminishes the osmotic potential between the root and soil solution, reducing water availability, primarily through the uptake of Na^+ and Cl^- (Tavakkoli et al. 2011, Abbasi et al. 2016), and ion toxicity arises from elevated concentrations of Na^+, Cl^-, or SO_4^{-2}. Excessive Na^+ uptake disrupts the equilibrium of essential ions like K^+, affecting crucial cellular functions necessary for plant growth (Rus et al. 2001). This disruption in the K^+/Na^+ ratio, essential for enzyme activation and protein synthesis, ultimately impedes plant growth and productivity (Ahmed et al. 2020), while SO_4^{-2} induces ion toxicity that affects nutrient uptake (Ullah and Khan 2021).

In terms of plant development, high salinity affects germination and vegetative growth, causing disruptions in nutrient balance and impedes nutrient uptake (Shrivastava and Kumar 2015, Sharma et al. 2016a). During the first stages, salinity reduces the endogenous levels of phytohormones and inhibits seedling establishment (Egamberdieva and Kucharova 2009). Although, in reproductive development processeses, such as microsporogenesis and stamen filament, elongation is inhibited, resulting in programmed cell death, ovule abortion, and the senescence of fertilized embryos (Shrivastava and Kumar 2015). Additionally, salinity adversely impacts photosynthesis by reducing leaf area, chlorophyll content, and stomatal conductance (Netondo et al. 2004).

Moreover, salinity induces secondary effects, such as the heightened uptake of chloride (Cl), which is often linked to a reduction in nitrate uptake by crops (Yu et al. 2016, Hasana and Mayake 2017, Kanagaraj and Desingh 2017). Another consequence is oxidative stress, because elevated salt levels impact chlorophyll content and diminish the activity of photosynthetic electron transport, resulting in a consequential disturbance in electron transfer (Morcillo and Manzanera 2021). This disturbance triggers a burst of reactive oxygen species (ROS). ROS burst directly harms plant tissues through toxic byproducts of aerobic metabolism, such as $O_2{\bullet}^-$, $OH{\bullet}$ and H_2O_2 (Apel and Hirt 2004, Voothuluru and Sharp 2012, Chawla et al. 2013).

ROS are generated in the chloroplast, mitochondria, and peroxisome. The increased production of ROS results in lipid peroxidation, causing damage to photosystem II, impeding photosynthesis and membrane permeability, causing damage to nucleic acids, inhibiting enzymes, and triggering programmed cell death (Sharma et al. 2012, Ullah and Khan 2021).

Above all these effects of salinization on plants, it is important to highlight that in agriculture, instances of salinization caused by anthropogenic sources are typically associated with the extensive use of fertilizers. The over use of these fertilizers leads to an excess of nutrients, which plays a significant role in inducing changes in the soil. Among mineral fertilizers, KCl has the most notable impact on soil salinity, as potassium is efficiently absorbed by plants as a macronutrient, while chloride, considered a micronutrient, remains in the soil (White and Broadley 2001, Manning 2010). Consequently, a high concentration of Cl^- in the soil, unrelated to NaCl, is plausible. Like Na^+, elevated levels of chloride ions can adversely affect plant growth and yield (Morcillo and Manzanera 2021). For instance, Cl^- disrupts photosynthesis and interferes with the absorption of other essential anions for plant growth, such as NO_3^- and PO_4^{2-} (Geilfus 2018).

Considering salinity to be an imminent global challenge given the salinization of several previously agricultural areas, a significant number of halophile microorganisms have been documented and evaluated for their contributions to advances in sustainable agriculture under salt stress. Research has demonstrated that inoculating plants with various growth-promoting halophilic microorganisms reduces Cl^- absorption and increases plant growth (Nadeem et al. 2014, Karlidag et al. 2010, Abd El-Azeem et al. 2012). In addition, these bacteria produce molecules capable of inducting synthetic metabolic pathways, reducing hydric stress and promoting biocontrol against pathogens. Furthermore, additional advantages of incorporating halophilic microorganisms and their derivatives into the plant rhizosphere will be discussed below.

Plant growth-promoting

Most of the studies involving applications of halophiles in agriculture focus on microbial cultures exhibiting plant-growth-promoting (PGP) characteristics in salt stress. This is attributed to the fact that the plant rhizosphere harbors symbiotic or non-symbiotic microorganisms that can exert positive effects on plants (Reang et al. 2022). This biota is diverse and modular, enhancing nutrient availability for plants.

Microbial cultures can present PGP either directly or indirectly. Direct promotion of growth occurs when plant-growth-regulators (PGRs) are released into the soil, such as auxins (indole acetic acid - IAA), cytokinins (CKs), gibberellins (GAs), polyamines (PAs) (Bashan et al. 2004, Hassan et al. 2019) or by the solubilization of certain nutrients such as phosphorus (P), iron (Fe), and nitrogen (N) fixation (El-Tarabily and Youssef 2010, Glick 2015). Microorganisms also have the capability to stimulate nutrient transport in plants (Bertrand et al. 2000) and may play a role in regulating stress hormone (such as ethylene) levels in plants.

As a stress hormone, the entire signaling pathway of ethylene is involved in the plant's response to salinity stress and could potentially mediated the stress signal transduction in plants (Tao et al. 2015). Halophilic bacteria with capabilities to produce the enzyme 1-aminocyclopropane-1-carboxylic acid (ACC) deaminase can mitigate the detrimental effects of salt stress on plant growth (Glick 2014, Acuna et al. 2019, Nagaraju et al. 2021). Plants inoculated with these positive isolates to ACC deaminase activity show increased tolerance to salt stress (Nishimura et al. 2007, Tiwari et al. 2016).

The presence of positive ACC deaminase activity in these isolates indicates their ability to irreversibly cleave ACC molecules, a crucial step in the ethylene biosynthetic pathway. This enzymatic action results in the production of α-ketobutyrate and ammonia, potentially PGP by reducing ethylene levels and alleviating stress under various stressful conditions (Orozco-Mosqueda et al. 2020, Shilev et al. 2020, Saraf et al. 2010).

On the other hand, indirect growth promotion occurs through the elimination of pathogens by the production of cyanide and siderophores, mentioned in detail in the section "Biocontrol". Some of these properties of growth promoters have been exploited in various applications, including biofertilizers, microbial rhizoremediation, and biopesticides and represent significant importance in agricultural practice (Reang et al. 2022).

Reang et al. (2022) isolated halophilic and halotolerant bacteria capable of tolerating salt concentrations up to 25%, with optimal growth occurring between 10–15% NaCl at 35°C. These bacteria demonstrated the ability to solubilize potash and phosphate, fix nitrogen, and produce the enzyme 1-aminocyclopropane-1-carboxylic acid (ACC) deaminase and indole acetic acid (IAA).

Not only isolates obtained from athalassohaline environments, but some halophilic microorganisms obtained from marine environments also produce plant growth regulators (PGRs). Certain bacterial strains isolated from mangrove roots and rhizosphere are mentioned as potentially capable of P-solubilization or N-fixation (Bashan et al. 2009, El-Tarabily and Youssef 2010, Castro et al. 2018, El-Tarabily et al. 2020), including IAA-producing *Azotobacter* spp. (Ravikumar et al. 2004, Kathiresan and Selvam 2006), and *Pseudoalteromonas maricaloris* with ACC-producing abilities (El-Tarabily and Youssef 2011).

Some authors have reported that halotolerant marine rhizosphere-competent actinobacteria are capable of promoting the growth and seed yields of *Salicornia bigelovii*, stimulating in planta levels of polyamines (PAs) and PGRs, and reducing the effects of ethylene (El-Tarabily et al. 2020, Mathew et al. 2020). El-Tarabily et al. (2021) also isolated actinobacteria from mangrove rhizosphere with high phosphorus solubilizing activity, such as *Streptomyces coelicoflavus*, and identified isolates like *Streptomyces polychromogenes*, *Streptomyces bacillaris*, and *Streptomyces ferrugineus* capable of producing IAA, PAs, and ACC, respectively. These characteristics highlight the significant potential of marine halophiles as PGP, positioning them as valuable sources of biofertilizers for enhancing plant growth.

Exopolysaccharides production

Marine microorganisms produce exopolysaccharides (EPS) as an important factor for their survival in extreme environments, such as high pressure, high temperatures, and high salinity (Schmid 2018). EPS are large carbohydrate polymers with high molecular weights, released by microorganisms during their growth and metabolic processes (Yang et al. 2021).

Various findings have suggested that the cryoprotective effect, water-holding capacity, and robust thermostability of marine EPS are attributed to their monosaccharide composition, high molecular weight, hydrophobicity, and polycharged characteristics (Qi et al. 2022).

EPS are crucial for the survival of halophiles in stress environments, because halophilic bacteria synthesize EPS to endure osmotic pressure (Gan et al. 2019). In addition to protecting themselves through EPS production, halophilic microorganisms could, consequently, shield their plant hosts against environmental fluctuations and other abiotic stresses such as drought, salinity, or heavy metal pollution.

Among the effects of plant-associated exopolysaccharides, EPS possesses significant potential for mitigating drought stress in plants (Morcillo and Manzanera 2021). This is attributed to their ability to enhance soil structure and water retention, facilitate microbial colonization and biofilm formation, and modulate the plant's response to water deficit. In addition, EPS helps plants to increase the time available to adapt their metabolism and become acclimatized to cope with heat and cold stresses.

In detail, EPS reduces Cl^- uptake and immobilizes cations, such as Ca^{2+}, Mg^{2+}, and K^+ ions from salts. This Ca^{2+} and K^+ retention by EPS might reduce the internal level of these cations in plants, producing positive impacts on plant growth (Ashraf et al. 2004, Geddie and Sutherland 1993). EPS can also adhere to soil particles, forming aggregates that are influenced by the nature and quantity of organic matter, as well as the specific type of soil particles (Chaney and Swift 1986, Or et al. 2007). The formation of aggregates occurs through a binding effect involving cation bridges, hydrogen bonding, van der Waals forces, and anion adsorption mechanisms (Chenu 1993). The highly interconnected structure of EPS enables it to contract and expand while maintaining saturation across diverse matric potentials (Or et al. 2007). All these effects collectively extend the timeframe for plants to undergo metabolic adjustments in response to drought stress.

Even with the enormous potential for the application of EPS-producing halophiles in agriculture, considering their widespread presence in saline environments, the exploration for new biologically active EPS from halophilic bacteria is still in its early stages (Qi et al. 2022). In this context, the marine environment could serve as a valuable locale for the bioprospecting of halophilic EPS producers. Many marine strains which can produce EPS were identified, such as *Pseudoalteromonas* spp., *Bacillus* spp., *Alteromonas* spp., and the *Vibrio* spp. (Gugliandolo and Maugeri 1998, Jouault et al. 2001, Guezennec 2002, Zanchetta et al. 2003).

Halophilic microorganisms in biocontrol

Halophiles associated with the plant rhizosphere can also play a fundamental role in plant defense by producing hydrolytic enzymes, siderophores, volatile compounds, like hydrogen cyanide (HCN), organic acids and other biomolecules derived from their unique metabolism. Some of these metabolism products are discussed below.

Enzymes

Halophilic microorganisms are known to produce various hydrolytic enzymes, including amylases, lipases, proteases, chitinases, and nucleases, which have diverse biotechnological applications (de Lourdes Moreno et al. 2013). The biocontrol potential of rhizobacterial enzymes against plant pathogens has been demonstrated in several studies, some of which showcase their ability to control fungal pathogens such as *Fusarium oxysporum* (Qaiser et al. 2015, Gowtham et al. 2016, Guleria et al. 2016, Passari et al. 2018, Samaras et al. 2018, Dukare and Paul 2021, el Kinany et al. 2021), as well as their role in controlling phytopathogens and indirectly promoting plant growth (Reinhold-Hurek et al. 2006, López et al. 2018). Among the biocidal biomolecules involved in these effects are chitinolytic enzymes (endochitinase, exochitinase, and chitobiase), along with other lytic enzymes, including proteinases, cellulases, amylases, endoglucanases, pectinases, and lipases.

Chitinases play a crucial role in biocontrolling fungal plant pathogens, particularly those with cell walls predominantly composed of chitin—their target substrate (Aloo et al. 2023). These enzymes catalyze the hydrolysis of β-1, 4-linkages in chitin, leading to the direct inhibition of hyphal growth in various fungal pathogens (Oyeleyeo and Normi 2018).

Nevertheless, for phytopathogen control, the enzymes cellulases and endoglucanases contribute to the degradation of cellulose and other cell wall polymers, making them significant in both phytopathogen control and the colonization of plant endospheres, fostering more intricate interactions with plants (Reinhold-Hurek et al. 2006, López et al. 2018).

Siderophores

Siderophores serve multiple functions in plants, including respiration, photosynthesis, bioremediation, plant-growth-promotion, phytoremediation of heavy metals, and bioprotection (Singh et al. 2022, Aloo et al. 2023). These biomolecules are Fe-chelating produced under iron (Fe)-deficient conditions (Trapet et al. 2016). Rhizosphere bacteria produce siderophores as a part of their regular metabolism when facing Fe limit, aiming to enhance Fe acquisition. These siderophores can form complexes by chelating ferric ions, plants obtain the bound Fe by degrading these complexes (Rajkumar et al. 2009).

Beyond their role in aiding plant Fe-acquisition in Fe-limited soils, rhizobacterial siderophores also contribute to plant bioprotection (Aloo et al. 2022). The primary mechanism underlying the control of phytopathogens by siderophores involves nutritional competition, restricting iron availability to pathogens and impeding their growth (Ahmed and Holmström 2015).

Volatile organic compounds

Volatile organic compounds (VOCs) are low-molecular-weight biomolecules, produced through the oxidation of glucose from various intermediates of microbial primary and secondary metabolism. These molecules play important functions in microbiome plant interactions (Santoro et al. 2011).

In plants, VOCs play a dual role by either positively or negatively modulating their growth, inducing systemic resistance, serving as a direct nutrient source, and influencing the production of secondary metabolites. VOCs exhibit direct inhibitory effects against plant pathogens, contributing to soil fungistasis and suppressiveness, and affecting various organisms like insects and nematodes as attractants or repellents (Schulz-Bohm et al. 2017). They also impact microbial growth and function as long-distance communication signals (infochemicals) in both air and aquatic environments (Vaselova et al. 2019).

Additionally, VOCs exhibit antimicrobial properties and play a crucial role in competitive microbial interactions within ecological niches, such as the plant rhizosphere, and in antagonistic relationships between plant-associated bacteria and pathogens (Vaselova et al. 2019). Even without physical contact, microorganisms can significantly alter plant root system development, plant physiology, hormonal pathways, and biomass production (Schulz-Bohm et al. 2017). In this context, they release various organic and inorganic volatile compounds, including alkenes, alcohols, aldehydes, organic acids, esters, ketones, terpenes, benzoids, pyrazines, as well as nitrogen and sulfur compounds (Schenkel et al. 2015).

Ammonia and hydrogen cyanide (HCN) are inorganic volatile compounds most common in rhizobacteria that inhibit various phytopathogens (Aloo et al. 2023). HCN is synthesized by some Pseudomonas, Chromobacterium, and Rhizobium species and formed from glycine by HCN synthase (Vaselova et al. 2019). Besides that, cyanides are known for their biocontrol activity against fungal diseases. Bano and Mussarat (2023) detected the production of hydrogen cyanide (HCN) in *Pseudomonas aeruginosa* strain, inhibiting the growth of *Fusarium oxysporum*, *Trichoderma herizum*, and *Alternaria alternata*. Radouane et al. (2023) observed the production of HCN by halophilic bacteria isolated from seawater, demonstrating the bioprotective potential of citrus fruit against *Penicillium digitatum* and *Penicillium italicum*, which are pathogenic fungi causing postharvest citrus fruit diseases.

Organic acids

The conversion of soil-unavailable mineral forms into forms accessible to plants is crucial in plant-microbe systems. In addition to enzymatic degradation, the solubilization of P, Zn, and K can also occur via organic acids (OAs), such as 2-ketogluconic acid, citric acid, succinic acid, and lactic acid (Muleta et al. 2013, Ramesh et al. 2014, Hussain et al. 2015). The production of these organic acids typically arises from the microbial metabolism's oxidation of glucose as a primary source of both carbon and energy (Macias-Benitez et al. 2020).

Various microorganisms are studied for their P solubilization abilities. Some research has shown that inoculating PGP strains capable of producing organic acids and dissolving inorganic phosphate can increase soil P levels and improve crop yields (Yazdani et al. 2009, Israr et al. 2016, El-Sayed and Hagab 2020).

Some strains of *Bacillus, Enterobacter, Achromobacter xylosoxidans* and *Delftia* sp. (Ma et al. 2011, Singh and Jha 2016, Sarkar et al. 2018, Roy and Roy 2019) are known to solubilize inorganic P and synthesize OAs such as the citric, gluconic, lactic, succinic, and propionic acids (Zheng et al. 2018). This acidification could be a mechanism for the solubilization of inorganic P (Mendes et al. 2020).

Kapadia et al. (2022) evaluated halophilic bacteria isolated from rhizosphere soil samples collected along the Arabian seacoasts of Dandi, Navsari, India. Isolates identified as *Bacillus halotolerans, Bacillus* sp., *Achromobacter* sp., and *Streptomyces* sp. demonstrated positive effects on PGP under high salt concentrations of 15–20 percent. When the multipotent isolate was inoculated on rice, it significantly enhanced root length by 50%, plant dry weight by 25%, and tiller formation by 41% compared to the control. The authors observed that salinity induced a combination of plant PGPRs in these microbes, including biofilm formation, siderophore production, increased phosphorus solubilization, and secretion of organic acids, which modulated phytohormones to help plants combat salt stress. The study highlights that, beyond the induction of IAA, ACC deaminase activity, siderophore production, and biofilm formation, the production of organic acids and phosphate solubilization plays a crucial role in validating plant growth-promoting capabilities.

Biosurfactants

Marine environments pose challenges for most life forms due to high osmotic pressure and salinity and become even more hostile due to nutrient scarcity and the presence of toxic pollutants (Kubicki et al. 2019). Despite these challenges, various niches harbor microorganisms specifically adapted to cope with these conditions, utilizing evolved mechanisms developed over millions of years (Olasanmi and Thring 2018). Therefore, marine microorganisms showcase distinctive metabolic and physiological characteristics, serving as a significant reservoir for biomolecules like biosurfactants.

Biosurfactants are microbial surfactants comprising amphiphilic compounds with distinct hydrophobic and hydrophilic groups. Their production could serve as one such survival mechanism in saline environments and can be produced abundantly by a range of microorganisms that are still relatively unexplored.

These extracellular molecules are produced by microbial cell surfaces during stationary growth and exhibit the ability to reduce surface and interfacial tensions in fluids (Fariq and Yasmin 2020). Hence, biosurfactants find applications as detergents, wetting and foaming agents, solubilizers, emulsifiers, and dispersants, being interesting tools in biomedicine, bioremediation, microbial oil recovery, waste management, food processing and agriculture (Fariq and Saeed 2016, Sharma 2016b).

Some marine bacteria have gained recognition as valuable reservoirs of biosurfactants with surface-active characteristics. Biosurfactants derived from marine microorganisms include a high structural diversity, encompassing a broad spectrum of fatty acid derivatives, lipoamino acids, lipopolysaccharides, lipopeptides, phospholipids and glycolipids (Sobrinho et al. 2013).

Specifically, glycolipid biosurfactants from marine microorganisms are known to exhibit antimicrobial potential. Strains such *Brevibacterium casei*, *Serratia marcescens*, *Streptomyces* sp., and *Aspergillus ustus*, are recognized for their capacity to act as biosurfactants with antimicrobial, anti-adhesive, and anti-biofilm properties against a wide range of human pathogens (Gudiña et al. 2016).

Beyond medicine, in agriculture, biosurfactants can be applied for plant pathogen elimination, providing support through the enhancement of beneficial microbe-plant interactions and protection from phytopathogens. Additionally, they contribute to soil improvement, stimulate effective foliar fertilizer uptake, and enhance nutrient bioavailability (Sachdev and Cameotra 2013, Kubicki et al. 2019), showing potential as tools for replacing aggressive surfactants in pesticide industries.

Traditional approaches are employed to screen the microbial community for biosurfactant production. However, there is a scarcity of research on halophilic and marine microorganisms, limiting access to biosurfactants from these diverse microbial populations. In this context, there is a necessity to investigate more innovative biosurfactants that have the potential to enhance drought resistance in plants.

Secondary metabolites

Microbial secondary metabolites are low-molecular-mass products that are not essential for the growth of the producing cultures (Ruiz et al. 2010). This is partly attributed to the ecological interactions of these microorganisms in their natural habitats, including both intra and interspecific competition. In instances where environmental resources are limited, some bacteria are eliminated by different species, prompting them to produce compounds that negatively impact their competitors (Aloo et al. 2023), thus the production of secondary metabolites by microorganisms may be directly linked to these interactions.

Due to their highly adaptable metabolism in challenging environments, halophilic microorganisms possess the capability to synthesize a diverse range of secondary metabolites with unique chemical structures. In agriculture, some of these biomolecules can be used as biopesticides (herbicide, fungicide, insecticide, and acaricide), disease resistance inducers, PGP (phytohormone production), abiotic (growth inhibition due to chemical application, salinity, and stress water) and biotic (pests and diseases) stress mitigators (Aloo et al. 2023).

Concerning the bioprospecting of metabolites from marine microbial halophiles applicable in agricultural systems, Ogaki et al. (2019) evaluated *Peniciullium* spp., and *Acremonium* sp. strains isolated from Antarctic deep-sea sediments that showed herbicidal and antifungal activities against the phytopathogen *Cladosporium cladosporioides*. A Nuclear Magnetic Resonance Spectroscopy (NMR) analysis showed the presence of aromatic compounds that could serve as natural antimicrobial agents.

In another screening assays with halophiles, Radouane et al. (2023) isolated halophilic bacteria from saline water of the Dead Sea, the Agadir Sea and the Rabat Sea. These bacteria were *Bacillus amyloliquefaciens*, *Bacillus subtilis*, *Bacillus velezensis*, *Bacillus vallismortis* and *Ochrobactrum thiophenivorans*. These strains presented antifungal activity against *Penicillium digitatum* and *Penicillium italicum* in citrus disease *in vitro*, probably by the liberation of bioactive metabolites in culture media that inhibited mycelial growth of fungal strains.

Bibi et al. (2018) also identified antimicrobial metabolites produced by halophilic bacterial strains isolated from rhizosphere and endophytic association with *Salsola imbricata* (mangrove), such as *Actinobacteria* spp., *Nocardiopsis dassonvillei*, *Pseudonocardia* sp., *Arthrobacter* sp., and *Streptomyces* sp. Some of the compounds produced by the strains, such as sulfamethoxypyridazine, sulfamerazine, and dimetridazole, have been previously reported for their diverse biological activities, including antibiotic properties. These same bacterial strains were also evaluated in screening assays and demonstrated antifungal activity against five different phytopathogens.

Once identified as promising and safe for human and animal health, all these biomolecules bioprospected from marine sources can serve as valuable tools in modern and sustainable agriculture. One of them is through the application of nanotechnology for crop protection. Nanotechnology has the potential to enhance both the competitiveness and sustainability of agriculture by encapsulating bioactives in nanoparticle systems (nanoformulations). It improves the application of these biological substances to pests, mitigating their rapid physical degradation and providing more stability (Koul 2019).

Nanoformulation technologies, such as nanoemulsification, green synthesis, and biosynthesis or biogenic synthesis, enable the controlled release of control agents and the delivery of biomolecules to specific targets (Schmaltz et al. 2023). This, in turn, enhances the delivery efficacy of the bioactive substance and imparts greater safety profiles to biopesticides, thereby reducing their toxicity to humans and the environment. This increased safety contributes to greater adoption by end-users (Koul 2019). All these techniques can connect the knowledge gained from research on microbial secondary metabolites, such as halophilic biomolecules, with practical applications in both small and large-scale farming.

New perspectives in the exploration of halophilic microorganisms

To bring industrial biotechnology to the same level of competitiveness as the chemical industry, the methods based on biological products must not only match their chemical counterparts in efficacy but also ensure energy and time advantages, minimal water usage, substrate economy, and a contamination-free continuous fermentation process (Liu et al. 2019). Halophiles, with their ability to thrive in harsh conditions where few other organisms survive, offer significant advantages in bioprocesses by reducing microbial contamination, especially in processes lacking sterility control. However, despite these production benefits, there are persistent challenges in biomolecule production by halophilic microorganisms that need to be addressed.

Bioprospecting poses one of the greatest challenges in utilizing halophiles in bioprocesses due to the high cost of bioprocessing, which is a significant drawback. Consequently, new approaches are being developed to tackle this challenge (Liu et al. 2019). Given the difficulties and costs associated with bioprocessing halophiles, significant genetic advancements have been made to enhance the performance of these microorganisms in industrial applications, such as in the production of polyhydroxyalkanoates (PHAs), ectoines, biosurfactants, and antioxidants.

In this regard, studies utilizing "omics" approaches can offer insights into ecology and may also be relevant for future genetic manipulations and bioprospecting assays, opening new possibilities for the utilization of halophilic microorganisms (Bolhuis et al. 2017, Tutuncu et al. 2019). Genome manipulation is currently a new strategy, primarily in industry, as emerging genomic tools, including genomics, metagenomics, transcriptomics, proteomics, and metabolomics, can reveal many biosynthetic capabilities of individual microorganisms or microbial communities in hypersaline environments (Wang et al. 2014).

Genome exploration has become an essential and widely adopted tool for understanding the genetic basis of secondary metabolite production by analyzing genetic profiles of biosynthetic gene clusters (BGCs) encoding essential enzymes in synthetic metabolic pathways. The use of Next-Generation Sequencing (NGS) has made the study of microbial diversity more accessible, allowing for the identification of hidden metabolic pathways and the genes involved in these activities (Corral et al. 2019).

Metagenomic analyses can uncover the composition and microbial biodiversity without the need for individual microorganism identification, while also assessing the potential of metabolic genes in the evaluated organisms. For example, Couto-Rodríguez and Montalvo-Rodríguez (2019) identified halophilic bacteria belonging to the Euryarchaeota, Bacteroidetes, and Proteobacteria groups in thalassohaline environments with a diversity of metabolic genes related to nitrogen fixation, ammonia oxidation, sulfate reduction, sulfur oxidation, and phosphate solubilization. Can-Herrera et al. (2021) discovered, also through metagenomic studies, the presence of *Bacillus subtilis* bacteria in estuary samples from Campeche, Mexico. The analyses also revealed that *B. subtilis* could be an efficient source of proteases, lipases, and cellulases from halophilic microbial communities in unexplored areas. Thus, bioinformatics studies and genetic analyses can also uncover the presence of key genes for the synthesis of compounds of interest.

Another noteworthy example pertains to polyhydroxyalkanoates (PHAs), which hold promise as bioplastic materials but are hindered by their high industrial production costs, constraining their commercial viability despite their environmentally friendly properties. Hou et al. (2013), through transcriptome analyses, detected the presence of an enzyme (β-ketothiolase) accountable for PHA accumulation in *Haloferax mediterranei* strains via two co-transcribed gene sets (referred to as bktBαβ and phaAαβ), indicating that such halophiles could serve as promising cell factories for PHA synthesis.

H. mediterranei stands out as one of the most well-known halophilic microorganisms within the Archaea domain, drawing attention for its potential in biotechnological applications (Matarredona et al. 2021). However, its utilization in fermentative processes can incur high costs due to its preference for defined or complex media and its ability to metabolize a diverse array of carbon sources. Hence, studies like these underscore the potential application of omics tools in pre-industrial process evaluations.

However, despite the numerous advancements in omics, further development of these available genetic tools is still necessary, as they have some drawbacks due to the lack of tangible results translating into known or even novel compounds. Nonetheless, they can still be considered cost- and timesaving compared to traditional methods of halophilic microorganism bioprocessing, which often require complex nutritional demands and production parameters.

Conclusion

The bioprospection of halophilic and halotolerant microorganisms in marine environments presents a fascinating avenue for scientific research. Some marine environments can exhibit harsh conditions for microbial survival, such as high salt concentrations, UV radiation, or extreme temperatures. As mentioned in this review, halophilic microorganisms, due to their unique characteristics, can produce various bioactive compounds that hold promising applications, particularly in agriculture. These biomolecules, including biosurfactants, exopolysaccharides, siderophores, pigments, enzymes, and other extracellular metabolites, offer a rich source for the identification of new natural products that can be applied in plant growth promotion, biocontrol activity, bioremediation, and protective biomolecules. Considering salinity as an imminent global challenge due to the salinization of several previously agricultural areas, a significant number of these halophilic biomolecules are interesting for the development of different bioproducts, which may contribute to a sustainable agriculture model under salt stress or even mesophilic conditions in future assays.

Acknowledgements

This work was carried out with support from the Postdoctoral scholarship financed by the Coordenação de Aperfeiçoamento de Pessoal de Nível Superior - Brasil (CAPES) - Financing Code 001.

References

Abaramak, G., Kırtel, O. and Toksoy Öner, E. (2020). Fructanogenic halophiles: A new perspective on extremophiles. pp. 123–13. *In*: Salwan, R. and Sharma, V. (eds.). Physiological and Biotechnological Aspects of Extremophiles. Elsevier.

Abbasi, H., Jamil, M., Haq, A., Ali, S., Ahmad, R. and MaliK, Z. (2016). Salt stress manifestation on plants, mechanism of salt tolerance and potassium role in alleviating it: a review. Zemdirbyste-Agricul. 103: 229–238.

Abd El-Azeem, S.A.M., Elwan, M.W.M., Sung, J.K. and Ok, Y.S. (2012). Alleviation of salt stress in eggplant (*Solanum melongena* L.) by plant-growth-promoting rhizobacteria. Commun. Soil Sci. Plant Anal. 43: 1303–1315.

Abdulrahman, I., Jamal, M.T., Pugazhendi, A., Dhavamani, J. and Satheesh, S. (2022). Antibiofilm activity of secondary metabolites from bacterial endophytes of Red Sea soft corals. Int. Biodeter. Biodegr. 173: 105462.

Acuna, J.J., Campos, M., Mora, M.L., Jaisi, D.P. and Jorquera, A. (2019). ACCD-producing rhizobacteria from an Andean Altiplano native plant (*Parastrephia quadrangularis*) and their potential to alleviate salt stress in wheat seedlings. Appl. Soil Ecol. 136: 184–190.

Ahmed, E. and Holmström, S.J.M. (2015). Microbe–mineral interactions: The impact of surface attachment on mineral weathering and element selectivity by microorganisms. Chem. Geo. 403: 13–23.

Ahmed, M., Hasanuzzaman, M., Raza, M.A., Malik, A. and Ahmad, S. (2020). Plant nutrients for crop growth, development and stress tolerance. pp. 43–92. *In*: Roychowdhury, R., Choudhury, S., Hasanuzzaman, M. and Srivastava, S. (eds.). Sustainable Agriculture in the Era of Climate Change. Springer International Publishing, Springer, Cham.

Aljohny, B. (2015). Halophilic bacterium—A review of new studies. Biosci. Biotechnol. Res. Asia 12(3): 2061–2069.

Al-Karablieh, N. (2017). Antimicrobial activity of bacillus persicus 24-DSM isolated from dead sea mud. Open Microbiol. J. 11(1): 372–383.

Aloo, B.N., Nyongesa, B.O., Were, J.O., Were, B.A. and Tumuhairwe, J.B. (2023). Rhizobacterial biomolecules for sustainable crop production and environmental management: plausible functions and molecular mechanisms. pp. 1–30. *In*: Kumar, A., Bilal, M., Ferreira, L.F. and Madhuree, K. (eds.). Microbial Biomolecules: Emerging Approach in Agriculture, Pharmaceuticals and Environment Management. Academic Press.

Apel, K. and Hirt, H. (2004). Reactive oxygen species: metabolism, oxidative stress and signal transduction. Annu. Rev. Plant Biol. 55: 373–399.

Ashraf, M., Hasnain, S., Berge, O. and Mahmood, T. (2004). Inoculating wheat seedlings with exopolysaccharide-producing bacteria restricts sodium uptake and stimulates plant growth under salt stress. Biol. Fertil. Soils. 40: 157–162.

Bano, N. and Musarrat, J. (2003). Characterization of a new *Pseudomonas aeruginosa* strain NJ-15 as a potential biocontrol agent. Curr. Microbiol. 46: 324–328.

Bashan, Y., Salazar, B., Puente, M.E., Bacilio, M. and Linderman, R. (2009). Enhanced establishment and growth of giant cardon cactus in an eroded field in the Sonoran Desert using native legume trees as nurse plants aided by plant growth-promoting microorganisms and compost. Biol. Fert. Soils 6: 585–594.

Bashan, Y., Holguin, G. and de-Bashan, L.E. (2004). *Azospirillum*-plant relationships: physiological, molecular, agricultural and environmental advances (1997–2003). Can. J. Microbiol. 50: 521–577.

Bertrand, H., Plassard, C., Pinochet, X., Touraine, B., Normand, P. and Cleyet-Marel, J.C. (2000). Stimulation of the ionic transport system in Brassica napus by a plant growth-promoting rhizobacterium (*Achromobacter* sp.). Can. J. Microbiol. 46: 229–236.

Bibi, F., Strobel, G.A., Naseer, M.I., Yasir, M., Khalaf Al-Ghamdi, A.A. and Azhar, I.E. (2018). Halophytes-associated endophytic and rhizospheric bacteria: Diversity, antagonism and metabolite production. Biocontrol Sci. Technol. 28(2): 192–213.

Bolhuis, H., Martín-Cuadrado, A.B., Rosselli, R., Pašić, L. and Rodriguez-Valera, F. (2017). Transcriptome analysis of *Haloquadratum walsbyi*: Vanity is but the surface. BMC Genomics 18(1).

Bowers, K.J., Mesbah, N.M. and Wiegel, J. (2009). Biodiversity of poly-extremophilic Bacteria: Does combining the extremes of high salt, alkaline pH and elevated temperature approach a physico-chemical boundary for life? Aquat. Biosyst. 5: 9.

Bremer, E. and Krämer, R. (2019). Responses of microorganisms to osmotic stress. Annu. Rev. Microbiol. 73: 313–334.

Bui, E.N. (2013). Soil salinity: a neglected factor in plant ecology and biogeography. J. Arid. Environ. 92: 14–25.

Can-Herrera, L.A., Gutierrez-Canul, C.D., Dzul-Cervantes, M.A.A., Pacheco-Salazar, O.F., Chi-Cortez, J.D. and Carbonell, L.S. (2023). Identification by molecular techniques of halophilic bacteria producing important enzymes from pristine area in Campeche, Mexico. Braz. J. Biol. 83: 1–9.

Castro, R.A., Dourado, M.N., de Almeida, J.R., Lacava, P.T., Nave, A., de Melo, I.S. et al. (2018). Mangrove endophyte promotes reforestation tree (*Acacia polyphylla*) growth. Braz. J. Microbiol. 49: 59–66.

Chaney, K. and Swift, R.S. (1986). Studies on aggregate stability. I. Reformation of soil aggregates. J. Soil Sci. 37: 329–335.

Chawla, S., Jain, S. and Jain, V. (2013). Salinity induced oxidative stress and antioxidant system in salt-tolerant and salt-sensitive cultivars of rice (*Oryza sativa* L.). J. Plant Biochem. Biotechnol. 22: 27–34.

Chenu, C. (1993). Clay- or sand-polysaccharide associations as models for the interface between micro-organisms and soil: Water related properties and microstructure. Geoderma. 56: 143–156.

Corral, P., Amoozegar, M.A. and Ventosa, A. (2019). Halophiles and their biomolecules: recent advances and future applications in biomedicine. Mar. Drugs 18: 33.

Couto-Rodríguez, R.L. and Montalvo-Rodríguez, R. (2019). Temporal analysis of the microbial community from the crystallizer ponds in Cabo Rojo, Puerto Rico, using metagenomics. Genes 10(6): 422.

Cui, H.L. and Dyall-Smith, M.L. (2021). Cultivation of halophilic archaea (class Halobacteria) from thalassohaline and athalassohaline environments. Mar. Life Sci. Technol. 3(2): 243–251.

Dang, H., Zhu, H., Wang, J. and Li, T. (2009). Extracellular hydrolytic enzyme screening of culturable heterotrophic bacteria from deep-sea sediments of the Southern Okinawa Trough. World J. Microbiol. Biotechnol. 25: 71–79.

de Lourdes Moreno, M., Pérez, D., García, M. and Mellado, E. (2013). Halophilic bacteria as a source of novel hydrolytic enzymes. Life. 3(1): 38–51.

Delabary, G.S., Silva, M.C., da Silva, C.S., Baratieri, L.Z., Melo, T.M., Stramosk, C.A. et al. (2020). Influence of temperature and culture media on growth and lipolytic activity of deep-sea *Halomonas sulfidaeris* LAMA 838 and *Marinobacter excellens* LAMA 842. Ocean Coastal Res. 68.

Dell'Anno, A. and Danovaro, R. (2005). Extracellular DNA plays a key role in deep-sea ecosystem functioning. Science 309: 2179.

Diba, H., Cohan, R.A., Salimian, M., Mirjani, R., Soleimani, M. and Khodabakhsh, F. (2021). Isolation and characterization of halophilic bacteria with the ability of heavy metal bioremediation and nanoparticle synthesis from Khara Salt Lake in Iran. Arch. Microbiol. 203(7): 3893–3903.

Dukare, A. and Paul, S. (2021). Biological control of *Fusarium* wilt and growth promotion in pigeon pea (*Cajanus cajan*) by antagonistic rhizobacteria, displaying multiple modes of pathogen inhibition. Rhizosphere 17: 100278.

Edbeib, M.F., Wahab, R.A. and Huyop, F. (2016). Halophiles: biology, adaptation and their role in decontamination of hypersaline environments. World J. Microbiol. Biotechnol. 32(8): 135.

Egamberdieva, D. and Kucharova, Z. (2009). Selection for root colonizing bacteria stimulating wheat growth in saline soils. Biol. Fertility Soil. 45: 563–571.

el Kinany, S., el Hassan, A., Symanczik, S., Ouzine, M., Oufli, M., Haggoud, A. et al. (2021). Characterization of plant growth-promoting mechanisms of rhizobacteria from the rhizosphere of date palms in tafilalet oases. J. Global Agri. Ecol. 12(3): 1–18.

El-Bondkly, E.A.M., El-Bondkly, A.A.M. and El-Bondkly, A.A.M. (2021). Marine endophytic fungal metabolites: A whole new world of pharmaceutical therapy exploration. Heliyon. 8;7(3): e06362.

El-Sayed, S.Y. and Hagab, R.H. (2020). Effect of organic acids and plant growth promoting rhizobacteria (PGPR) on biochemical content and productivity of wheat under saline soil conditions. Middle East J. 9(2): 227–242.

El-Tarabily, K. and Youssef, T. (2010). Enhancement of morphological, anatomical and physiological characteristics of seedlings of the mangrove *Avicennia marina* inoculated with a native phosphate-solubilizing isolate of *Oceanobacillus picturae* under greenhouse conditions. Plant Soil. 332: 147–162.

El-Tarabily, K. and Youssef, T. (2011). Improved growth performance of the mangrove *Avicennia marina* seedlings using a 1-aminocyclopropane-1-carboxylic acid deaminase-producing isolate of Pseudoalteromonas maricaloris. Plant Growth Regul. 65: 473–483.

El-Tarabily, K.A., Sham, A., Elbadawi, A.A., Hassan, A.H., Alhosani, B.K.K., El-Esawi, M.A. et al. (2021). A consortium of rhizosphere-competent Actinobacteria exhibiting multiple plant growth-promoting traits improves the growth of *Avicennia marina* in the United Arab Emirates. Front. Mar. Sci. 8: 715123.

El-Tarabily, K.A., ElBaghdady, K.Z., AlKhajeh, A.S., Ayyash, M.M., Aljneibi, R.S., El-Keblawy, A. et al. (2020). Polyamine-producing actinobacteria enhance biomass production and seed yield in Salicornia bigelovii. Biol. Fertil. Soils 56: 499–519.

Fariq, A. and Saeed, A. (2016). Production and biomedical applications of probiotic biosurfactants. Curr. Microbiol. 72: 489–495.

Fariq, A. and Yasmin, A. (2020). Production, characterization and bioactivities of biosurfactants from newly isolated strictly halophilic bacteria. Process Biochem. 98: 1–10.

Galinski, E.A. (1995). Osmoadaptation in bacteria. Adv. Microb. Physiol. 37: 272–328.

Gan, L., Li, X., Wang, H., Peng, B. and Tian, Y. (2019). Structural characterization and functional evaluation of a novel exopolysaccharide from the moderate halophile *Gracilibacillus* sp. SCU50. Int. J. Biol. Macromol. 154: 1140–1148.

Geddie, J.L. and Sutherland, I.W. (1993). Uptake of metals by bacterial polysaccharides. J. Appl. Bacteriol. 74: 467–472.

Geilfus, C.M. (2018). Chloride: from nutrient to toxicant. Plant Cell Physiol. 59: 877–886.

Glick, B.R. (2014). Bacteria with ACC deaminase can promote plant growth and help to feed the world. Microbiol. Res. 169: 30–39.

Glick, B.R. (2015). Introduction to plant growth promoting bacteria. pp. 1–28. *In*: Glick, B.R. (ed.). Beneficial Plant-Bacterial Interactions. Springer International Publishing, Switzerland.

Gowtham, H.G., Hariprasad, P., Nayak, S.C. and Niranjana, S.R. (2016). Application of rhizobacteria antagonistic to *Fusarium oxysporum* f. sp. *lycopersici* for the management of Fusarium wilt in tomato. Rhizosphere 2: 72–74.

Gudiña, E., Teixeira, J. and Rodrigues, L. (2016). Biosurfactants produced by marine microorganisms with therapeutic applications. Mar. Drugs 14(2): 38.

Guezennec, J. (2002). Deep-sea hydrothermal vents: A new source of innovative bacterial exopolysaccharides of biotechnological interest. J. Ind. Microbiol. Biot. 29: 204–208.

Gugliandolo, C. and Maugeri, T.L. (1998). Temporal variations in heterotrophic mesophilic bacteria from a marine shallow hydrothermal vent off the Island of Vulcano (Eolian Islands, Italy) Microb. Ecol. 36: 13–22.

Guleria, S., Walia, A., Chauhan, A. and Shirkot, C.K. (2016). Molecular characterization of alkaline protease of *Bacillus amyloliquefaciens* SP1 involved in biocontrol of *Fusarium oxysporum*. Int. J. Food Microbiol. 232: 134–143.

Gunde-Cimerman, N., Ramos, J. and Plemenitaš, A. (2009). Halotolerant and halophilic fungi. Mycol. Res. 113: 1231–1241.

Hamidi, M., Mirzaei, R., Delattre, C., Khanaki, K., Pierre, G., Gardarin, C. et al. (2019). Characterization of a new exopolysaccharide produced by *Halorubrum* sp. TBZ112 and evaluation of its anti-proliferative effect on gastric cancer cells. Biotech. 9(1): 1.

Hasana, R. and Mayake, H. (2017). Salinity stress alters nutrient uptake and causes the damage of root and leaf anatomy in maize. KnE Life Sci. 3: 219–225.

Hassan, M.K., McInroy, J.A. and Kloepper, J.W. (2019). The interactions of rhizodeposits with plant growth-promoting rhizobacteria in the rhizosphere: a review. Agriculture 9: 1–42.

Henciya, S., Vengateshwaran, T.D., Gokul, M.S., Dahms, H.U. and James, R.A. (2020). Antibacterial activity of halophilic bacteria against drug-resistant microbes associated with diabetic foot infections. Curr. Microbiol. 77(11): 3711–3723.

Hou, J. and Cui, H.L. (2018). *In vitro* antioxidant, antihemolytic and anticancer activity of the carotenoids from halophilic archaea. Curr. Microbiol. 75: 266–271.

Hou, J., Feng, B., Han, J., Liu, H., Zhao, D., Zhou, J. et al. (2013). Haloarchaeal-type β-ketothiolases involved in Poly(3-Hydroxybutyrate-co-3-Hydroxyvalerate) synthesis in *Haloferax mediterranei*. App. Environ. Microbiol. 79(17): 5104–5111.

Hussain, S.M., Ahmad, N., Jabeen, F., Javid, A., Aslam, N., Hussain, M. et al. (2015). Effects of citric acid and phytase supplementation on nutrient digestibility and growth performance of *Cirrhinus mrigala* fingerlings fed on corn gluten (30%) meal based diets. Int. J. Biosciences 6(7): 82–91.

Israr, D., Mustafa, G., Khan, K.S., Shahzad, M., Ahmad, N. and Masood, S. (2016). Interactive effects of phosphorus and *Pseudomonas putida* on chickpea (*Cicer arietinum* L.) growth, nutrient uptake, antioxidant enzymes and organic acids exudation. Plant Physiol. Biochem. 108: 304–312.

Jouault, S.C., Chevolot, L., Helley, D., Ratiskol, J., Bros, A., Sinquin, C. et al. (2001). Characterization, chemical modifications and *in vitro* anticoagulant properties of an exopolysaccharide produced by *Alteromonas infernus*. BBA-Gen Subjects 1528: 141–151.

Kanagaraj, G. and Desingh, R. (2017). Salinity influences physiological traits of seven Sesame (*Sesamum indicum* L.) varieties. J. Sci. Agri. 1: 188–196.

Kannan, K., Pitchiah, S., Joseph, J.G., Ganapathy, D., Sundarrajan, S. and Ramakrishna, S. (2022). Marine archaeal extracellular polymeric substances from *Halococcus* AMS12, their characterization, and biological properties. J. Mar. Sci. Eng. 10(11): 1788.

Kapadia, C., Patel, N., Rana, A., Vaidya, H., Alfarraj, S., Ansari, M.J. et al. (2022). Evaluation of plant growth-promoting and salinity ameliorating potential of halophilic bacteria isolated from saline soil. Front. Plant Sci. 13: 946217.

Karlidag, H., Esitken, A., Yildirim, E., Donmez, M.F. and Turan, M. (2010). Effects of plant growth promoting bacteria on yield, growth, leaf water content, membrane permeability and ionic composition of strawberry under saline conditions. J. Plant Nutr. 34: 34–45.

Kathiresan, K. and Selvam, M.M. (2006). Evaluation of beneficial bacteria from mangrove soil. Bot. Mar. 49: 86–88.

Koul, O. (2019). Nanobiopesticides: An introduction. pp. 1–15. *In*: Koul, O. (ed.). Nano-Biopesticides Today and Future Perspectives. Elsevier.

Krishnamurthy, M., Jayaraman Uthaya, C., Thangavel, M., Annadurai, V., Rajendran, R. and Gurusamy, A. (2020). Optimization, compositional analysis, and characterization of exopolysaccharides produced by multi-metal resistant *Bacillus cereus* KMS3-1. Carbohydrate Polymers 227: 115369.

Kubicki, S., Bollinger, A., Katzke, N., Jaeger, K.E., Loeschcke, A. and Thies, S. (2019). Marine biosurfactants: Biosynthesis, structural diversity and biotechnological applications. Mar. Drugs 17(7): 408.

Kushner, D.J. and Kamekura, M. (1998). Physiology of halophilic eubacteria. Halophilic Bact. 1: 109–138.

Liu, C., Baffoe, D.K., Zhan, Y., Zhang, M., Li, Y. and Zhang, G. (2019). Halophile, an essential platform for bioproduction. J. Microbiologic Methods 166: 105704.

López, J.L., Alvarez, F., Príncipe, A., Salas, M.E., Lozano, M.J., Draghi, W.O. et al. (2018). Isolation, taxonomic analysis and phenotypic characterization of bacterial endophytes present in alfalfa (*Medicago sativa*) seeds. J. Biotechnol. 267: 5562.

Loukas, A., Kappas, I. and Abatzopoulos, T.J. (2018). HaloDom: A new database of halophiles across all life domains. J. Biol. Res. Thessaloniki. 25(1).

Ma, Y., Galinski, E.A., Grant, W.D., Oren, A. and Ventosa, A. (2010). Halophiles: Life in saline environments. Appl. Environ. Microbiol. 76: 6971–6981.

Ma, Y., Prasad, M.N.V., Rajkumar, M. and Freitas, H. (2011). Plant growth promoting rhizobacteria and endophytes accelerate phytoremediation of metalliferous soils. Biotechnol. Adv. 29: 248–258.

Macias-Benitez, S., Garcia-Martinez, A.M., Caballero Jimenez, P., Gonzalez, J.M., Tejada Moral, M. and Parrado Rubio, J. (2020). Rhizospheric organic acids as biostimulants: Monitoring feedbacks on soil microorganisms and biochemical properties. Front. Plant Sci. 11: 633.

Manning, D.A.C. (2010). Mineral sources of potassium for plant nutrition. A review. Agron. Sustain. Dev. 30: 281–294.

Martin, D.D., Ciulla, R.A. and Roberts, M.F. (1999). Osmoadaptation in archaea. Appl. Environ. Microbiol. 65: 1815–1825.

Matarredona, L., Camacho, M., Zafrilla, B., Bravo-Barrales, G., Esclapez, J. and Bonete, M.J. (2021). The survival of *Haloferax mediterranei* under stressful conditions. Microorganisms 9(2): 336.

Mathew, B.T., Torky, Y., Amin, A., Mourad, A.I., Ayyash, M.M., El-Keblawy, A. et al. (2020). Halotolerant marine rhizosphere-competent actinobacteria promote *Salicornia bigelovii* growth and seed production using seawater irrigation. Front. Microbiol. 11: 552.

Mendes, G., Murta, H.M., Valadares, R.V., Silveira, W.B., da Silva, I.R. and Costa, M.D. (2020). Oxalic acid is more efficient than sulfuric acid for rock phosphate solubilization. Miner. Eng. 155: 106458.

Montoya, L., Vizioli, C., Rodríguez, N., Rastoll, M.J., Amils, R. and Marin, I. (2013). Microbial community composition of Tirez lagoon (Spain), a highly sulfated athalassohaline environment. Aquat. Biosyst. 9(1).

Mora-Ruiz, M. del R., Font-Verdera, F., Orfila, A., Rita, J. and Rosselló-Móra, R. (2016). Endophytic microbial diversity of the halophyte *Arthrocnemum macrostachyumacross* plant compartments. FEMS Microbiol. Ecol. 92(9): fiw145.

Morcillo, R. and Manzanera, M. (2021). The effects of plant-associated bacterial exopolysaccharides on plant abiotic stress tolerance. Metabolites 11(6): 337.

Muleta, D., Assefa, F., Börjesson, E. and Granhall, U. (2013). Phosphate-solubilising rhizobacteria associated with *Coffea arabica* L. in natural coffee forests of southwestern Ethiopia. J. Saudi Soc. Agri. Sci. 12(1): 73–84.

Muńoz, J.A., Pérez-Esteban, B., Esteban, M., de la Escalera, S., Gómez, M.A., Martínez-Toledo, M.V. and González-López, J. (2001). Growth of moderately halophilic bacteria isolated from sea water using phenol as the sole carbon source. Folia Microbiologica 46(4): 297–302.

Nadeem, S.M., Ahmad, M., Zahir, Z.A., Javaid, A. and Ashraf, M. (2014). The role of mycorrhizae and plant growth promoting rhizobacteria (PGPR) in improving crop productivity under stressful environments. Biotechnol. Adv. 32: 429–448.

Nagaraju, Y., Mahadevaswamy, Naik, N.M., Gowdar, S.B., Narayanarao, K. and Satyanarayanarao, K. (2021). ACC deaminase-positive halophilic bacterial isolates with multiple plant growth-promoting traits improve the growth and yield of Chickpea (*Cicer arietinum* L.) under salinity stress. Front. Agron. 3: 681007.

Naghoni, A., Emtiazi, G., Amoozegar, M.A., Cretoiu, M.S., Stal, L.J., Etemadifar, Z. et al. (2017). Microbial diversity in the hypersaline Lake Meyghan, Iran. Sci. Reports 7(1).

Nazareth, S., Gonsalves, V. and Nayak, S. (2011). A first record of obligate halophilic aspergilli from the dead sea. Indian J. Microbiol. 52(1): 22–27.

Netondo, G.W., Onyango, J.C. and Beck, E. (2004). Sorghum and salinity: II. Gas exchange and chlorophyll fluorescence of sorghum under salt stress. Crop Sci. 44: 806–811.

Netondo, G.W., Onyango, J.C. and Beck, E. (2004). Sorghum and salinity: I. Response of growth, water relations and ion accumulation to NaCl salinity. Crop Sci. 44: 797–805.

Nishimura, T., Vertes, A.A., Shinoda, Y., Inui, M. and Yukawa, H. (2007). Anaerobic growth of *Corynebacterium glutamicum* using nitrate as a terminal electron acceptor. Appl. Microb. Biotech. 75: 889–897.

Nissenbaum, A. (1975). The microbiology and biogeochemistry of the Dead Sea. Microb. Ecol. 2: 139–161.

Ogaki, M.B., Coelho, L.C., Vieira, R., Neto, A.A., Zani, C.L., Alves, T.M. et al. (2019). Cultivable fungi present in deep-sea sediments of Antarctica: Taxonomy, diversity and bioprospecting of bioactive compounds. Extremophiles 24(2): 227–238.

Olasanmi, I. and Thring, R. (2018). The role of biosurfactants in the continued drive for environmental sustainability. Sustainability 10: 4817.

Or, D., Phutane, S. and Dechesne, A. (2007). Extracellular polymeric substances affecting pore-scale hydrologic conditions for bacterial activity in unsaturated soils. Vadose Zone J. 6: 298–305.

Orellana, R., Macaya, C., Bravo, G., Dorochesi, F., Cumsille, A., Valencia, R. et al. (2018). Living at the frontiers of life: Extremophiles in Chile and their potential for bioremediation. Front. Microbiol. 9: 2309.

Oren, A. (2002). Diversity of halophilic microorganisms: Environments, phylogeny, physiology and applications. J. Industrial Microbiol. Biotech. 28(1): 56–63.

Oren, A. (2008). Microbial life at high salt concentrations: phylogenetic and metabolic diversity. Saline Syst. 4: 2.

Orozco-Mosquedaa, M.C., Glick, B.R. and Santoyoc, G. (2020). ACC deaminase in plant growth-promoting bacteria (PGPB): An efficient mechanism to counter salt stress in crops. Microbiol. Res. 235: 126439.

Oyeleyeo, A. and Normi, Y.M. (2018). Chitinase: Diversity, limitations and trends in engineering for suitable applications. Bioscience Reports 38(4).

Passari, A.K., Lalsiamthari, P.C., Zothanpuia, V.V., Mishra, V.K., Yadav, M.K. et al. (2018). Biocontrol of Fusarium wilt of *Capsicum annuum* by rhizospheric bacteria isolated from turmeric endowed with plant growth promotion and disease suppression potential. European J. Plant Pathol. 150(4): 831846.

Plemenitaš, A., Lenassi, M., Konte, T., Kejžar, A., Zajc, J., Gostinčar, C. et al. (2014). Adaptation to high salt concentrations in Halotolerant/Halophilic fungi: A molecular perspective. Front. Microbiol. 5: 199.

Qaiser, J., Seong, L.Y., Deok, J.H., Suk, P.Y. and Yong, K.K. (2015). Isolation and biocontrol potential of *Bacillus amyloliquefaciens* Y1 against fungal plant pathogens. Korean J. Soil Sci. Fertilizer. 48(5): 485491.

Qi, M., Zheng, C., Wu, W., Yu, G. and Wang, P. (2022). Exopolysaccharides from marine microbes: Source, structure and application. Mar: Drugs 20(8): 512.

Radouane, N., Adadi, H., Ezrari, S., Kenfaoui, J., Belabess, Z., Mokrini, F. et al. (2023). Exploring the bioprotective potential of halophilic bacteria against major postharvest fungal pathogens of citrus fruit *Penicillium digitatum* and *Penicillium italicum*. Horticulturae 9(8): 922.

Rahman, A.K.M.M., Ahmed, K.M., Butler, A.P. and Hoque, M.A. (2018). Influence of surface geology and micro-scale land use on the shallow subsurface salinity in deltaic coastal areas: a case from southwest Bangladesh. Environ. Earth Sci. 77: 423.

Rajkumar, M., Ae, N. and Freitas, H. (2009). Endophytic bacteria and their potential to enhance heavy metal phytoextraction. Chemosphere 77(2): 153160.

Ramesh, A., Sharma, S.K., Sharma, M.P., Yadav, N. and Joshi, O.P. (2014). Inoculation of zinc solubilizing *Bacillus aryabhattai* strains for improved growth, mobilization and biofortification of zinc in soybean and wheat cultivated in Vertisols of central India. Appl. Soil Ecol. 1(73): 87–96.

Ravikumar, S., Kathiresan, K., Ignatiammal, S.T., Selvam, M.B. and Shanthy, S. (2004). Nitrogen-fixing azotobacters from mangrove habitat and their utility as marine biofertilizers. J. Exp. Mar. Biol. Ecol. 312: 5–17.

Reang, L., Bhatt, S., Tomar, R.S. et al. (2022). Plant growth promoting characteristics of halophilic and halotolerant bacteria isolated from coastal regions of Saurashtra Gujarat. Sci. Rep. 12: 4699.

Reinhold-Hurek, B., Maes, T., Gemmer, S., Van Montagu, M. and Hurek, T. (2006). An endoglucanase is involved in infection of rice roots by the not-cellulose-metabolizing endophyte *Azoarcus* sp. Strain BH72. Mol. Plant Microbe Interact. 19(2): 181188.

Renn, D., Shepard, L., Vancea, A., Karan, R., Arold, S.T. and Rueping, M. (2021). Novel enzymes from the red sea brine pools: Current state and potential. Front. Microbiol. 12: 732856.

Roberts, M.F. (2004). Osmoadaptation and osmoregulation in archaea: update. Front Biosci. 9: 1999–2019.

Roberts, M.F. (2005). Organic compatible solutes of halotolerant and halophilic microorganisms. Saline Syst. 1(1): 5.

Rodriguez-Valera, F., Ruiz-Berraquero, F. and Ramos-Cormenzana, A. (1981). Characteristics of the heterotrophic bacterial populations in hypersaline environments of different salt concentrations. Microb. Ecol. 7: 235–243.

Rojas-Tapias, D., Moreno-Galván, A., Pardo-Díaz, S., Obando, M., Rivera, D. and Bonilla, R. (2012). Effect of inoculation with plant growth-promoting bacteria (PGPB) on amelioration of saline stress in maize (*Zea mays*). Appl. Soil Ecol. 61: 264–272.

Roy, S. and Roy, M. (2019). Characterization of plant growth promoting feature of a neutromesophilic, facultatively chemolithoautotrophic, sulphur oxidizing bacterium *Delftia* sp. strain SR4 isolated from coal mine spoil. Int. J. Phytoremediation. 21: 531–540.

Ruiz, B., Chávez, A., Forero, A., García-Huante, Y., Romero, A., Sánchez, M. et al. (2010). Production of microbial secondary metabolites: Regulation by the carbon source. Crit. Rev. Microbiol. 36(2): 146–167.

Rus, A., Yokoi, S., Sharkhuu, A., Reddy, M., Lee, B.H., Matsumoto, T.K. et al. (2001). AtHKT1 is a salt tolerance determinant that controls Na entry into plant roots. Proc. Natl. Acad. Sci. 98: 14150–14155.

Sabet, S. (2012). Halophilic viruses. pp. 81–116. *In*: Vreeland, R.H. (ed.). Advances in Understanding the Biology of Halophilic Microorganisms. Springer, Netherlands.

Sachdev, D.P. and Cameotra, S.S. (2013). Biosurfactants in agriculture. Appl. Microbiol. Biotechnol. 97(3): 1005–1016.

Samaras, A., Efthimiou, K., Roumeliotis, E. and Karaoglanidis, G.S. (2018). Biocontrol potential and plant-growth promoting effects of *Bacillus amyloliquefaciens* MBI 600 against *Fusarium oxysporum* f. sp. *radicislycopersici* on tomato. Acta Horticulturae 1207: 139146.

Santhaseelan, H., Dinakaran, V.T., Dahms, H.U., Ahamed, J.M., Murugaiah, S.G., Krishnan, M. et al. (2022). Recent antimicrobial responses of halophilic microbes in clinical pathogens. Microorganisms 10(2): 417.

Santoro, M.V., Zygadlo, J., Giordano, W. and Banchio, E. (2011). Volatile organic compounds from rhizobacteria increase biosynthesis of essential oils and growth parameters in peppermint (*Mentha piperita*). Plant Physiol. Biochem. 49(10): 11771182.

Saraf, M., Jha, C.K. and Patel, D. (2010). The role of ACC deaminase producing PGPR in sustainable agriculture. pp. 365–385. *In*: Maheshwari, D.K. (ed.). Plant Growth and Health Promoting Bacteria. Springer.

Sarkar, A., Ghosh, P.K., Pramanik, K., Mitra, S., Soren, T., Pandey, S. et al. (2018). A halotolerant Enterobacter sp. displaying ACC deaminase activity promotes rice seedling growth under salt stress. Res. Microbiol. 169: 20–32.

Satbhai, A., Kasodekar, A., Pachuau, L. and Bharambe, N. (2015). Isolation of halophiles from the Dead Sea and exploring their potential biotechnological applications. Int. J. Curr. Microbiol. App. Sci. 2: 1–17.

Schenkel, D., Lemfack, M., Piechulla, B. and Splivallo, R. (2015). A meta-analysis approach for assessing the diversity and specificity of belowground root and microbial volatiles. Front. Plant Sci. 6: 707.

Schmaltz, S., Silva, M.A., Ninaus, R.G., Guedes, J.V.C., Zabot, G.L., Tres, M.V. et al. (2023). Biomolecules in modern and sustainable agriculture. Biotech. 13(2).

Schmid, J. (2018). Recent insights in microbial exopolysaccharide biosynthesis and engineering strategies. Curr. Opin. Biotechnol. 53: 130–136.

Schulz-Bohm, K., Martín-Sánchez, L. and Garbeva, P. (2017). Microbial volatiles: Small molecules with an important role in intra- and inter-kingdom interactions. Front. Microbiol. 8: 2484.

Seckbach, J., Oren, A. and Stan-Lotter, H. (2013). Polyextremophiles: Life Under Multiple Forms of Stress. Springer, Heidelberg, Germany.

Sharma, A., Vaishnav, A., Jamali, H., Srivastava, A.K., Saxena, A.K. and Srivastava, A.K. (2016a). Halophilic bacteria: Potential bioinoculants for sustainable agriculture and environment management under salt stress. pp. 297–325. *In*: Choudhary, D., Varma, A. and Tuteja, N. (eds.). Plant-Microbe Interaction: An Approach to Sustainable Agriculture. Springer, Singapore.

Sharma, D. (2016b). Classification and Properties of Biosurfactants, Biosurfactants in Food. Springer.

Sharma, P., Jha, A.B., Dubey, R.S. and Pessarakli, M. (2012). Reactive oxygen species, oxidative damage and antioxidative defense mechanism in plants under stressful conditions. J. Botany. 217037.

Shilev, S. (2020). Plant-growth-promoting bacteria mitigating soil salinity stress in plants. Appl. Sci. 10: 7326.

Shrivastava, P. and Kumar, R. (2015). Soil salinity: A serious environmental issue and plant growth promoting bacteria as one of the tools for its alleviation. Saudi J. Biol. Sci. 22(2): 123–131.

Singh, P., Chauhan, P.K., Upadhyay, S.K., Singh, R.K., Dwivedi, P., Wang, J. et al. (2022). Mechanistic insights and potential use of siderophores producing microbes in rhizosphere for mitigation of stress in plants grown in degraded land. Front. Microbiol. 13: 898979.

Singh, R.P. and Jha, P.N. (2016). A halotolerant bacterium *Bacillus licheniformis* HSW-16 augments induced systemic tolerance to salt stress in wheat plant (*Triticum aestivum*). Front. Plant Sci. 7: 1890.

Śliżewska, W., Struszczyk-Świta, K. and Marchut-Mikołajczyk, O. (2022). Metabolic potential of halophilic filamentous fungi—current perspective. Int. J. Mol. Sci. 23(8): 4189.

Sobrinho, H.B., Luna, J.M., Rufino, R.D., Porto, A.L.F. and Sarubbo, L.A. (2013). Biosurfactants: classification, properties and environmental applications, Recent Dev. Biotechnol. 11: 1–29.

Soliman, G.S.H. and Trüper, H.G. (1982). *Halobacterium pharaonis* sp. nov., a new, extremely haloalkaliphilic archaebacterium with low magnesium requirement. Zentralbl Bakteriol. Hyg. Abt. I. Orig. C3: 318–329.

Sorokin, D.Y., Kuenen, J.G. and Muyzer, G. (2011). The microbial sulfur cycle at extremely haloalkaline conditions of soda lakes. Front Microbiol. 2: 44.

Sorokin, D.Y., Toshchakov, S.V., Kolganova, T. and Kublanov, I.V. (2015). Halo(natrono)archaea isolated from hypersaline lakes utilize cellulose and chitin as growth substrates. Front. Microbiol. 6: 942.

Tao, J.J., Chen, H.W., Ma, B., Zhang, W.K., Chen, S.Y. and Zhang, J.S. (2015). The role of ethylene in plants under salinity stress. Front. Plant Sci. 6: 1059.

Tavakkoli, E., Fatehi, F., Coventry, S., Rengasamy, P. and McDonald, G.K. (2011). Additive effects of Na+ and Cl− ions on barley growth under salinity stress. J. Exp. Bot. 62: 2189–2203.

Thombre, R.S., Joshi, V.S. and Oke, R.S. (2017). Halophiles: Pharmaceutical potential and biotechnological applications. pp. 111–139. *In*: Thangadurai, D. and Sangeetha, J. (eds.). Industrial Biotechnology. Apple Academic Press.

Tiwari, S., Lata, C., Chauhan, P.S. and Nautiyal, C.S. (2016). Pseudomonas putida attunes morphophysiological, biochemical and molecular responses in Cicer arietinum L. during drought stress and recovery. Plant Physiol. Biochem. 99: 108–117.

Trapet, P., Avoscan, L., Klinguer, A., Pateyron, S., Citerne, S., Chervin, C. et al. (2016). The Pseudomonas fluorescens siderophore pyoverdine weakens Arabidopsis thaliana defense in favor of growth in iron deficient conditions. Plant Physiol. 171(1): 675693.

Truitt, C. and Deole, R. (2021). Viruses of extremely halophilic prokaryotes. In Bacteriophages in Therapeutics. *In*: Bhardwaj, S.B. (ed.). Bacteriophages in Therapeutics. IntechOpen.

Tutuncu, H.E., Balci, N., Tuter, M. and Karaguler, N.G. (2019). Recombinant production and characterization of a novel esterase from a hypersaline lake, Acıgöl, by metagenomic approach. Extremophiles 23(5): 507–520.

Ullah, A., Bano, A. and Khan, N. (2021). Climate change and salinity effects on crops and chemical communication between plants and plant growth-promoting microorganisms under stress. Front. Sustain. Food Syst. 5: 618092.

Vavourakis, C.D., Ghai, R., Rodriguez-Valera, F., Sorokin, D.Y., Tringe, S.G., Hugenholtz, P. et al. (2016). Metagenomic insights into the uncultured diversity and physiology of microbes in four hypersaline soda lake brines. Front. Microbiol. 7: 211.

Vavourakis, C.D., Ghai, R., Rodriguez-Valera, F., Sorokin, D.Y., Tringe, S., Hugenholtz, P. et al. (2016). Metagenomic insights into the uncultured diversity and physiology of microbes in four hypersaline soda lake brines. Front. Microbiol. 7: 211.

Ventosa, A., Nieto, J.J. and Oren, A. (1998). Biology of moderately halophilic aerobic bacteria. Microbiol. Mol. Biol. Rev. 62: 504–544.

Ventosa, A. and Arahal, D.R. (2009). Physico-chemical characteristics of hypersaline environments and their biodiversity. In Extremophiles; CRC Press: Boca Raton, USA.

Ventosa, A., Oren, A. and Ma, Y. (2011). Halophiles and Hypersaline Environments, Springer: Berlin/ Heidelberg, Germany.

Veselova, M.A., Plyuta, V.A. and Khmel, I.A. (2019). Volatile compounds of bacterial origin: Structure, biosynthesis and biological activity. Microbiol. 88(3): 261–274.

Voothuluru, P. and Sharp, R.E. (2012). Apoplastic hydrogen peroxide in the growth zone of the maize primary root under water stress. Increased levels are specific to the apical region of growth maintenance. J. Experi. Botany. 64: 1223–1233.

Wang, H., Fewer, D.P., Holm, L., Rouhiainen, L. and Sivonen, K. (2014). Atlas of nonribosomal peptide and polyketide biosynthetic pathways reveals common occurrence of nonmodular enzymes. Proc. Nat. Acad. Sci. 111(25): 9259–9264.

Weinisch, L., Kühner, S., Roth, R., Grimm, M., Roth, T., Netz, D.J.A. et al. (2018). Identification of osmoadaptive strategies in the halophile, heterotrophic ciliate *Schmidingerothrix salinarum*. PLOS Biol. 16(1): e2003892.

White, P.J. and Broadley, M.R. (2001). Chloride in soils and its uptake and movement within the plant: A review. Ann. Bot. 44: 967–988.

Wielen, P.W.J.J., Bolhuis, H., Borin, S., Daffonchio, D., Corselli, C., Giuliano, L. et al. (2005). Bio deep scientific party: the enigma of prokaryotic life in deep hypersaline anoxic basins. Science 307: 121–123.

Wu, J.H., McGenity, T.J., Rettberg, P., Simões, M.F., Li, W.J. and Antunes, A. (2022). The archaeal class Halobacteria and astrobiology: Knowledge gaps and research opportunities. Front. Microbiol. 13: 1023625.

Yang, L., Wei, Z., Li, S., Xiao, R., Xu, Q., Ran, Y. et al. (2021). Plant secondary metabolite, daphnetin reduces extracellular polysaccharides production and virulence factors of *Ralstonia solanacearum*. Pestic. Biochem. Phys. 179: 104948.

Yazdani, M., Bahmanyar, M.A., Pirdashti, H. and Esmaili, M.A. (2009). Effect of phosphate solubilization microorganisms (PSM) and plant growth promoting rhizobacteria (PGPR) on yield and yield components of corn (*Zea mays* L.). World Acad. Sci. Eng. Technol. 49: 90–92.

Youssef, N.H., Ashlock-Savage, K.N. and Elshahed, M.S. (2012). Phylogenetic diversities and community structure of members of the extremely halophilic Archaea (order Halobacteriales) in multiple saline sediment habitats. Appl. Environ. Microbiol. 78: 1332–1344.

Yu, Y., Xu, T., Li, X., Tang, J., Ma, D., Li, Z. et al. (2016). NaCl-induced changes of ion homeostasis and nitrogen metabolism in two sweet potato (*Ipomoea batatas* L.) cultivars exhibit different salt tolerance at adventitious root stage. Environ. Exp. Bot. 129: 23–36.

Zajc, J., Kogej, T., Galinski, E.A., Ramos, J. and Gunde-Cimerman, N. (2014). Osmoadaptation strategy of the most halophilic fungus, *Wallemia ichthyophaga*, growing optimally at salinities above 15% NaCl. Appl. Environ. Microbiol. 80(1): 247–256.

Zanchetta, P., Lagarde, N. and Guezennec, J. (2003). A new bone-healing material: A hyaluronic acid like bacterial exopolysaccharide. Calcif. Tissue Int. 72: 74–79.

Zheng, B.X., Ibrahim, M., Zhang, D.P., Bi, Q.F., Li, H.Z., Zhou, G.W. et al. (2018). Identification and characterization of inorganic-phosphate-solubilizing bacteria from agricultural fields with a rapid isolation method. AMB Express. 8: 47.

Ziko, L., Adel, M., Malash, M.N. and Siam, R. (2019). Insights into red sea brine pool specialized metabolism gene clusters encoding potential metabolites for biotechnological applications and extremophile survival. Mar. Drugs 17: 273.

Chapter 4

Marine Microbes for Bioremediation of Fishing Industry Wastes towards the Blue Economy

Sihono,[1] *Endar Marraskuranto,*[1] *Nurhayati,*[1] *Muhamad Darmawan*[2] and *Hari Eko Irianto*[1,3,*]

Introduction

The fisheries and aquaculture industries represent the world's most important economic sectors and play a crucial role in providing the livelihood and food of millions of people, especially in developing countries (FAO 2021). The total world fisheries and aquaculture production reported by the Food and Agriculture Organization (FAO) reached 171 million tons in 2016 (Al Khawli et al. 2019) and increased to 177.8 million tons in 2019, of which 88% was intended for human consumption (USD 406 billion). The production was slightly higher in 2020, estimated at 178 million tons, contributed by capture fisheries at about 90 million tons (51%), and aquaculture at 88 million tons (49%). The export value of fisheries and aquaculture commodities in 2019 was USD 161 billion, around 11% of the export value of agricultural products and an increase of 7% annually (FAO 2021, FAO 2022, Khiari 2022).

[1] Research Center for Marine and Land Bioindustry, National Research and Innovation Agency of the Republic of Indonesia, Lombok, Indonesia.

[2] Research and Development Center for Merine and Fisheries Product Processing and Biotechnology, Ministry of Marine Affairs and Fisheries of the Republic of Indonesia, Jakarta, Indonesia.

[3] Faculty of Food Technology and Health, Sahid University, Jakarta, Indonesia.

* Corresponding author: hariekoirianto1960@gmail.com

The production and export value of fisheries and aquaculture is predicted to increase annually, which will encourage the rise of worldwide income, the development of technology in processing and value chains, and the increase in per capita fish consumption. However, 50–70% of the global production of fisheries and aquaculture is considered waste (Olsen et al. 2014, Khiari 2022). Over 20 million tons of discards from worldwide fisheries are produced annually, which is equal to 25% of the marine catch total production (Racioppo et al. 2021). Ideia et al. (2020) reported that more than 60% of the catch is discarded, only 40% being used for human consumption. Al Khawli et al. (2019) estimated that 30–85% of the total weight of the various catches is made up of fish by-products. Fish waste is typically burned, landfilled, thrown into the sea, or even left abandoned, all of which have detrimental effects on biodiversity, human health, and the environment (Racioppo et al. 2021).

The increase in fisheries production and consumption has consequently increased the amount of fish waste across the world. Waste generated from the fisheries industry and aquaculture exhibits serious global issues because of its impact on society, economy, and the environment. This is related to energy consumption across the life cycle stages, from fisheries production to waste generation (Abeliotis et al. 2015, Khiari 2022). The social ramifications of fish waste are diverse and can impact various facets of society. Fish waste reduces the accessibility of nutritious food supplies, contributing to food insecurity, especially in communities that heavily depend on seafood as their main protein source. Fish waste could also worsen social inequalities, as marginalized communities frequently endure the consequences of food insecurity and financial difficulties due to diminished access to fish resources (Gustavsson et al. 2011). Communities dependent on the seafood industry for their livelihoods may encounter social and cultural disruptions when fish waste results in economic decline and the loss of traditional practices (Pascual-Fernandez et al. 2020).

The economic impact of fish waste is significant, influencing multiple sectors of the economy. Fish waste leads to direct financial loss for fishermen, seafood processors, and businesses engaged in the seafood supply chain, thereby diminishing their total earnings and profitability (Kruijssen et al. 2020). The instability caused by fish waste in terms of availability and pricing can disrupt seafood markets, resulting in price fluctuation and decreased market efficiency (Racioppo et al. 2021). Fish waste has the potential to result in unemployment and financial uncertainty for individuals employed in the seafood sector, affecting their livelihoods and the economic stability of coastal communities heavily dependent on fishing (Coppola et al. 2021).

The environmental impact of fish waste are significant and can affect marine ecosystems, biodiversity, and natural resources. Fish waste contributes to the overexploitation of marine resources, resulting in the depletion of fish populations and degradation of ecosystems (Herbert-Read et al. 2022). Fish waste has the potential to change marine habitats and ecosystems, impacting biodiversity, coral reefs, and other delicate marine environments (Nama et al. 2023). Fish waste frequently entails the disposal of non-target species and bycatch, resulting in unintentional mortality and ecological imbalances within marine ecosystems (Gilman et al. 2020).

Food waste is "composed of raw or cooked food materials and includes food loss, before, during or after meal preparation in the household, as well as food discarded in the process of manufacturing, distribution, retail, and food service activities" (European Commission 2010). In terms of the fisheries industry and aquaculture, waste is divided into two categories: fish loss and fish waste. The characteristics of fish are perishable materials, therefore, fish loss and fish waste are very critical. It is estimated that annually, up to 35% of the world's productivity in fisheries and aquaculture is lost or wasted (FAO 2022). More than 40% of physical losses occur in developing countries (Kruijssen et al. 2020, Racioppo et al. 2021).

Fish loss and fish waste can occur along value chains and stages, from catch or production to processing and final consumption. Fish loss is not only the discarded fish or part of fish (physical loss), but also loss in the nutrition, quality, and market force (De Silva 2011). In developing countries, there is a high rate of fish loss and waste during loading and unloading, handling, processing, transportation, and storage. Moreover, they have constraints in facilities and infrastructures, especially in cold value chains. The problem in fisheries and aquaculture is caused not only by fish loss and fish waste but also by the waste of the fisheries industry. The characteristic of the fisheries industry is high water consumption; hence, it contributes to high wastage of water (Riberio and Naval 2019). On the other hand, developed countries also face the issue of fish loss and waste. A considerable amount of primary fish and seafood production is lost because of discard rates ranging from 9–15% of marine catches. Additionally, a significant portion of purchased fish and seafood is wasted by households (Gustavsson et al. 2011). Some key points regarding factors that affect fish loss and waste are market imperfections like demand fluctuations and price volatility. This can result in the disposal or wastage of fish due to excess or unsold stock. Consumer preferences for specific fish species or cuts result in waste as retailers discard less popular or visually unappealing fish parts. Furthermore, retail practices such as excessive stocking and strict expiration date policies can also lead to avoidable waste (Godfray et al. 2010). Fish loss and waste in developed countries have environmental consequences, including the squandering of resources such as fuel, water, and energy used in fishing and processing operations, as well as the release of greenhouse gases from decomposing organic matter (Wabnitz et al. 2018).

Comprehensive waste management in the fisheries industry and aquaculture is urgently needed to reduce serious global impacts on the society, economy, and environment. Relating to the issue, the term "Blue Economy" refers to the concept adopted by developed economies in the oceans and marine resources. It aims to maintain the sustainability of the environment in the oceans and coastal regions while simultaneously encouraging economic growth, social inclusion, and the preservation or improvement of livelihoods. The concept of blue economy is to manage the various aspects of oceanic sustainability, such as pollution, healthy ecosystems, and sustainable fisheries (Vazquez et al. 2020, European Commission 2020, World Bank 2017).

An aspect of the blue economy is managing pollution from fisheries waste or discards. Following the global population and industrialization growth, increasing pollution from the fisheries industry and aquaculture waste has been a serious issue for the water ecosystem, not only in inland waters but also in oceans. Bioremediation using marine microbes is a technique to reduce fisheries waste or discards that is in-line with the blue economy concept. The technique employs efficient technology reducing costs and is environmentally friendly. Recently, there have been many worldwide reports that industrial waste is being biodegraded using enzymes derived from marine microbes. Therefore, this chapter is focused on exploring the sources of potential marine microbes and their utilization and application for fisheries waste bioremediation.

The characterization of wastes from the fishing industry

The present research reported that fish contributed to 15.3% of the world's production of animal sources for human consumption in 2018 (FAO 2020, Boyd et al. 2022). Fish provide around 17% of the worldwide intake of animal protein, making it an essential component of human nutrition (FAO 2020, Kruijssen et al. 2020). The global consumption of aquatic food increased by 3% annually from 1960 to 2019 (FAO 2022). Following the population growth that causes the consumption of aquatic products to increase annually, the development of fishing technologies has been the primary driver of the tremendous expansion in fisheries and aquaculture production. As a result, there has been a notable global increase in the production of fishery waste, raising serious issues for the environment and the economy (Coppola et al. 2021).

The amount of fish processed prior to being placed in the market is almost 70% of the total fish captured (Hou et al. 2016, Coppola et al. 2021). The process results in huge wastage (20–80%) depending on the processing level (scaling, gutting, or filleting) and species, because each species has a certain shape, size, and composition. The process produces discards that mainly include bones (9–15%), fins and skin (1–3%), heads (9–12%), trimmings (15–20%), scales (5%), and viscera (12–18%) (Rustad et al. 2011, Arnaud et al. 2018). The same problems are also faced by aquaculture processing. The discard generated from filleting aquaculture plants is about 60% (trimmings, heads, viscera, and frames) of the total fish weight (Vazquez et al. 2020).

Waste from fisheries can be classified into two categories: fish by-products and fish waste/discards. Fish by-products refer to the left-outs that are not generally regarded as ordinary marketable products (fillet, eviscerated, beheaded fish, or round), but can be recycled after treatment or converted to industrial or edible products. The term by-product refers to something that can be utilized. Today, the raw materials, whether edible or inedible, left over after the primary product are referred to as a by-product (Penarubia et al. 2020, Rustad et al. 2011, Rustad 2003). Waste refers to the products that have to be destroyed, composted, or burned; they cannot be converted into value-added products or feed (Rustad et al. 2011).

Based on their characteristics, fish by-products can be divided into two categories: those that are highly enzyme-soluble and readily broken down, such as viscera and blood, and those that are more stable, including bones, heads, and skin (Khawli et al. 2019). In terms of their nutritional content and functional qualities, prior research has demonstrated the significance of fish by-products (50–80% moisture, 15–30% crude protein, and 0–25% fat) (Ghaly et al. 2013, Saranya et al. 2018, Nawaz et al. 2020). Research has demonstrated the significance of fishery by-products from a food perspective (Jayathilakan et al. 2012, Mahboob 2014).

The fishing industry is divided into several categories, i.e., recreational, subsistence, and commercial (FAO 2014). Aquaculture, fish processing, fish products, and fish marketing are among the commercial fishing industries. Each of these commercial fisheries has its industrial effluent characteristics. Generally, fishing industry waste has certain characteristics, including high levels of oil, fat, grease (FOG), biochemical oxygen demand (BOD), and nitrogen content (Shahidul Islam et al. 2004, Divya et al. 2015b, Aanand et al. 2017, Agyarkwa et al. 2022). Wastewater from the fishing industry is also contaminated with various microbes, pathogenic bacteria, viruses, and other microflora, contributing to the heavy organic content of seafood processing wastes (Divya et al. 2015b, Sule et al. 2016, Aanand et al. 2017, Eunice and Voke 2022). Therefore, wastewater treatment is necessary to both minimize the unpleasant aromas created during food preparation and to enable water to be discharged without causing pollution.

Martinez-Porchas et al. (2014) revealed concerns about aquaculture effluents causing eutrophication, hyper nutrification, and nitrification modifications. These are among the most prevalent issues. This means that aquaculture effluents contain contaminants consisting of nitrogen (nitrite, nitrate, ammonia), phosphorus (PO_4), carbon, and total suspended solids. Meanwhile, effluents from the fish canning industry have different characteristics, including a high volume of water with high salinity and contaminants in several forms, both dissolved, colloidal, and undissolved particles (Cristóvão et al. 2015).

Marine microbes

The marine ecosystem is a significant reservoir of biodiversity, sustenance, power, and medicinal substances. According to Overmann and Lepleux (2016), almost 90% of the biomass in the ocean is made up of microorganisms. The magnitude and importance of the marine environment may be quantified by its coverage of approximately 70% of the Earth's surface (Overmann and Lepleux 2016). Various environmental factors, including temperature, pH, pressure, and salt, contribute to the proliferation of marine species. Marine microbes have the ability to adapt, both physiologically and biochemically, to the physical, chemical, and biological aspects of the ocean environment. The marine environment is subject to a variety of sporadic changes, such as variations in sea level rise, sea surface temperature, pH of the surrounding water, light, and UV light patterns, and tropical storms (Neale et al. 2023). Marine environments are home to extreme halophiles that can withstand high salt concentrations (Bowers et al. 2009). Additionally, according to Wai et al. (1999), marine bacteria can adapt to environments deficient in nutrients and endure lengthy periods of fasting. Microorganisms that are psychrophilic and barophilic are favored by the lower temperatures and higher pressures found in marine environments (Delong and Yayanos 1987). Because of their variety of metabolic processes, marine microbes make up approximately 90% of the total biomass found in oceans and are crucial to the cycling of nutrients. Due to their constant exposure to extreme pH, pressure, temperature, salinity, and oxygen concentrations at varying depths, they are frequently categorized as extremophiles (Dash et al. 2013).

Source of marine microbial diversity

Microbial communities inhabit and proliferate in many habitats. Microorganisms inhabit different depths of the water, spanning from the surface to the ocean bottom. These depths include diverse ecosystems including coastal zones, coral reefs, and deep open ocean waters.

Coastal zones

Coastal regions are dynamic ecosystems where the interactions between land and sea promotes a profusion of microbial life. The research undertaken by Farrer et al. (2022) highlights the importance of estuaries, salt marshes, and mangrove forests as areas with high levels of microbial diversity. The diverse microbial communities in these settings are well-suited to different circumstances, impacted as they are by tidal changes and human activities (Coombs 2019, Pozo-Solar et al. 2022). Their contributions to nutrient cycling and ecosystem health have been thoroughly investigated, revealing their crucial significance in coastal ecology (Rodrigues-Filho et al. 2023). These dynamic settings function as focal points of microbial variety, making a substantial contribution to the process of nutrient cycling and the general well-being of the ecosystem (Yue et al. 2023). According to Liang et al. (2023) bacteria constitute the most abundant microbial community in coastal zones, followed by archaea and fungi. Moreover, based on the report by Haider et al. (2016) coastal areas and ports exhibit distinct bacterial abundances, with *Flavobacteriia* more prevalent in harbors and *Gammaproteobacteria* more prevalent at sea. This discrepancy may stem from increased organic waste in harbors, favoring *Flavobacteriia* in nutrient-rich environments.

Coral reefs

Coral reefs, often referred to as the ocean's rainforests, are habitats for remarkable marine organism diversity. The microbial communities within this dynamic ecosystem engage in complex interactions with corals and other creatures, significantly influencing the nutrient cycles and the stability of these fragile ecosystems (Maher et al. 2022). The study conducted by Ricci et al. (2019) has shown the crucial significance of microbial communities in maintaining the health and resilience of corals. The presence of symbiotic connections, namely with algae and certain microbial species, is essential for the functioning and stability of coral ecosystems. Research on the variety of microorganisms present in coral reefs provides valuable information about how they react to environmental pressures and their possible use in biotechnology (Kanisan et al. 2023). Microbial populations in coral reef ecosystems can change due to environmental disturbances, such as climate change and attacks by pathogenic bacteria (Liu et al. 2024). Coral reefs are one of the most important ecosystems in the ocean, providing high primary productivity that support about a third of the marine biodiversity, including bacteria, viruses. archaea, and fungi. Liu et al. (2024), Li et al. (2013) report that common bacterial species found in coral reefs include *Gammaproteobacteria, Alphaproteobacteria, Firmicutes, Bacteroidetes,* and *Actinobacteria.*

Deep open ocean

The deep open ocean remains a mysterious and extreme environment, yet it harbors diverse microbial life forms. Zhang et al. (2022) have researched that the richness and diversity of microbial community is highest in seamount sediments, followed by that in the cold seep, and the lowest in trenches. Gaining knowledge about the variety and roles of microorganisms in the open ocean is essential for understanding the basic ecological mechanisms and tackling environmental issues, such as pollution and climate change (Liao et al. 2022). Investigations have revealed the presence of distinct extremophiles flourishing in hydrothermal vents, cold seeps, and abyssal plains. These extremophiles have specific adaptations and biochemical capacities, which provide potential for innovative biotechnological uses (Raddadi et al. 2015). Exploring deep-sea microbial communities enhances our comprehension of life in extreme habitats and its possible ramifications for several industries (Liao et al. 2022). The deep sea, characterized by intense pressure, darkness, and frigid temperatures, harbors extremophilic microorganisms thriving around hydrothermal vents and cold seeps (Jin et al. 2019). These extremophiles exhibit extraordinary adaptations and metabolic capacities, which have significant promise for biotechnological uses and provide valuable insights into the adaptability of life in severe conditions (Rekadwad et al. 2023). Research conducted by Dewapriya and Kim (2014) estimates that the ocean biomass is made up of about 3.6×10^{29} bacterial cells. However, the reported population is far smaller because the isolation techniques employed in their production failed. The majority of marine bacteria cannot be isolated using traditional culture-based techniques, hence a metagenomic approach and other culture-independent techniques are needed to uncover these novel groupings of bacteria that beneath the ocean's depths are yet unknown.

Types of marine microbes

Various environmental factors, including temperature, pH, pressure, and salt, contribute to the proliferation of marine species. Microbes in the ocean are highly diverse, but some of the most common and important types include bacteria, viruses, archaea, and fungi.

Marine bacteria

Marine bacterial communities often demonstrate significant variety, with individual samples potentially containing more than 20,000 species (Khan et al. 2023). Distinct bacterial populations populate various marine environments, such as coastal surface waters, deep open ocean waters, and sediments (Khan et al. 2023). Marine microorganisms play a crucial role in maintaining and regulating biological and chemical processes within the marine environment. They are essential for the remineralization of organic materials through nitrogen and carbon fixation. The foundation of the ocean food webs is also because to these bacteria.

Fisheries waste breaks down mostly due to the action of marine microbes. Single-stage co-fermentation is a procedure that has been demonstrated to be an efficient means of breaking down unprocessed crustacean shell waste by some species of bacteria, including *Bacillus subtilis*, *Priestia megaterium*, and *Bacillus amyloliquefaciens* (Vakkachan et al. 2023). According to Vakkachan et al. (2023), these bacteria have high deproteinization (> 90%) and demineralization efficiency (> 90%) due to their concurrent chitinolytic and proteolytic activity. This process of degradation which is environmentally benign is a pollution control and waste management method for crustacean shell debris.

The microbial communities in deep-sea sediments exhibit diversity and variation, such as carbon, nitrogen, and sulfur cycling, across seamount, trench, and cold seep ecosystems (Zhang et al. 2022). Several studies have reported that *Pseudomonas* sp. and *Ochrobatrum* sp., isolated from coastal sediment, are responsible for degrading fluorene and naphthalene resulting from oil spill form fishing vessels. Furthermore, Kimata et al. (2004) found that *Pseudomonas aeruginosa* is commonly found in coastal surface waters. This bacterium produces proteases that function in deproteinizing shrimp and crab shell waste (Oh et al. 2000).

Marine viruses

Marine viruses represent the predominant biological entities in aquatic environments. In 2019, research identified approximately 200,000 distinct marine viruses (De Jesus 2019). These viruses were categorized into five primary groupings, mostly according to their distinct environments. Marine viruses are acknowledged as significant contributors in marine ecosystems, exerting influence on several processes, including the well-being of marine organisms and global biogeochemical cycles. Their role as the primary factors behind fish infections was initially acknowledged in the early 1960s. Subsequently, it has been established that they are accountable for diseases in several marine organisms, ranging from bacteria to protists, molluscs, crustaceans, fish, and mammals (Middelboe and Brussaard 2017). The valorization of marine waste can be achieved through various alternative methods, including hydrolysis, ensilaging, fermentation, and gelation. The technologies have been meticulously optimized to extract and generate highly concentrated marine oils, functional protein food components, pharmaceutical-grade biopolymers, and textiles (Rudovica et al. 2021).

Marine fungi

Marine fungus are fungal species that inhabit marine or estuarine environments. Although they do not belong to the same taxonomic group, they do inhabit the same environment. Obligate marine fungus are organisms that exclusively develop in the marine environment, either fully or partially immersed in seawater. Facultative marine fungus often inhabits terrestrial or freshwater environments but have the ability to survive or produce spores in a marine ecosystem. A total of 444 marine fungal species have been documented, consisting of seven genera and ten species of basidiomycetes, as well as 177 genera and 360 species of ascomycetes (Anon. 2023). As of October 2023, there have been a recorded total of 1,947 species of marine fungus (Anon. 2023). Distinct maritime settings harbor diverse fungal populations. Fungi inhabit a wide range of environments, including deep ocean areas, coastal waterways, mangrove swamps, and estuaries with low salt levels. Marine fungi are essential for the transfer of energy and the recycling of nutrients in marine ecosystems, since they facilitate the breakdown and cycling of dissolved organic matter (Gonçalves et al. 2022).

Archaea

Archaea is a taxonomic domain consisting of unicellular bacteria. These creatures are prokaryotes, which indicates that they do not possess a distinct nucleus and organelles enclosed by a membrane. Originally, Archaea were categorized as bacteria and referred to as archaebacteria, however, this classification is now obsolete. Archaea are renowned for their capacity to thrive in harsh settings characterized by excessive acidity, temperature, alkalinity, or salinity. They may be found in diverse habitats such as hot springs, hydrothermal vents, and glaciers (Nagrale and Gawande 2018). Archaea have a notable range of variation in their physical structure, chemical processes, and global spread. They can perform several ecological functions, including carbon fixation, nitrogen cycling, organic compound turnover, and the maintenance of microbial symbiotic and syntrophic communities. Regarding global diversity, it is estimated that there are around 0.8–1.6 million operational taxonomic units (OTUs) of prokaryotes worldwide (Louca et al. 2019). OTUs are used as a measure of microbial richness (Louca et al. 2019). Archaea's ability to thrive in diverse habitats and their substantial contributions to Earth's ecosystems are facilitated by their genetic variety.

Marine microbes demonstrate higher adaptability and tolerance to extreme environments, such as high temperatures, high pressure, pH variations, and environments with high salt concentrations, compared to their counterparts on land (Kumar and Sharma 2019). Since they can quickly adapt to a variety of harsh environments, marine microorganisms are a good choice for bioremediation. Bioremediation harnesses the innate capacity of marine microbes to naturally transform hazardous chemical components in waste into less harmful forms. Implementing this strategy can contribute to the sustainable and environmentally friendly management of fishing industry wastewater, thereby reducing its negative impact on aquatic ecosystems.

Utilization and application of marine microbes for bioremediation

Reducing the release of pollutants from the fishing industry wastewater into the environment is crucial for maintaining ecological balance, sustainability, and human health. A technique known as "bioremediation" uses biological processes to reduce, degrade, modify, eliminate, immobilize, detoxify, mineralize, or change contaminant concentrations into a state that is neither poisonous nor harmful. The type of pollutant, such as agrochemicals, chlorinated compounds, dyes, greenhouse gases, heavy metals, hydrocarbons, radioactive waste, and plastic trash, determine the best treatment strategy for contaminated water (Ojha et al. 2021). As mentioned previously, the fishing industry waste has certain characteristics, including high levels of oil, fat, grease (FOG), biochemical oxygen demand (BOD), nitrogen, and phosphorus (Aanand et al. 2017, Agyarkwa et al. 2022, Islam et al. 2004, Martinez-Porchas et al. 2014). Nevertheless, the habitats that receive aquaculture effluents are unable to fully absorb and integrate those nutrients into biogeochemical cycles and food chains. As a result, significant nutrient accumulation tends to occur, leading to ecological imbalances (Martinez-Porchas et al. 2014). Bioremediation is a waste treatment method that utilizes microorganisms and other living organisms to eliminate toxins, pollutants, and impurities from soil, water, and other environments (Hlihor et al. 2017, Zhang et al. 2021). Microbial bioremediation has the potential to be a quick, affordable, and environmentally friendly, and a human-health-safe method of destroying or immobilizing pollutants (Gadd 2000, Gheewala and Annachhatre 1997, Heitzer and Sayler 1993). Bacteria, virus, fungi, and archaea are important microorganisms frequently employed in bioremediation.

Bioremediation can be used to clean up a wide range of pollutants, including organic compounds, heavy metals, and various pollutants found in different industrial wastewaters. Some of the most common types of pollutants that can be cleaned up using bioremediation include (Aanand et al. 2017, Jasmin et al. 2020, Kashem et al. 2023, Macaulay and Rees 2014, Roy et al. 2023):

1. *Organic compounds*: bioremediation can be used to degrade various organic compounds, such as hydrocarbons, pesticides, and nitrogen compounds, into simpler molecules or mineralized forms like carbon dioxide, water, and methane.
2. *Heavy metals*: certain strains of microorganisms can biosorb heavy metals and other pollutants, removing them from the wastewater. For example, certain strains of *Rhizobia* and *Azotobacter* can biosorb heavy metals like lead, cadmium, and mercury.
3. *Nitrogen compounds*: bioremediation can be used to remove nitrogen compounds, such as ammonia and nitrate from wastewater. Nitrogen-oxidizing bacteria like *Nitrosomonas* and *Nitrobacter* can convert nitrogen compounds into nitrate; while denitrifying bacteria like *Azotobacter* and *Pseudomonas* can reduce nitrate to nitrogen gas. Several microalgae species are also used as bioremediation agents to remove nitrogen compound.
4. *Phosphorus*: certain bacteria, such as *Acinetobacter* sp. and *Bacillus* sp., are capable of removing phosphorus from wastewater.

Since the fishing industry has specific characteristics, many studies have been conducted focusing on the use of marine microbes as bioremediation agents for fishing industry wastewater (Aanand et al. 2017, Azad et al. 2001, Bibbins-Martínez et al. 2023, Divya et al. 2015a, Mohamad et al. 2017, Palmeiro-Sánchez et al. 2021, Prasertsan et al. 1993, Samani et al. 2016, Selvi et al. 2014, Urano et al. 2002, Zokaeifar et al. 2014). Even marine invertebrates such as sponges have also been used as research materials for bioremediation of fish farming waste (Longo et al. 2022).

Table 1 shows microbes that have been used in fishing industry wastewater bioremediation.

Table 1. Listed microbes used in fishing industry wastewater bioremediation.

No	Species	Microbe groups	Function	Reference
1	*Pseudomonas* sp.	Bacteria	Nitrate reducing strain; hydrocarbons degradation	Aanand et al. (2017), Jayashree et al. (2012)
2	*Brevibacillus* sp.	Bacteria	Nitrate reducing strain	Aanand et al. (2017)
3	*Rhodocyclusgelatinosus*	Bacteria	COD removal	Prasertsan et al. (1993)
4	*Rhodovulumsulfidophilum*	Bacteria	COD reduction	Azad et al. (2001)
5	*Bacillus* sp.	Bacteria	Ammonical nitrogen reduction	
6	Combined *Nitrosomonas* and *Nitrobacter*	Bacteria	Ammonia reduction	Selvi et al. (2014)
7	*Bacillus cereus* and *Bacillus* sp.	Bacteria	BOD, COD, ammonia, nitrate, and total nitrogen reduction	Divya et al. (2015a)
8	*Nitrobacter*, yeast, *Bacillus subtilis*	Consortium bacteria and yeast	Total nitrogen and total phosphorus removal	Mohamad et al. (2017)
9	*Bacillus* sp.	Bacteria	Ammonical nitrogen reduction	Samani et al. (2016)
10	Mixture of *Bacillus* sp.	Bacteria	Total ammonical nitrogen, nitrite and nitrate reduction	Zokaeifar et al. (2014)
11	*Trichosporonovoides*	Yeast	TOC reduction	Urano et al. (2002)
12	*Hymeniacidonperlevis*	Sponge	Bacterial filter and removal	Longo et al. (2022)
13	Mixed microbial consortium	Not specifically mentioned	Organic pollutants conversion	Palmeiro-Sánchez et al. (2021)

It has been described in the previous section that the characteristics of fisheries industry effluents depend on the type of industry, but in general fishing industry wastewater has certain characteristics, including high levels of biochemical oxygen demand (BOD), fat, oil and grease (FOG), and nitrogen (Aanand et al. 2017, Agyarkwa et al. 2022, Divya et al. 2015a, Islam et al. 2004). This paragraph is limited by the types of wastewaters generated from two types of fishing industries, namely aquaculture and fish canning industry.

In their study, Cristóvão et al. (2015) combined physical and biological methods in handling fish-canning industry wastewater, including sedimentation, chemical coagulation-flocculation, aerobic biological degradation, reverse osmosis, and ultraviolet disinfection. In one of the stages of waste treatment, they employed a biological aerobic reactor operated at various hydraulic retention times (HRT) to assess the organic matter removal efficiency. The activated sludge process involved the aeration of wastewater to provide oxygen for aerobic microorganisms' metabolic processes, leading to the formation of activated sludge and the subsequent reduction of organic content in the wastewater. The study evaluated the effectiveness of this biological treatment in removing organic matter from fish-canning wastewater, with the highest DOC removal occurring at the reactor outlet. Overall, the findings of the study indicate high removal efficiencies for various contaminants, making the treated effluent suitable for recycling or reuse in the industrial process, with a unit cost of 0.85 €/m^3 for obtaining a treated effluent to be reused in the process.

In relation to aquaculture effluents, bioremediation is still focused on wastewater treatment containing organic matter that is harmful to the environment and waters surrounding aquaculture areas (Jasmin et al. 2020). Although aquaculture sludge (solid waste) is also part of aquaculture effluent, bioremediation application on sludge is still not well-documented (Jasmin et al. 2020). Beneficial microbial organisms are employed in well-known bioremediation procedures to treat contaminated water or trash. Since bioremediation doesn't use chemicals, it is seen as an environmentally beneficial way to manage organic waste. For example, microorganisms use phosphate and nitrogen as sources of nutrition. By using the inorganic substances present, such as fixing the inorganic carbon (CO_2), autotrophs are able to generate their own food.

Autotrophic bacteria include ammonia-oxidizing and nitrite-oxidizing bacteria, which are vital to the nitrification and denitrification processes (Jasmin et al. 2020). To multiply, autotrophic bacteria absorb and convert soluble nitrogen and phosphorus that are available in the biological system. In the meantime, non-living organic matter is destroyed or rendered immobile by heterotrophic bacteria in order to manufacture carbon for their cell walls. In response, these cells catalyze the oxidation of these substances by acting as donors of electrons. Heterotrophic bacteria can shift ammonia nitrogen into non-toxic compounds known as microbial mass, but they do not play a significant role in the nitrification and denitrification processes like autotrophic bacteria do. Since bioremediation works on a nearly identical principle to that of treating wastewater and culture water, it holds great promise for utilizing beneficial bacteria to break down the toxins in the sludge. However, a number of important variables, like the type of bacteria present, the amount of toxins present, the surrounding environment, and others, affect how well bioremediation works (Jasmin et al. 2020).

Mechanisms of the utilization of marine in fishing industry wastewater bioremediation

Bioremediation involves two mechanisms: biosorption and bioaccumulation (Ojha et al. 2021, Bibbins-Martínez et al. 2023, Roy et al. 2023). Biosorption is a rapid and reversible passive adsorption method. Biosorption can occur in both living and dead biomass. Since the biomass obtained from industrial waste can be repeatedly generated and reused, this technology has low-cost requirements. By absorbing poisons from their surroundings, marine organisms can increase their concentration or accumulation of toxins through a process known as bioaccumulation. Included are both external and intracellular processes. For bioaccumulation, only live biomass is allowed to occur. Since this method operates in the presence of living cells, where reuse is restricted, it is expensive.

The exact roles of microbes in bioremediation include the degradation of environmental contaminants into less toxic forms, detoxification, and mineralization of wastes into inorganic compounds. Some specific examples of the roles of microbes in bioremediation are (Aanand et al. 2017, Al-Gheethi et al. 2018).

- *Detoxification and mineralization*: microorganisms are used to detoxify and mineralize wastes, converting them into inorganic compounds such as carbon dioxide, water, and methane;
- *Degradation of pollutants*: microbes are capable of degrading various pollutants present in the wastewater, such as ammonia, nitrogen compounds, fat, oil, grease, and other organic materials;
- *Bioaugmentation*: where competent microflora is introduced to the contaminated site to speed up the *bioremediation process*;
- *Biodegradation of specific pollutants*: specific microbes, such as *Pseudomonas* sp., *Brevibacillus* sp., *Arthrobacter* sp., and *Rhodococcus* sp. are capable of degrading various pollutants like petroleum, pesticides, hydrocarbons, and nitrogen compounds.

Applications in various waste management scenarios within the fishing industry

The seafood industry produces large volumes of liquid waste or effluent that will pollute the surrounding environment if it is not managed right at the beginning. The effluent contains very high concentrations of biochemical oxygen demand (BOD) and chemical oxygen demand (COD), which can cause real environmental problems. Proper effluent management will produce outputs that have great potential benefits, such as food additives, nutrasetical ingredients, nutritional supplements, flavor ingredients, fertilizers, plant biostimulants, animal and fish feed (Venugopal and Sasidharan 2021).

There are several methods applied for waste management of the fishing or seafood industry, including physico-chemical, biological, membrane, and others: for example: physico-chemical methods which include neutralization (pH adjustment), coagulation, flocculation, chemical precipitation, chemical oxidation (ozonation, UV radiation, halogenation, electrolysis, polymeric adsorption, and others). Coagulation and flocculation serve to reduce fat, oil, and grease (FOG) and BOD value; ozonation, UV radiation and halogenation serve as disinfectants for wastewater; pH adjustment serves to remove the charge on colloidal particles, thus facilitating the coagulation and flocculation process (Venugopal and Sasidharan 2021).

Biological methods cover the following processes (Venugopal and Sasidharan 2021).

Aerobic treatment processes:

- It activated sludge systems, aerobic lagoons, trickling filters, and rotating disc contactors are common aerobic processes.
- In activated sludge treatment, microorganisms degrade organic materials in wastewater using excess dissolved oxygen and nutrients.
- This process reduces soluble BOD_5, fats, and nutrients but is energy-intensive.

Anaerobic Digestion (AD):

- AD is a green technology that involves hydrolysis, fermentation, and methane production.
- The parameters include hydraulic retention time, organic loading rate, pH, temperature, and pretreatment.
- Anaerobic systems (e.g., UASBR, AF, AFBR) offer advantages like low sludge yield, methane-rich biogas, and lower costs.

Biological treatments for seafood wastewater:

- Aerobic processes are suitable for removing organics from seafood wastewater.
- A laboratory-scale aerobic bioreactor treated high saline fish processing wastewater effectively.
- Anaerobic treatment followed by aerobic treatment is ideal for fish effluents.

Specific examples:

- Corsino et al. (2016) used an AD system with simultaneous nitritation–denitritation at salinities up to 5 percent.
- An activated sludge process combined with anaerobic ammonium oxidation achieved COD reduction and ammonium removal.

Mesophilic anaerobic treatment:

- Sludge from salmon farming effluents was treated in stirred tank reactors at 35°C.
- Dilution with tap water stabilized COD and produced methane.

Bio-Electrochemical Process:

- This is a green technique that generates electricity from wastewater.
- Jayashree et al. (2012) used an up-flow microbial fuel cell (MFC) to treat seafood processing waste.
- At an organic loading rate of 0.6 g per day, the MFC achieved up to 95% reduction in COD.

Hydrodynamic cavitation:

- This is a new technology used to treat fish processing wastewater.
- It involves using an orifice plate in the reactor to create a cavitation effect.
- Intensification with H_2O_2 and titanium oxide led to about 70% COD reduction in 120 minutes.
- The treatment also improved the biodegradability of the effluent.

Overall, biological treatments offer effective solutions for wastewater treatment, considering both aerobic and anaerobic processes. They also offer promising solutions for sustainable wastewater treatment and energy production.

Highlighting the effectiveness and benefits of utilizing marine microbes

Bioremediation offers several advantages over other methods of pollution cleanup. Some of the key advantages include (Divya et al. 2015a, Aanand et al. 2017, Bibbins-Martínez et al. 2022, Roy et al. 2023):

Eco-friendly approach: bioremediation is an environmentally friendly technology that uses natural processes to degrade and remove pollutants from the environment, making it a sustainable and green alternative to traditional cleanup methods.

Minimizes environmental impact: by utilizing microorganisms to degrade contaminants into less toxic forms, bioremediation helps minimize the environmental impact of pollutants, especially in cases such as seafood processing plant wastewater, which contains high organic load and contamination.

Cost-effectiveness: bioremediation can be a cost-effective solution for pollution cleanup, as it often requires less equipment and energy compared to traditional methods. It can also be applied *in situ*, reducing the need for extensive site excavation and transportation of contaminated materials.

Versatility: bioremediation can be applied to a wide range of contaminants, including organic compounds, heavy metals, and various pollutants found in different industrial wastewater. This versatility makes it a valuable tool for addressing diverse pollution problems.

Long-term effectiveness: unlike some traditional methods that may only contain or move pollutants to a different location, bioremediation can lead to the complete degradation and removal of contaminants, providing a long-term solution to pollution cleanup.

Challenges

Since 1961, the average annual growth rate of the world's aquatic food consumption (apart from algae) has been 3.0 percent, while the population has grown at a pace of 1.6 percent (FAO 2022). This growth is accompanied by an increase in the number of global fishing industries that are able to meet global human protein needs. The excesses of the growth of the fishing industry that occur are an increase in the amount of fishery waste produced. This can be dangerous, especially for environmental conditions. Bioremediation techniques are expected to be able to overcome environmental problems resulting from an increase in the amount of fishery waste produced. The use of marine microbes as a tool for the bioremediation process is one of the effective alternatives to be able to overcome environmental problems that occur. This is in line with the blue economy concept developed, namely, to be able to create sustainable and environmentally friendly conditions.

Bioremediations have many advantages when compared to traditional remediation techniques. In traditional remediation technique, contaminated items are physically removed or treated. Typically, these techniques involve chemical treatment, soil washing, incineration, and excavation (Reynolds 2023). There is a considerable risk of secondary pollution and environmental disturbance associated with the mechanical extraction and transportation of contaminated materials. Due to the substantial usage of heavy machinery and the need to dispose of polluted materials, traditional methods are frequently more expensive. These techniques do not support ecological restoration, and they frequently require the landfill disposal of polluted materials, which exacerbates long-term environmental problems. Conventional remediation techniques sometimes overlook the potential long-term ecological advantages of bioremediation in favour of immediate containment (Tudi et al. 2021).

However, several challenges must be faced to develop the bioremediation process using marine microbes to overcome the environmental issues especially caused by fishery industry pollution. Despite the number of marine microbes that can be used for the bioremediation process, exploration of various other types of marine microbial sources that can be used for the bioremediation process is still needed. Determination of the right types of bacteria, viruses, and fungi originating from the sea to overcome various types of environmental problems, especially those derived from fishery industry waste, must still be improved. Exploration of these various types of marine microbial sources requires considerable cost and time. Another challenge that must be faced in the use of marine microbes for the bioremediation process is the need for appropriate and efficient technology to obtain, develop, and apply the use of marine microbes in the bioremediation process. The need for a big investment becomes its own challenge in the future. In addition, the application of the use of marine microbes in the bioremediation process to overcome the problem of environmental pollution originating from the fishing industry will not run well without support from the government. The implementation of appropriate government policies to overcome environmental problems by utilizing marine microbes in the bioremediation process is expected to support the realization of the blue economy concept so that sustainability and environmentally friendly schemes in the field of fisheries industry can still be implemented.

The challenges that must be faced in the development of the use of marine microbes in the bioremediation process to overcome these environmental problems can be overcome. Good and effective cooperation between various parties such as academics, researchers, government, private sector, and other parties can realize environmental sustainability in the context of the blue economy.

Authors' contribution

All authors have an equal contribution in the preparation of the manuscript. Hari Eko Irianto is designated as the corresponding author.

References

Aanand, S., Divya, M., Deepak, T., Padmavathi, P. and Manimekalai, D. (2017). Review on seafood processing plant wastewater bioremediation-A potential tool for waste management the user has requested enhancement of the downloaded file Review on seafood processing plant wastewater bioremediation—A potential tool for waste management. International Journal of Applied Research 3(7): 01–04. https://www.researchgate.net/publication/318098015.

Abeliotis, K., Lasaridi, K., Costarelli, V. and Chroni, C. (2015). The implications of food waste generation on climate change: The case of Greece. Sustainable Production and Consumption 3: 8–14.

Agyarkwa, M.A., Azaglo, G.S.K., Kokofu, H.K., Appah-Sampong, E.K., Nerquaye-Tetteh, E.N., Appoh, E. et al. (2022). Surveillance of WHO priority gram-negative pathogenic bacteria in effluents from two seafood processing facilities in Tema, Ghana, 2021 and 2022: A descriptive study. Int. J. Environ. Res. Public Health 19(17): 10823. doi: 10.3390/ijerph191710823. PMID: 36078530; PMCID: PMC9518130.

Al-Gheethi, A.A., Efaq, A.N., Bala, J.D., Norli, I., Abdel-Monem, M.O. and Kadir, M.O. (2018). Removal of pathogenic bacteria from sewage-treated effluent and biosolids for agricultural purposes. Applied Water Science 8: 74. https://doi.org/10.1007/s13201-018-0698-6.

Al Khawli, F.A., Ferrer, E., Berrada, H., Barba, F.J., Pateiro, M., Domínguez, R. et al. (2019). Review - Innovative green technologies of intensification for valorization of seafood and their by-Products. Marine Drugs 17(689): 1–21. doi:10.3390/md17120689.

Anonymous. (2023). Marine fungi. Mushroom Research Foundation: Thailand. https://marinefungi.org/.

Arnaud, C., de Lamballerie, M. and Pottier, L. (2018). Effect of high-pressure processing on the preservation of frozen and re-thawed sliced cod (*Gadus morhua*) and salmon (*Salmo salar)* fillets. High Pressure Research 38(1): 62–79. https://doi.org/10.1080/08957959.2017.1399372.

Azad, S.A., Vikineswary, S., Ramachandran, K.B. and Chong, V.C. (2001). Growth and production of biomass of Rhodovulumsulfidophilum in sardine processing wastewater. Letters in Applied Microbiology 33: 264–268.

Bibbins-Martínez, M., Juárez-Hernández, J., López-Domínguez, J.Y., Nava-Galicia, S.B., Martínez Tozcano, L.J., Juárez-Atonal, R. et al. (2023). Potential application of fungal biosorption and/or bioaccumulation for the bioremediation of wastewater contamination: A review. J. Environ. Biol. 44: 135–145.

Bowers, K.J., Mesbah, N.M. and Wiegel, J. (2009). Biodiversity of poly-extremophilic Bacteria: does combining the extremes of high salt, alkaline pH and elevated temperature approach a physicochemical boundary for life? Saline Syst. 5(1): 9.

Boyd, C.E., McNevin, A.A. and Davis, R.P. (2022). The contribution of fisheries and aquaculture to the global protein supply. Food Security 14: 805–827. https://doi.org/10.1007/s12571-021-01246-9.

Coombs, J.M. (2019). Microbial communities in salt marsh systems and their responses to anthropogenic pollutants. *In*: Hurst, C. (eds.). Understanding Terrestrial Microbial Communities. Advances in Environmental Microbiology, vol 6. Springer, Cham. https://doi.org/10.1007/978-3-030-10777-2_8.

Coppola, D., Lauritano, C., Esposito, F.P., Riccio, G., Rizzo, C. and de Pascale, D. (2021). Review - Fish Waste: From problem to valuable resource. Marine Drugs 19(116): 1–39. https://doi.org/10.3390/md19020116.

Corsino, S.F., Capodici, M., Morici, C., Torregrossa, M. and Viviani, G. (2016). Simultaneous nitritation–denitritation for the treatment of high-strength nitrogen in hypersaline wastewater by aerobic granular sludge. Water Research 88: 329–336. https://doi.org/10.1016/j.watres.2015.10.041.

Cristóvão, R.O., Botelho, C.M., Martins, R.J.E., Loureiro, J.M. and Boaventura, R.A.R. (2015). Fish canning industry wastewater treatment for water reuse—a case study. Journal of Cleaner Production 87: 603–612. https://doi.org/10.1016/j.jclepro.2014.10.076.

Dash, H.R., Mangwani, N., Chakraborty, J., Kumari, S. and Das, S. (2013). Marine bacteria: potential candidates for enhanced bioremediation. Appl. Microbiol. Biotechnol. 97(2): 561–571.

De Jesus, E.I.G. (2019). Hundreds of thousands of marine viruses discovered in world's oceans. Nature. Doi: 10.1038/d41586-019-01329-w.

Delong, E.F. and Yayanos, A.A. (1987). Properties of the glucose-transport system in some deep-sea bacteria. Appl. Environ. Microbiol. 53: 527–532.

De Silva, D.A.M. (2011). Value chain of fish and fishery products: origin, functions and application in developed and developing country markets. Food and Agriculture Organization.

Dewapriya, P. and Kim, S.K. (2014). Marine microorganisms: an emerging avenue in modern nutraceuticals and functional foods. Food Res. Int. 56: 115–125.

Divya, M., Aanand, S., Srinivasan, A. and Ahilan, B. (2015a). Bioremediation-An eco-friendly tool for effluent treatment: A review. International Journal of Applied Research 1(12): 530–537.

Divya, M., Aanand, S., Srinivasan, A., Ahilan, B. and Uma, A. (2015b). Bioremediation of seafood processing plant effluents using indigenous bacterial isolates. International Journal of Advanced Biotechnology and Research 6(3): 443–449.

Eunice, O.E. and Voke, U. (2022). Bacteriological and physicochemical analysis of fishpond effluents in warri and its environs, Nigeria. Journal of Advances in Microbiology 22(8): 92–100. https://doi.org/10.9734/jamb/2022/v22i830493.

European Commission. (2010). The blue economy reports. The reuse policy of European Commission documents is regulated by Decision 2011/833/EU (OJ L 330, 14.12.2011, p. 39).

European Commission. (2020). The EU blue economy report 2020. Publication office of European Union. Luxembourg.

Farrer, E.C., Van Bael, S.A., Clay, K. and Smith, M.K.K.H. (2022). Plant-microbial symbioses in coastal systems: their ecological importance and role in coastal restoration. pp. 1805–1822. *In*: Estuaries and Coasts (Vol. 45, Issue 7). Springer. https://doi.org/10.1007/s12237-022-01052-2.

Food and Agriculture Organization. (2014). https: //www.fao.org/faoterm/collection/fisheries/en/. Accessed in January 5, 2024.

Food and Agriculture Organization. (2020). The state of world fisheries and aquaculture 2020. Sustainability in action. Rome. https://doi.org/10.4060/ca9229en.

Food and Agriculture Organization. (2021). Trade in fisheries and aquaculture products: a major international commodity. FAO Support to The WTO Negotiations at the 12TH Ministerial Conference.

Food and Agriculture Organization. (2022). The state of world fisheries and aquaculture 2022. Towards Blue Transformation. Rome, FAO. https://doi.org/10.4060/cc0461en.

Gadd, G.M. (2000). Bioremedial potential of microbial mechanisms of metal mobilization and immobilization. Curr. Opi. Biotech 11: 271–279.

Ghaly, A.E., Ramakrishnan, V.V., Brooks, M.S., Budge, S.M. and Dave, D. (2013). Fish processing wastes as a potential source of proteins, amino acids and oils: a critical review. J. Microb. Biochem. Technol. 5(4): 107–129. http://dx.doi.org/10.4172/1948-5948.1000110.

Gheewala, S.H. and Annachatre, A.P. (1997). Biodegradation of aniline. Water Sci Technol., pp. 36–53.

Gilman, E., Perez Roda, A., Huntington, T., Kennelly, S.J., Suuronen, P., Chaloupka, M. et al. (2020). Benchmarking global fisheries discards. Sci. Rep. 10: 14017. https://doi.org/10.1038/s41598-020-71021-x.

Godfray, H.C.J., Beddington, J.R., Crute, I.R., Haddad, L., Lawrence, D., Muir, J.F. et al. (2010). Food security: the challenge of feeding 9 billion people. Science 327(5967): 812–818.

Gonçalves, M.F.M., Esteves, A.C. and Alves, A. (2022). Marine fungi: Opportunities and challenges. Encyclopedia 2(1): 559–577. https://doi.org/10.3390/encyclopedia2010037.

Gustavsson, J., Cederberg, C., Sonesson, U., van Otterdijk, R. and Meybeck, A. (2011). Global food losses and food waste: extent, causes and prevention. Food and Agriculture Organization of the United Nations.

Haider, Md. N., Nishimura, M. and Kogure, K. (2016). Bacterial community structure and diversity of closely located coastal areas. Open Journal of Marine Science 6: 423–439.

Heitzer, A. and Sayler, G.S. (1993). Monitoring efficacy of bioremediation. Trends Biotech. 11: 334–343.

Herbert-Read, J.E., Thornton, A., Amon, D.J., Birchenough, S.N.R., Cˆ ot´ e, I.M., Dias, M.P. et al. (2022). A global horizon scan of issues impacting marine and coastal biodiversity conservation. Nat. Ecol. Evol. https://doi.org/10.1038/s41559-022- 01812-0.

Hou, Y., Shavandi, A., Carne, A., Bekhit, A.A., Bun Ng, T., Cheung, R.C.F. et al. (2016). Marine shells: potential opportunities for extraction of functional and health-promoting materials. Critical Reviews in Environmental Science and Technology, DOI: 10.1080/10643389.2016.1202669.

Hlihor, R.M., Gavrilescu, M., Tavares, T., Favier, L. and Olivieri, G. (2017). Bioremediation: An overview on current practices, advances, and new perspectives in environmental pollution treatment. BioMed Research International, 1–2.

Ideia, P., Pinto, J., Ferreira, R., Figueiredo, L., Spínola, V. and Castilho, P.C. (2020). Fish processing industry residues: A review of valuable products extraction and characterization methods. Waste and Biomass Valorization 11: 3223–3246. https://doi.org/10.1007/s12649-019-00739-1.

Islam, M.S., Khan, S. and Tanaka, M. (2004). Waste loading in shrimp and fish processing effluents: Potential source of hazards to the coastal and nearshore environments. Marine Pollution Bulletin 49(1–2): 103–110. https://doi.org/10.1016/j.marpolbul.2004.01.018.

Jasmin, M.Y., Syukri, F., Kamarudin, M.S. and Karim, M. (2020). Potential of bioremediation in treating aquaculture sludge: Review article. Aquaculture 519: 1–5. https://doi.org/10.1016/j.aquaculture.2019.734905.

Jayathilakan, K., Sultana, K., Radhakrishna, K. and Bawa, A.S. (2012). Utilization of byproducts and waste materials from meat, poultry and fish processing industries: a review. J. Food Sci. Technol. 49(3): 278–293. doi: 10.1007/s13197-011-0290-7.

Jayashree, R., Evany Nithya, S., Prasanna, P.R. and Krishnaraju, M. (2012). Biodegradation capability of bacterial species isolated from oil contaminated soil. J. Acad. Indus. Res. 1(3): 140.

Jin, M., Gai, Y., Guo, X., Hou, Y. and Zeng, R. (2019). Properties and applications of extremozymes from deep-sea extremophilic microorganisms: A mini review. Marine Drugs 17(12). https://doi.org/10.3390/md17120656.

Kanisan, D.P., Quek, Z.B.R., Oh, R.M., Afiq-Rosli, L., Lee, J.N., Huang, D. et al. (2023). Diversity and distribution of microbial communities associated with reef corals of the Malay Peninsula. Microbial Ecology 85(1): 37–48. https://doi.org/10.1007/s00248-022-01958-1.

Kashem, A.H.M., Das, P., Abdul Quadir, M., Khan, S., Thaher, M.I., Al ghasal, G. et al. (2023). Microalgal bioremediation of brackish aquaculture wastewater. Science of the Total Environment 873: 1–10. http://dx.doi.org/10.1016/j.scitotenv.2023.162384.

Khan, I., Ullah, R., Ali, S. and Shah, M. (2023). Bacterial diversity in the marine environment and the cutting-edge tools for isolation, identification and characterization of the marine microbiome. pp. 225–243. In book: Marine Biotechnology: Applications in Food, Drugs and Energy. DOI: 10.1007/978-981-99-0624-6_11.

Khawli, F., Ferrer, E., Berrada, H., Barba, F.J., Pateiro, M., Domínguez, R. et al. (2019). Review - Innovative green technologies of intensification for valorization of seafood and their by-products. Marine Drugs 17(689): 1–21. doi:10.3390/md17120689.

Khiari, Z. (2022). Sustainable upcycling of fisheries and aquaculture wastes using fish-derived cold-adapted proteases. Frontiers in Nutrition 9: 1–9. doi: 10.3389/fnut.2022.875697.

Kimata, N., Nishino, T., Suzuki, S. and Kogure, K. (2004). *Pseudomonas aeruginosa* isolated from marine environments in Tokyo Bay. Microbial Ecology 47: 41–47. DOI: 10.1007/s00248-003-1032-9.

Kruijssen, F., Tedesco, I., Ward, A., Pincus, L., Love, D. and Thorne-Lyman, A.L. (2020). Loss and waste in fish value chains: A review of the evidence from low and middle-income countries. Global Food Security 26: 1–13. https://doi.org/10.1016/j.gfs.2020.100434.

Kumar, A. and Sharma, S. (2019). *M*icrobes and Enzymes in Soil Health and Bioremediation (Springer Nature Singapore, Ed.; Vol. 16). http://www.springer.com/series/14379.

Li, J., Chen, Q., Zhang, S., Huang, H., Yang, J. and Tian, X.P. (2013). Highly heterogeneous bacterial communities associated with the South China Sea reef corals *Porites lutea, Galaxeafascicularis* and *Acropora millepora*. PLoS ONE 8: e71301. doi: 10.1371/journal.pone.0071301.

Liang, S., Li, H., Wu, H., Yan, B. and Song, A. (2023). Microorganisms in coastal wetland sediments: a review on microbial community structure, functional gene, and environmental potential. Front. Microbiol. 14. https://doi.org/10.3389/fmicb.2023.1163896.

Liao, J., Weller, D.L., Kovac, J. and Huang, Y. (2022). Editorial: Functional diversity of aquatic microorganisms and their roles in water quality. Editorial. https://doi.org/10.3389/frwa.2022.879155.

Liao, J., Huang, Y., Kovac, J. and Weller, D.L. (eds.). (2022). Functional diversity of aquatic microorganisms and their roles in water quality. Frontiers Media SA. https://doi.org/10.3389/978-2-88976-370-2.

Liu, M., Yin, F., Zhai, W., Tian, P., Zhou, Y., Jia, Z. et al. (2024). Diversity of culturable bacteria from the coral reef areas in the South China sea and their agar-degrading abilities. Microorganisms 12(1): 187. https://doi.org/10.3390/microorganisms12010187.

Longo, C., Pierri, C., Mercurio, M., Trani, R., Cardone, F., Carbonara, P. et al. (2022). Bioremediation capabilities of *Hymeniacidonperlevis* (*Porifera, Demospongiae*) in a land-based experimental fish farm. Journal of Marine Science and Engineering 10(7): 874. https://doi.org/10.3390/jmse10070874.

Louca, S., Mazel, F., Doebeli, M. and Parfrey, L.W. (2019). A census-based estimate of earth's bacterial and archaeal diversity. PLoS Biology 17(2). https://doi.org/10.1371/journal.pbio.3000106.

Macaulay, B.M. and Rees, D. (2014). Bioremediation of oil spills: A review of challenges for research advancement. Annals of Environmental Science 8: 9–37.

Mahboob, S. (2014). Isolation and characterization of collagen from fish waste material- skin, scales and fins of *Catlacatla* and *Cirrhinusmrigala*. J. Food Sci. Technol. Doi 10.1007/s13197-014-1520-6.

Maher, R.L., Epstein, H.E. and Vega Thurber, R. (2022). Dynamics of bacterial communities on coral reefs: implications for conservation. *In*: van Oppen, M.J.H. and Aranda Lastra, M. (eds.). Coral Reef Conservation and Restoration in the Omics Age. Coral Reefs of the World, vol 15. Springer, Cham. https://doi.org/10.1007/978-3-031-07055-6_7.

Martinez-Porchas, M., Martinez-Cordova, L.R., Lopez-Elias, J.A. and Porchas-Cornejo, M.A. (2014). Bioremediation of aquaculture effluents. pp. 541–555. *In*: Das, S. (ed.). Microbial Biodegradation and Bioremediation. Elsevier Inc. London.

Middelboe, M. and Brussaard, C.P.D. (2017). Marine viruses: Key players in marine ecosystems. In Viruses (Vol. 9, Issue 10). MDPI AG. https://doi.org/10.3390/v9100302.

Mohamad, K.A., Mohd, S.Y., Sarah, R.S., Mohd, H.Z. and Rasyidah, A. (2017). Total nitrogen and total phosphorus removal from brackish aquaculture wastewater using effective microorganism. AIP Conference Proceedings 1885(1): 020127. https://doi.org/10.1063/1.5002321.

Nagrale, D.T. and Gawande, S.P. (2018). Archaea: Ecology, application. *In*: Sharma, S. and Varma, A. (eds.). Microbial Resource Conservation. Soil Biology, vol 54. Springer, Cham. https://doi.org/10.1007/978-3-319-96971-8_16.

Nama, S., Shanmughan, A., Bhusan, B., Nayak, Bhushan, S. and Ramteke, K. (2023). Impacts of marine debris on coral reef ecosystem: A review for conservation and ecological monitoring of the coral reef ecosystem. Marine Pollution Bulletin, vol. 189.

Nawaz, A., Li, E., Irshad, S., Xiong, Z., Xiong, H., Shahbaz, H.M. et al. (2020). Valorization of fisheries by-products: Challenges and technical concerns to food industry, Trends in Food Science & Technology. doi: https://doi.org/10.1016/j.tifs.2020.02.022.

Neale, P.J., Williamson, C.E., Banaszak, A.T., Häder, D.P., Hylander, S., Ossola, R. et al. (2023). The response of aquatic ecosystems to the interactive effects of stratospheric ozone depletion, UV radiation, and climate change. Photochem. Photobiol. Sci. 22: 1093–1127. https://doi.org/10.1007/s43630-023-00370-z.

Oh, Yi-Su, Shih, Ing-Lung, Tzeng, Yew-Min and Wang, San-Lang. (2000). Protease produced by Pseudomonas aeruginosa K-187 and itsapplication in the deproteinization of shrimp and crab shell wastes. Enzyme and Microbial Technology 27: 3–10. https://doi.org/10.1016/S0141-0229(99)00172-6.

Ojha, N., Karn, R., Abbas, S. and Bhugrad, S. (2021). Bioremediation of industrial wastewater: A review. IOP Conf. Series: Earth and Environmental Science 796. doi:10.1088/1755-1315/796/1/012012.

Olsen, R.L., Toppe, J. and Karunasagar, I. (2014). Challenges and realistic opportunities in the use of by-products from processing of fish and shellfish. Trends in Food Science & Technology 36: 144–151. http://dx.doi.org/10.1016/j.tifs.2014.01.007.

Overmann, J. and Lepleux, C. (2016). Marine bacteria and archaea: diversity, adaptations, and culturability. pp. 21–55. In book: The Marine Microbiome: An Untapped Source of Biodiversity and Biotechnological Potential. Doi: 10.1007/978-3-319-33000-6_2.

Palmeiro-Sánchez, T., Campos, J.L. and Mosquera-Corral, A. (2021). Bioconversion of organic pollutants in fish-canning wastewater into volatile fatty acids and polyhydroxyalkanoate. International Journal of Environmental Research and Public Health 18(19). https://doi.org/10.3390/ijerph181910176.

Pascual-Fernández, J.J., Pita, C. and Bavinck, M. (eds.). (2020). Small-scale Fisheries in Europe: Status, Resilience and Governance (Vol. 23). Springer International Publishing. https://doi.org/10.1007/978-3-030-37371-9.

Penarubia, O.R., Toppe, J. and James, D. (2020). Fish waste management: turning waste into healthy feed with antimicrobial properties. Asian Fisheries Science 33: 11–15. https://doi.org/10.33997/j.afs.2020.33.S1.002.

Pozo-Solar, F., Cornejo-D´Ottone, M., Orellana, R., Yepsen, D.V. and Bassi, N. (2022). Dissolved greenhouse gases and benthic microbial communities in coastal wetlands of the Chilean coast semiarid region. PLOS ONE 17(9): e0271208. https://doi.org/10.1371/journal.pone.0271208.

Prasertsan, P., Choorit, W. and Suwanno, S. (1993). Optimization for growth of Rhodocyclusgelatinosus in seafood processing effluents. World Journal of Microbiology and Biotechnology 9(5): 593–596. https://doi.org/10.1007/BF00386302.

Racioppo, A., Speranza, B., Campaniello, D., Sinigaglia, M., Corbo, M.R. and Bevilacqua, A. (2021). Review - fish loss/waste and low-value fish challenges: state of art, advances, and perspectives. Foods 10: 1–25. https://doi.org/10.3390/foods10112725.

Raddadi, N., Cherif, A., Daffonchio, D., Neifar, M. and Fava, F. (2015). Biotechnological applications of extremophiles, extremozymes and extremolytes. pp. 7907–7913. *In*: Applied Microbiology and Biotechnology (Vol. 99, Issue 19). Springer Verlag. https://doi.org/10.1007/s00253-015-6874-9.

Reynolds, N. (2023). Bioremediation vs. traditional remediation methods: a comparative analysis. Journal of Biodiversity, Bioprospecting and Development (Vol 9:05). 2 pp. DOI: 10.37421/2376-0214.2023.9.60.

Rekadwad, B.N., Li, W.J., Gonzalez, J.M., PunchappadyDevasya, R., AnanthapadmanabhaBhagwath, A., Urana, R. et al. (2023). Extremophiles: the species that evolve and survive under hostile conditions. In 3 Biotech (Vol. 13, Issue 9). Springer Science and Business Media Deutschland GmbH. https://doi.org/10.1007/s13205-023-03733-6.

Ricci, F., Rossetto Marcelino, V., Blackall, L.L., Kühl, M., Medina, M. and Verbruggen, H. (2019). Beneath the surface: Community assembly and functions of the coral skeleton microbiome. In Microbiome (Vol. 7, Issue 1). BioMed Central Ltd. https://doi.org/10.1186/s40168-019-0762-y.

Riberio, F.H.M. and Naval, L.P. (2019). Reuse alternatives for effluents from the fish processing industry through multi-criteria analysis. Journal of Cleaner Production 227: 336–345. https://doi.org/10.1016/j.jclepro.2019.04.110.

Rodrigues-Filho, J.L., Macêdo, R.L., Sarmento, H., Pimenta, V.R.A., Alonso, C., Teixeira, C.R. et al. (2023). From ecological functions to ecosystem services: linking coastal lagoons biodiversity with human well-being. pp. 2611–2653. In Hydrobiologia (Vol. 850, Issues 12–13). Springer Science and Business Media Deutschland GmbH. https://doi.org/10.1007/s10750-023-05171-0.

Roy, R., Samanta, S., Pandit, S., Naaz, T., Banerjee, S., Rawat, J.M. et al. (2023). An overview of bacteria-mediated heavy metal bioremediation strategies. Applied Biochemistry and Biotechnology. https://doi.org/10.1007/s12010-023-04614-7.

Rudovica, V., Rotter, A., Gaudêncio, S.P., Novoveská, L., Akgül, F., Akslen-Hoel, L.K. et al. (2021). Valorization of marine waste: use of industrial by-products and beach wrack towards the production of high added-value products. In Frontiers in Marine Science (Vol. 8). Frontiers Media S.A. https://doi.org/10.3389/fmars.2021.723333.

Rustad, T. (2003). Utilisation of marine by-products. Electron. J. Environ. Agric. Food Chem. 2(4): 458–463.

Rustad, T., Storrø, I. and Slizyte, R. (2011). Review - Possibilities for the utilisation of marine by-products. International Journal of Food Science and Technology 46: 2001–2014. doi:10.1111/j.1365-2621.2011.02736. x.

Samani, M.N., Jafaryan, H., Gholipour, H., Harsij, M. and Farhangi, M. (2016). Effect of different concentration of profitable Bacillus on bioremediation of common carp (*Cyprinus carpio*) pond discharge. Iranian Journal of Aquatic Animal Health 2(2): 44–54.

Saranya, R., Jayapriy, A.J. and Selvi, A.T. (2018). Purification, characterization, molecular modeling and docking study of fish waste protease. International Journal of Biological Macromolecules. doi: 10.1016/j.ijbiomac.2018.06.119.

Selvi, V., Sathiyamoorthy, M. and Karthikeya, D.R.C. (2014). Nitrification of fish processing wastewater using mixed cultures of nitrosomonas and nitrobactor for ammonia degradation (Phase-I). Global Journal of Science Frontier Research: G Bio-Tech & Genetics 14(1).

Shahidul Islam, Md., Khan, S. and Tanaka, M. (2004). Waste loading in shrimp and fish processing effluents: potential source of hazards to the coastal and nearshore environments. Marine Pollution Bulletin 49(1–2): 103–110. https://doi.org/10.1016/j.marpolbul.2004.01.018.

Sule, I.O., Agbabiaka, T.O., Ahmed, R.N., Saliu, B.K. and Olayinka, K.J. (2016). Bacteriological and physicochemical analysis of wastewater from fishponds. Ethiopian Journal of Environmental Studies & Management 9(2): 167–178. http://dx.doi.org/10.4314/ejesm.v9i2.5.

Tudi, M., Ruan, H.D., Wang, L., Lyu J., Sadler, R., Connell, D. et al. (2021). Agriculture development, pesticide application and its impact on the environment. Int. J. Environ. Res. Public 18: 1112. 1–24. https://doi.org/ 10.3390/ijerph18031112.

Urano, N., Sasaki, E., Ueno, R., Namba, H. and Shida, Y. (2002). Bioremediation of fish cannery wastewater with yeasts isolated from a drainage canal. Marine Biotechnology 4: 559–564.

Vakkachan, P., Gopakumar, S.T. and Janardhanan, R.K. (2023). Degradation of marine crustacean shell wastes through single-stage co-fermentation using proteolytic and chitinolytic bacteria. Environ. Sci. Pollut. Res. https://doi.org/10.1007/s11356-023-30355-5.

Vazquez, J.A., Amado, I.R., Sotelo, C.G., Sanz, N., Pérez-Martín, R.I. and Valcárcel, J. (2020). Production, characterization, and bioactivity of fish protein hydrolysates from aquaculture turbot (*Scophthalmus maximus*) wastes. Biomolecules 10(310): 1–13. doi:10.3390/biom10020310.

Venugopal, V. and Sasidharan, A. (2021). Seafood industry effluents: Environmental hazards, treatment and resource recovery. Journal of Environmental Chemical Engineering 9: 1–17. https://doi.org/10.1016/j.jece.2020.104758.

Wabnitz, C.C.C., Lam, V.W.Y., Reygondeau, G., Teh, L.C.L., Al-Abdulrazzak, D., Khalfallah, M. et al. (2018). Climate change impacts on marine biodiversity, fisheries and society in the Arabian Gulf. PLoS ONE 13(5): e0194537. https://doi.org/ 10.1371/journal.pone.0194537.

Wai, S.N., Mizunoe, Y. and Yoshida, S. (1999). How *Vibrio cholerae* survive during starvation. FEMS Microbiol Lett. 180: 123–131.

World Bank. (2017). The Potential of the Blue Economy: Increasing Long-term Benefits of the Sustainable Use of Marine Resources for Small Island Developing States and Coastal Least Developed Countries. World Bank, Washington DC.

Yue, Y., Rong, H., Yang, Z., Pan, X., Chen, Y. and Yang, M. (2023). Microbial diversity and functional profiling in coastal tidal flat sediment with pollution of nutrients and potentially toxic elements. Journal of Soils and Sediments 23(7): 2935–2950. https://doi.org/10.1007/s11368-023-03511-0.

Zhang, X., Das, S., Li, A., Ma, Q. and Tan, L. (2021). Editorial: Marine microbes for contaminant bioremediation. In Frontiers in Microbiology (Vol. 12). Frontiers Media S.A. https://doi.org/10.3389/fmicb.2021.762968.

Zhang, X., Wu, K., Han, Z., Chen, Z., Liu, Z., Sun, Z. et al. (2022). Front. Microbiol, Sec. Aquatic Microbiology 12. https://doi.org/10.3389/fmicb.2022.1029564.

Zokaeifar, H., Babaei, N., Saad, C.R., Kamarudin, M.S., Sijam, K. and Balcazar, J.L. (2014). Administration of *Bacillus subtilis* strains in the rearing water enhances the water quality, growth performance, immune response, and resistance against *Vibrio harveyi* infection in juvenile white shrimp, Litopenaeusvannamei. Fish & Shellfish Immunology 36(1): 68–74. https://doi.org/https://doi.org/10.1016/j.fsi.2013.10.007.

Chapter 5

Omics Translation

Marine-Derived Pharmaceuticals for the Blue Bio-economy

*Parthkumar Prajapati, Bhavika Parekh, Rimpal Bhanderi, Chandrashekar Mootapally** and *Neelam Nathani**

Introduction

Our understanding of biological systems at the molecular level has been completely transformed by the interdisciplinary field of omics, which primarily combines genomics, transcriptomics, proteomics, and metabolomics. Researchers unravel the complex workings of living things using this broad range of "omics" technologies, which shed insight on their genetic make-up, gene expression patterns, protein profiles and metabolite signatures. We can bridge the gap between basic research and clinical practice through translation omics and a potent use of these technologies to open the door for customized medicine and tailored treatments. Translational omics provides the potential for more accurate and effective healthcare by unraveling the molecular causes of diseases and adjusting interventions to a person's particular biological profile. As the field of omics continues to advance, it becomes increasingly essential for improving our understanding of complex biological systems and their practical applications in medicine and beyond (Collins and Varmus 2015). The field of omics sciences in marine biotechnology involves characterization and quantification of all biological molecules that are included in the structure, function and organization of cell, tissue or organism. The inevitable methods in multi omics in marine biotechnology include mass spectrometry and gel-based technique for proteins,

School of Applied Sciences and Technology (SAST), Gujarat Technological University, Chandkheda-382424, Ahmedabad.

* Corresponding authors: ap_chandrashekar@gtu.edu.in, chandu.avi@gmail.com; ap_neelam@gtu.edu.in, neelam.nathani13@gmail.com

sequencing and microarray for mRNA and DNA, and Nuclear Magnetic Resonance (NMR) for metabolites. All these techniques are used in marine biotechnology to derive pharmaceutical products from marine natural products (D'Ayala et al. 2008, McEneff et al. 2013, Tang et al. 2012, Xie et al. 2017). Omics is a decorative strategy in pharmacology and bioinformatics, which inspects the disease and effect of drugs on the whole genome, proteome or metabolome.

Marine biotechnology is the application of biotechnological techniques and principles to study the marine environment. The marine environment includes marine organisms, their genetic materials and compounds produced from marine organisms. Marine biotechnology represents a rapidly expanding field with immense potential to address critical global challenges. This field uses the distinctive biochemical and genetic resources of marine animals to create new applications in a variety of fields, including pharmaceuticals, agriculture, environmental conservation, and energy generation (Das and Patel 2020). It draws its inspiration from the abundance of biodiversity found in our oceans. The potential of marine biotechnology to offer sustainable solutions for various industries grows along with our understanding of complex marine environments. The relevance of the discipline is highlighted by the rise in interest in marine resource discovery and sustainable use, making it a vital contributor to both scientific knowledge and economic progress (Rotter et al. 2021).

The marine environment is a treasure trove of bio-compounds that have captivated the attention of scientists and researchers worldwide. The oceans provide a source of bioactive compounds with enormous promise for medication discovery, biotechnology, and numerous industrial applications (Bhatnagar and Kim 2010, Suleria et al. 2016). These compounds are produced by a varied variety of creatures, including coral reefs, microbes, fish, and sponges. These bio-compounds, which include unique enzymes, bioactive peptides, antimicrobial compounds, and natural products, have shown extraordinary qualities that can be used for the treatment of diseases, the production of sustainable agricultural methods, and the development of novel materials. As we delve deeper into marine biodiversity, we find a plethora of bio-compounds. Each one of these offers the potential to advance human health as marine-derived pharmaceuticals and sources for human wellbeing while advancing our knowledge of the natural world.

Marine-derived pharmaceuticals have emerged as a fascinating frontier in drug discovery, offering a wealth of bioactive compounds that hold the potential to combat a wide range of diseases. Researchers have extracted substances with distinctive chemical structures and powerful biological activities from the depths of the oceans. In the effort to create new medicines, these substances—which include antiviral medications, anti-inflammatory chemicals, and possible cancer therapeutics—are invaluable. These pharmacological prospects are abundant in the marine environment, and their use in medicine is a testimonial of the unrealized promise of marine biodiversity (Calado et al. 2018). As we continue to broaden our pharmacological toolkit through our investigation of marine species and their bioactive molecules, we give fresh promise to treating some of the most difficult diseases in the world.

In this chapter, we embark on a captivating journey into the world of 'Omics Translation,' where cutting-edge genomics, transcriptomics, proteomics, and metabolomics converge to unlock the untapped potential of marine-derived pharmaceuticals. Exploring the oceans' depths, we explore blue bio-economy—a burgeoning sector that combines the promise of new drug discovery with the sustainable exploitation of marine resources. We decipher the complex molecular profiles of marine organisms and the bioactive substances they contain, providing a glimpse into a future in which these molecules power invention, sustainable economic growth, and improvements in healthcare. We explore the relationship between science, business, and environmental protection through an interdisciplinary lens, highlighting how "Omics Translation" is poised to completely alter how we think about the development of pharmaceuticals and the blue bio-economy.

Translational omics

Omics is a field of study that identifies every component of a cell's biological makeup. Any biomolecule that is present in a cell and may be used in a molecular process in healthy, functional cells and tissues can be identified, described, and quantified with the use of omics technology. Various omics technologies fall under this category, like lipid omics, epigenomics, transcriptomics, genomics, and proteomics, which utilize the examination of DNA, metabolites, proteins, RNA, chromosomes, and functional and non-functional genes, among other things.

Metabolomics

Metabolomics has opened a window into the metabolic intricacies of marine organisms, revealing their adaptation strategies, bioactive compounds, and responses to environmental changes. Whether it is for drug discovery, biodiversity studies, or monitoring ecosystem health, metabolomics in marine organisms is a transformative tool that holds the key to unlocking the secrets of the world's oceans. Its applications not only advance scientific understanding but also offer valuable contributions to the fields of biotechnology, medicine (Stuart et al. 2020), and environmental conservation (Bayona et al. 2022). As we continue to unveil the metabolic mysteries of marine life, we gain the potential to harness the ocean's biochemical wealth for the benefit of both science and society. Metabolomics covers the investigation of metabolites, bodily fluids, and biomolecules found within cells or tissues. To better understand cellular biology and metabolic processes, it can be explored by merging it with computational biology (Reverter et al. 2020).

Two types of metabolites can be found in microorganisms or plants: they are primary and secondary metabolites. Secondary metabolites, such as antibiotics and pigments, are not engaged in the growth or development of plants or other organisms. The metabolome, which may be thought of as the result of one product serving as the substrate for another, connects metabolomics. NMR, mass spectroscopy, shotgun sequencing, and other techniques are used in metabolomics, and many statistical methods can be used for data interpretation. The sample's compounds must first be separated in order to conduct the metabolomics experiment. Mass spectroscopy requires the separation of compounds because they are mixed and heavy; however if NMR and shotgun sequencing techniques have been employed by the performer, this step can be bypassed. Separation aids in the detection of chemicals in the detector during Mass spectroscopy. Techniques such as high-performance liquid chromatography, liquid chromatography, or gas chromatography can be used to separate the mixtures of the chemicals (Alseekh et al. 2021, Emwas et al. 2019, Tang 2011).

Transcriptomics

The study of transcriptome, which includes the analysis of the RNA sets present in the cell, is thought to be best accomplished through the use of transcriptomics as a research method. Unlocking the potential of marine resources requires the critical application of transcriptomics, a potent molecular biology technique. Researchers can learn a great deal about the molecular mechanisms behind the adaptation of various marine organisms, including fish, corals, and algae, to environmental stressors, their responses to climate change, and their production of bioactive compounds, by examining the gene expression patterns of these organisms (Gleason 2019, Kong et al. 2020, Li et al. 2020, Mazurais et al. 2011, Seneca and Palumbi 2015). The discovery of novel pharmaceuticals and biotechnological applications derived from the rich genetic diversity of the oceans, as well as the development of resilient aquaculture practices and sustainable management of marine ecosystems, can benefit from this information. The level of RNA expression, RNA degradation, RNA trafficking, RNA functions, and RNA location are all included in transcriptome studies. It covers every type of RNA, including tRNA, mRNA, rRNA, miRNA, and siRNA, among others. Information in any organism typically transfers from DNA to RNA through transcription and is subsequently translated from RNA to protein, and gives us an idea about tissue specific or cell specific expression of protein in any cell. The assumption is that the expression level of RNA should be tissue- or cell-specific which gives rise to single cell transcriptomics. This conveys information about the degree of expression in a certain cell. Furthermore, transcriptomics allows for a comprehensive understanding of the intricate ecological interactions within marine communities, shedding light on how these resources can be harnessed responsibly for the benefit of both scientific knowledge and human well-being.

Proteomics

Proteomics is the study of proteins that are found in living things. They perform essential tasks for all living things, including the replication of DNA, the development of muscles and the synthesis of enzymes, hormones, and antibodies. Proteomics is essential to the discovery and utilization of marine resources. Researchers can learn more about the molecular underpinnings of a variety of biological processes, such as growth, reproduction, and responses to environmental stressors by analyzing the intricate protein profiles found in marine organisms (Rodrigues et al. 2012, Romero et al. 2019, Tomanek 2014). This knowledge can be crucial in understanding how these organisms adapt and survive. Proteomics also makes it easier to identify useful proteins that have uses in the food and pharmaceutical sectors (Dyshlovoy et al. 2017), like enzymes for processing seafood or bioactive substances with potential medical uses. This information not only improves our comprehension of marine ecosystems but also creates opportunities for the development of novel biotechnologies, sustainable management techniques, and the ethical use of marine resources, all of which eventually advance science and improve human welfare. Additionally, functional proteomics is a branch of proteomics that focuses on understanding the biological purpose of unidentified proteins and identifying the molecular basis of cellular activity (Wagner et al. 2005). Complexity in proteomics is much higher than in genomics and transcriptomics due to the fact that the function of any protein differs from cell to cell and from time to time. Distinct genes are expressed in different cell types, which means that the basic set of proteins produced by cells must be identified. There are different techniques of proteomics such as 2D Gel electrophoresis (2D-PAGE), Mass Spectrometry (MS), Liquid Chromatography (LC), Gel electrophoresis (SDS-PAGE), Protein microarrays, Shotgun Proteomics, Quantitative Proteomics, Protein sequencing, Phosphoproteomics, Protein interaction analysis, Top-down Proteomics, Targeted Proteomics, Post-Translational Modification Analysis, etc., These techniques collectively allow scientists to comprehensively study the proteome of cells, tissues, and organisms, providing insights into biological processes, disease mechanisms, and therapeutic targets (Bovee et al. 2011, Campos et al. 2016, Campos and de Almeida 2016, Inoue et al. 1995, Ji et al. 2014, Kwon et al. 2020, Leprêtre et al. 2020, Powell et al. 2005, Rivera-Pérez and Hernández-Saavedra 2021, Romeu et al. 2021, Thiyagarajan et al. 2009, Vilasi et al. 2013).

Genomics

A revolutionary tool for the sustainable management and use of marine resources is emerging: genomics. Through the process of deciphering the genetic code, genomics allows researchers to gain an understanding of the complex genetic diversity, adaptability, and evolutionary history of a wide range of marine organisms, from economically significant fish species to microorganisms essential to the health of ecosystems. This information is crucial for fisheries management since it aids in the creation of plans that guarantee the long-term wellbeing of marine populations. Additionally, genomics makes it easier to identify genes linked to desired aquaculture traits, improving breeding programs for higher yield and disease resistance. It looks at the collective characterization and quantification of all the genes in any given organism. A gene carries the instructions needed to make a certain protein or set of proteins. The ribosome receives instructions from the mRNA, and the tRNA uses those instructions to make proteins. Analysis and sequencing of genomes are aspects of genomics. Functional genomics, structural genomics, epigenomics, and metagenomics are only a few of the numerous types of available genome investigation tools (Balch et al. 2010, Reid and Whitehead 2016, Trindade et al. 2015, Wyche et al. 2017). Functional genomics helps with the annotation of unidentified genes, whereas structural genomics assists in characterizing the three-dimensional structure of the proteins encoded. Epigenomics is the study of epigenetic modifications of genetic material, such as DNA methylation and histone modifications. Metagenomics is a tool for analyzing microbiome from any habitat. It also plays a pivotal role in environmental monitoring, aiding in the assessment of the impact of pollution, climate change, and habitat degradation on marine ecosystems (Coelho et al. 2013, M. et al. 2012, Raimundo et al. 2021, Techtmann and Hazen 2016). In essence, genomics empowers us to unlock the potential of marine resources while safeguarding the delicate balance of our oceans and their biodiversity.

Lipidomics

Lipids play pivotal roles in cellular membranes and lipid particles like lipoproteins contribute to cellular barriers, membrane structures, signaling pathways and other vital functions. Lipidomics refers to the comprehensive characterization of lipid molecules involved in protein expression related to lipid metabolism and function. This field has evolved significantly due to advancements in mass spectrometry techniques, including liquid chromatography coupled with mass spectrometry, shotgun lipidomics, MS imaging, and ion mobility MS (Han 2016, Han et al. 2012). The lipidomics workflow encompasses sample preparation methods such as shotgun and LC-based lipidomics, followed by MS data acquisition involving ionization and MS analysis, leading to data generation. Subsequently, bioinformatics tools and spectral data analysis are employed for data processing (Seppänen-Laakso and Orešič 2009). Through lipidomics, researchers can identify various bioactive compounds, including antibiotics, antitumor agents, antifungals, and anti-inflammatory compounds, among others (Yang and Han 2016). In a study focusing on seagrasses, researchers utilized *Ulva armoricana* and *S. chordalis* to assess their antiproliferative properties. They employed GC-MS for compound separation and detection, subsequently evaluating anti-proliferative activity. Notably, phytol was identified as a precursor for industrially synthesizing vitamins E and K. Additionally, detected phytosterols exhibited potential in reducing cholesterol absorption, functioning as anti-inflammatory agents and contributing to cardiovascular disease treatment (Kendel et al. 2015). The recent advancements in MS-based lipidomics offer a unique opportunity to explore diverse polar lipid molecular species in marine organisms. Integrating high-throughput techniques with reliable bioinformatics pipelines is crucial for efficiently managing data and avoiding common omics challenges. Standardizing protocols for extraction, structural identification, and deposit vouchers of screened taxa is essential for result reliability and reproducibility. These efforts will drive further innovation and understanding in lipidomics, enabling the discovery of new bioactive compounds and applications in marine research and beyond.

Metagenomics

Metagenomics offers a powerful approach to unveil the composition and functional potential of microbial communities directly from environmental samples. This eliminates the need for laborious isolation and cultivation techniques, providing a comprehensive picture of microbial diversity and their roles within various ecosystems (Simon and Daniel 2011). Marine metagenomics focuses on deciphering the genetic content and functional capabilities of microorganisms residing in marine environments, encompassing seas, oceans, and estuaries (Sunagawa et al. 2015). Marine bioprospecting leverages this approach to identify novel organisms and genes from marine sources, ultimately leading to the discovery of diverse bioactive compounds with promising applications. A recent study exemplifies the power of this combined approach. Researchers employed metagenomic and single-cell genome sequencing to investigate filamentous marine cyanobacteria. This strategy facilitated the identification of a putative biosynthetic gene cluster responsible for apratoxin, a potent cytotoxin with potential therapeutic applications. Furthermore, this work enabled the isolation of a complete biosynthetic gene cluster from *Lyngbya bouilloni*, a marine cyanobacterium renowned for its production of various bioactive compounds (He et al. 2012). Another study employed metagenomic analysis of the marine sponge *Discodermia calyx*. Functional screening identified two clones capable of producing red pigments, subsequently characterized as porphyrins. Structural elucidation revealed the major pigment to be Zn-porphyrin 3, a molecule with potential applications in treating haematological disorders and anaemia, as well as environmental monitoring (Grindberg et al. 2011). These findings highlight the significant role of marine metagenomics in various fields, including drug discovery, ecosystem monitoring, and the development of novel biotechnologies. Advancements in sequencing technologies, bioinformatics tools, and interdisciplinary collaborations are continuously furthering our understanding of marine ecosystems and unlocking the immense potential of marine organisms for societal benefit.

Multi-omics

Marine ecosystems represent a tapestry of life teeming with diverse organisms interacting in a dynamic web. To fully comprehend the intricate workings of this environment, researchers are increasingly turning to multi-omics approaches. This powerful strategy integrates data from various "omics" disciplines, including genomics, proteomics, transcriptomics, metabolomics, and metagenomics to paint a holistic picture of biological systems. Within the realm of marine science, multi-omics offers an unprecedented level of understanding regarding the interplay between organisms, their functional capacities, and their responses to environmental stimuli. This approach allows researchers to move beyond studying individual components in isolation and delve into the complex network of interactions that govern the health and functionality of marine ecosystems. One compelling example of the power of multi-omics comes from a study utilizing a combination of metagenomics and metabolomics. Researchers focused on the ascidian model organism, *Didemnum molle*. By employing metagenomic sequencing, they were able to identify a novel precursor molecule within the sample, based on its primary sequence as a cyanobactin. Leveraging this information, they proceeded to extract and analyze the sample using Liquid Chromatography-Fourier Transform Mass Spectrometry (LC-FT-MS). This analysis led to the successful identification of a mature peptide product named Minimide. This newly discovered compound exhibits promising bioactivities, including potential for anti-microbial, anti-cancer, anti-inflammatory, and even neuroprotective effects (Donia et al. 2011). The true strength of multi-omics lies in its ability to integrate datasets from these diverse disciplines. By bridging the gap between genetic blueprint, protein expression, metabolic activity, and microbial community composition, researchers can gain a deeper understanding of complex biological processes within marine organisms. This integrated approach holds immense promise for the discovery of novel bioactive compounds with significant applications in medicine, environmental management, and beyond. Ultimately, the combined application of multi-omics approaches paves the way for a system-level understanding of marine ecosystems, empowering us to unlock their full potential and ensure their continued health and resilience.

Glycomics

Glycomics, a burgeoning field, delves into the intricate world of glycans within marine organisms. These sugar molecules, with their remarkable structural diversity, play a pivotal role in deciphering the structure, interactions, and functions of marine life. From algae and seaweeds to diverse marine invertebrates, each species harbors a unique repertoire of glycans. These biomolecules orchestrate a symphony of vital processes, including immunity, defense mechanisms, cellular signaling and structural support. Recent studies have yielded exciting results. Researchers successfully isolated chondroitin sulphate (CHS) with a novel structure from the skin of the Halaelurus burger shark. This unique CHS demonstrated the remarkable ability to improve insulin resistance. Employing sophisticated techniques like Nuclear Magnetic Resonance (NMR) and Fourier-Transform Infrared Spectroscopy (FTIR), scientists were able to characterize the isolated compound. Furthermore, in vivo experiments provided compelling evidence – the CHS significantly reduced body weight, blood glucose levels, and insulin levels in animal models (Hu et al. 2023). Another study unveils a novel *Mameliella alba* strain (LZ-28) isolated from a toxic marine dinoflagellate. Researchers employed comparative genomics method to reveal its close genetic relationship with other marine algae-associated *M. alba* strains, suggesting a crucial role in these ecosystems. Notably, LZ-28 demonstrated exceptional production of bio flocculants, which were further optimized for enhanced yield. Bio flocculants hold promise in the pharmaceutical field, with applications in drug delivery and wastewater treatment (Ren et al. 2022). This research underscores the power of omics in exploring microbe interactions within marine environments, paving the way for the discovery of novel biomolecules with significant pharmaceutical potential. The future of marine glycomics is brimming with potential. As research progresses, we can anticipate the discovery of novel polysaccharides with a plethora of bioactivities, including anticoagulant, anticancer, and antitumor properties. This field holds immense promise for unravelling the complex and fascinating world of marine glycans, paving the way for advancements in various sectors. By harnessing the potential of these marine resources, we can ensure sustainable development while promoting human health through groundbreaking discoveries.

Phenomics

Marine phenomics has emerged as a powerful tool for dissecting the intricate interplay between genotype, environment, and phenotype in marine organisms. By comprehensively analyzing observable characteristics, it sheds light on the remarkable phenotypic diversity exhibited by these creatures. This approach holds immense value in deciphering the intricate mechanisms underlying adaptation and responses to the unique challenges posed by marine environments. Marine phenomics not only elucidates the functional underpinnings of various traits but also sheds light on evolutionary relationships between species. Furthermore, it fosters a deeper understanding of the complex interplay between genotype, phenotype, and the environment. A recent study exemplifies the power of marine phenomics. Researchers employed this approach to identify a novel fatty acid transporter in diatoms. Disruption of this transporter resulted in altered levels of several crucial fatty acids, including the medically important omega-3 polyunsaturated fatty acids (PUFAs). Through a targeted screening strategy utilizing a collection of mutagenized diatoms, the study successfully identified temperature-sensitive mutants, effectively unravelling genes associated with stress responses and potentially novel proteins. By analyzing these mutant phenotypes, scientists gained valuable insights into the intricate fatty acid biosynthetic pathways, paving the way for future investigations into the manipulation of these pathways for the production of essential fatty acids with therapeutic potential (Johansson et al. 2019). In conclusion, marine phenomics offers a transformative approach for understanding the remarkable adaptability of marine organisms. By bridging the gap between genes and observable traits, this field holds immense promise for advancements in marine biology, biotechnology, and ultimately, the development of novel therapeutic strategies.

Epigenomics

Epigenetics, a subdiscipline of genomics, delves into the fascinating realm of epigenetic modifications across the entire genome of an organism. These modifications, encompassing mechanisms like DNA methylation, histone modifications, and RNA regulation, influence gene expression and play a crucial role in various biological processes, including disease development. Marine epigenetics specifically investigates how marine organisms respond to a dynamic interplay of environmental cues, stress factors, and ecological influences. This field explores the fascinating phenomenon of environmentally induced epigenetic modifications. By elucidating how such modifications occur during these changes, marine epigenetics sheds light on how they influence phenotypic variation, adaptation, and evolution within marine ecosystems. Understanding these epigenetic mechanisms holds immense significance. It allows researchers to decipher how marine organisms adjust to environmental pressures, potentially leading to the discovery of novel strategies for mitigating the effects of climate change and pollution on these vital ecosystems. Additionally, by unravelling the link between the environment and epigenetic modifications, marine epigenetics may pave the way for the development of novel tools for marine conservation and sustainable resource management.

Potential marine resources for bioprospection

Marine environments offer a treasure trove of potential resources for biotechnological research that often surpass what terrestrial biology can provide. A wide variety of life can be found in the vast and mainly uncharted depths of the world's oceans; many of these species have evolved unusual and fascinating biochemical adaptations in order to survive. These modifications can produce useful chemicals and enzymes that have uses in biotechnology, such as in the manufacturing of drugs, cosmetics, and food. The marine ecosystem is full of new biological wonders from extremophiles that thrive in high-pressure, high-temperature environments to organisms that can bioluminesce and decompose. Many of these can solve urgent global issues. Furthermore, a vast range of molecules and microorganisms with potential uses in drug development, biomaterials, and bioenergy are accessible in the oceans (Dalmaso et al. 2015, de la Calle 2017, Fonseca et al. 2023, Penesyan et al. 2010, Poli et al. 2017, Suleria et al. 2015). Marine life is an endless source of inspiration for cutting-edge biotechnological research due to its exceptional resilience and adaptability.

Marine ecosystems exhibit several unique characteristics compared to terrestrial ecosystems. Marine ecosystems contain fourteen distinct animal phyla and numerous species stemming from their ancient origins, diverse taxonomic levels, and extensive genetic diversity. They differ primarily in their physical environment, with marine life dwelling in water and terrestrial organisms inhabiting land. The broader marine ecosystem harbors a higher species diversity due to its varied niches, dispersal capacities, and prolific fecundity, influencing genetic structure and evolution. Unlike terrestrial ecosystems, marine ecosystems feature phytoplankton as primary producers and emphasize the dominance of grazers over primary producers. Carnivores play a significant role in structuring the marine ecosystem, but overexploitation threatens biodiversity and ecosystem functions. This is in contrast to terrestrial ecosystems, dominated by herbivores and human consumption of primary energy production. The ocean offers a greater variety of species at higher trophic levels, with around 400 marine species harvested for food compared to just ten from land. However, marine ecosystems are highly vulnerable to pollution, as all pollutants ultimately enter the sea, impacting marine diversity, food chains, and the quality of marine food.

Algae

Marine algae, also known as seaweeds, have gained significant attention in recent years due to their potential pharmaceutical properties. Their abundance of bioactive compounds, many of which have qualities that can be used for pharmaceutical and medical purposes, is one of their most promising features. Fucoidans, for instance, are abundant in brown algae like *Fucus vesiculosus* and have shown to have anti-inflammatory, anticoagulant, and anti-cancer properties (Cumashi et al. 2007). Sulfated polysaccharides with antiviral and immunomodulatory qualities can be found in red algae, such as *Gracilaria* sp. and *Porphyra* sp., which is why researchers are interested in using them to create antiviral medications (Alencar et al. 2019, Bhatia et al. 2013). Furthermore, lectins and polyphenols, which are abundant in green algae and include species like *Ulva lactuca*, may find use in the development of anti-inflammatory and anti-cancer medications (Erpel et al. 2020, Fitton et al. 2019). Moreover, marine algae offer an eco-friendly alternative to terrestrial sources for pharmaceutical production. They grow rapidly, do not require arable land or freshwater, and their cultivation has a minimal environmental footprint. This aligns with the growing need for sustainable and renewable sources of pharmaceutical compounds. As highlighted in research by (Holdt and Kraan 2011), the sustainable cultivation and harvesting of seaweeds, as well as the development of extraction methods can provide a consistent source of high-quality bioactive compounds for pharmaceutical applications. The potential pharmaceutical properties of marine algae not only underscore their significance as a valuable resource but also their role in addressing global health challenges and the development of innovative drugs. Fernández-Acero et al. (2019) have used *Nannochloropsis gaditana* which is a non-flagellated microalga generally used for biofuels and aquiculture. Using proteomics technology, they have documented around 7500 peptides and amongst, have annotated the 655 proteins and identified 'prohibitin' protein, which is generally used for diseases involving cellular replication control.

Seagrass

Seagrass are the only flowering plants found in the marine environment. There are almost sixty species which belong to four families named Posidoniaceae, Zosteraceae, Hydrocharitaceae, Cymodoceaceae. It is observed that seagrass evolved from terrestrial plants which recolonized the ocean seventy to one hundred million years ago. Seagrasses provide shelter and food to animals, invertebrate fish, crabs, turtle, etc. Seagrasses, often overshadowed by their marine counterparts like algae and coral reefs, have recently emerged as a promising source of pharmaceutical compounds with diverse potential applications. These submerged flowering plants, found in coastal and marine environments worldwide, have garnered attention for their ability to produce secondary metabolites and bioactive molecules. Some seagrasses, such as the Mediterranean *Posidonia oceanica*, are rich in phenolic compounds and polyphenols with demonstrated antioxidant properties, offering potential therapeutic applications in combating oxidative stress-related diseases (Kesraoui et al. 2011). Additionally, research has indicated that seagrass species contain anti-inflammatory and antimicrobial compounds as demonstrated in a study by (El-Rafie et al. 2016). The study highlights the anti-inflammatory potential of the seagrass *Halophila stipulacea* through the inhibition of proinflammatory cytokines.

Another study, employed a GC-MS-based metabolomic approach to analyze Seagrass *Halodule pinifolia.* They successfully identified ninety-eight metabolites in wild species, while 125 metabolites were identified in suspension callus culture. This study aimed to compare the metabolites in wild and suspension callus culture. Utilizing the Kyoto Encyclopedia of Genes and Genomes (KEGG) pathway, they comprehensively annotated the functions of all the identified metabolites. Notably, it was observed that these metabolites exhibited significant antimicrobial activities against *Pseudomonas aeruginosa PAO1*. Furthermore, a preliminary assessment of antibiofilm activity demonstrated that the methanolic extract of *H. pinifolia* showed remarkable biofilm formation inhibition, reaching 96%

compared to the control. Eight bioactive compounds, including 4-hydroxybenzoic acid, rosmarinic acid, 4-methoxybenzoic acid, p-coumaric acid, protocatechuic acid, caffeic acid, naringenin, and vanillic acid, were profiled. Among these compounds, 4-methoxybenzoic acid (4-MBA) displayed the highest bacterial growth inhibition and acts as a promising lead molecule with minimal inhibitory concentration (MIC) (Danaraj et al. 2020).

Plankton

Marine plankton, comprising a diverse array of microscopic organisms, including phytoplankton and zooplankton, have recently come under scrutiny for their potential pharmaceutical properties. Diatoms and dinoflagellates are examples of phytoplankton, which are recognized for their capacity to synthesize bioactive substances such as polyunsaturated fatty acids, polysaccharides, and toxins that have potential uses in drug development. These substances show a variety of intriguing medicinal qualities, such as potential anti-cancer and neuroprotective qualities as well as antioxidant and anti-inflammatory effects (Assunão et al. 2017, Terracciano et al. 2018). Studies that have been carried out, highlight the potential neuroprotective qualities of polyunsaturated fatty acids derived from diatoms and have unearthed a wealth of biomedical compounds within planktonic communities, with potential pharmaceutical applications. For example, they have identified cyclopeptides sourced from cyanobacteria (blue-green plankton), such as *Anabaena* sp. and *Microcystis* sp., which possess the ability to function as cancer-fighting agents. These cyanobacteria produce hydrophilic cyclic heptapeptides that have the potential to induce cellular damage upon uptake via organic anion-transporting polypeptides. Significantly, researchers have observed an elevated presence of organic anion-transporting polypeptides in cancerous cells compared to normal tissues, indicating the promising potential of microcystin as a candidate for cancer treatment (Dunstan et al. 1993, Pradhan and Ki 2022, Sainis et al. 2010). Leveraging omics technology, we can pinpoint the specific genes responsible for cyclopeptide production, enabling the inducement of these peptides through metabolic engineering—a pivotal step towards their utilization in cancer therapy.

Prospecting planktonic organisms can be used to find new compounds that are bioresources. It is clear that a lot of successful stories have already been published, which piques this organism's interest in the subject matter (Abida et al. 2013). By bioprospecting this biodiversity, new approaches to many of the world's problems of today may be found. For example, using phytoplankton as a feedstock could help produce biofuels that can reduce pollution (Johnson 2009). This microbe may be useful in the synthesis of medicinal and nutraceutical substances. Through the use of omics technology to mine new bioactive compounds, we can obtain a variety of pharmaceutical compounds that can be produced in large quantities and may be effective against organisms that are resistant to multiple drugs.

Marine venom

Marine venom is becoming known as a potential source of medicinal compounds. It is commonly linked to the stinging cells of animals such as jellyfish, sea anemones, fish, sea snakes, cone snails, etc. In order to repel predators or immobilize prey, these venomous organisms have developed an arsenal of bioactive molecules, many of which have medicinal qualities that can be used in therapeutic settings. Cone snails, for example, are producers of a wide variety of conopeptides, some of which have shown to have strong analgesic effects. Ziconotide, a synthetic conopeptide derived from the venom of the cone snail *Conusmagus*, is used as an analgesic (painkiller) to treat severe chronic pain by blocking the N-type calcium channel (Li et al. 2020, Safavi-Hemami et al. 2019). Furthermore, there are many substances in marine venom that may have anti-cancer effects. According to recent research *Heteractis crispa* sea anemone has been found to produce venom peptides that can specifically target cancer cells while sparing healthy ones (Fedorov et al. 2010). These venom peptides have demonstrated potential in preventing the spread and growth of tumors. For instance, a study reported that the peptide ShK-186, which is derived from the venom of the sea anemone *Stichodactyla helianthus*, has shown specific targeting of potassium channels in cancer cells, making it a viable candidate for anti-cancer therapy (Chang et al. 2015). Proteomics and mass spectrometry profiling have also been used to characterize the venom of *Montivipera bulgardaghica* and *Montivipera raddei*. The cytotoxicity screening of crude venom and isolated serine proteases of *Montivipera bulgardaghica* and *M. raddei* revealed high activity against A-549 human lung carcinoma cells (Nalbantsoy et al. 2017).

The convergence of cutting-edge technologies, encompassing genomics, proteomics, transcriptomics, and metabolomics offers a promising avenue for discovering novel drug candidates. This multifaceted approach facilitates the exploration of a wide array of post-translational modifications within marine venomous substances. Furthermore, the incorporation of advanced techniques like liquid chromatography-nuclear magnetic resonance (LC-NMR) plays a pivotal role in characterizing the presence of small molecules within venom.

Benthos

Marine benthos—a vast array of organisms that live on or in the ocean floor—has come to light as a valuable source of possible medicinal compounds. Soft corals, sponges, and tunicates are examples of these sedentary or slowly moving organisms that are known to produce a wide range of secondary metabolites and bioactive molecules with substantial pharmaceutical potential. Marine sponges are known to produce bioactive molecules, including polyketides and alkaloids, which exhibit anti-inflammatory effects. One of the studies underscores the pharmaceutical promise of these marine sponges, and corals emphasizing their potential applications in anti-inflammatory drug development (Li et al. 2021). Moreover, marine benthic organisms have shown strong anticancer qualities, containing substances that can specifically target and stop the growth of cancer cells. Studies have revealed that

Table 1. List of few potential therapeutic candidates screened from marine resources through omics approaches.

Location	Organism	Approach	Techniques	Isolated pharmaceutical compounds	Application	References
Coastal habitats, Qingdao China	Seaweed derived fungus *Aspergillus* sp. *340*	MB	NMR- MS-QTOF-HPLC	Furanone derivatives, sclerotiorumins D–F	Anti-Phytopathogenic compounds	(Huang et al. 2020)
Orkney island, Scotland	Marine bryozoan *Bugula neritina* and synthesized by the symbiotic bacterium *Candidatus Endobugulasertula*		LC- HRMS-NMR	Novel secondary metabolites	anti-cancer, anti-tumour, anti-viral and antibacterial properties	(Lindquist 2002)
Scottish and Antarctic sediments	85 isolates belonging to the phyla Firmicutes and Actinobacteria of genera Bacillus and Streptomyces		HR-MS/MS	Novel bioactive compounds	anticancer compound retimycin A, antifungal compound thanamycin, malyngamide C with a combination of antifungal and cytotoxic properties	(Purves et al. 2016)
Arac¸a´ Beach, São Sebastião, SP, Brazil	Streptomyces sp. BRB081		LC-MS/ MS	Surugamides, Desferrioxiamine E (nocardamine), diketopiperazines, Bre vianamide F	cathepsin B inhibiting proprieties, antibacterial, antiviral, cytotoxic, Cardiovascular dysfunction	(Tangerina et al. 2021)
Great Barrier Reef, Queensland, Australia	*Carteriospongiafoliascens, Cliona orientalis, Coscinodermamatthewsi, Irciniaramosa, Pericharaxheteroaphis,* and *Stylissafabelliformis*		UHPLC-LC-MS/ GS-MS	docosatetraenoic acid and didodecyl 3,3′-thiodipropionate (DLTDP)	antibacterial, antibiofilm, and anti-inflammatory activities, antioxidant	(Zhang et al. 2022)
Estuary Ria de Aveiro, Portugal,	*Emericellopsiscladophorae*		UPLC- QToF–MS/MS, and genome sequencing (Illumina HiSeq)	9,12,13-Trihydroxyoctadec-10-enoic acid, hymeglusin Novo-biocin A, daunomycinone, isoreserpin, flavopiridol, 2’-O-Galloylhyperin	Antifungal, antibacterial, anti-cancer, Anti-inflammatory	(Ameen et al. 2022)
Scottish Association for Marine Science, Oban, UK	43 specimens of macroalgae and microalgae, including seven macroalgal specimens, 14 Chlorophytes, 14 Haptophytes, three Ochrophytes, one Rhodophyte, and four diatomaceous		PCR-HPLC-LC-MS	Novel metabolite compounds	Inflammation, type 1 diabetes, metabolic disorders	(Hughes et al. 2021)

Table 1 contd. ...

...Table 1 contd.

Location	**Organism**	**Approach**	**Techniques**	**Isolated pharmaceutical compounds**	**Application**	**References**
Rongcheng, Shandong Province, China	*Pseudoalteromonasagarivorans*	TC & GN	RNA sequencing, Illumina sequencing	Compounds like Protease, Phospholipase, Collagenase	Can be used for treatment of dermal ulcers, cancer, cystic fibrosis, inflammation	(Liu et al. 2019)
Pacific Oyster	26 samples of *Crassostrea gigas*	TC	Transcriptome sequencing and annotation	Novel neuropeptide precursor	neurotransmitters, neuromodulators, neurohormones, and growth factors	(Réalis-Doyelle et al. 2021)
Jeju island coast of South Korea	*Charonialampassauliae*		RNA extraction Illumina sequencing	Conotoxins	Neurotoxins, new class of analgesics	(Hwang et al. 2021)
Le Dellec area in Brest Bay	*Crassostrea gigas*	PT and GN	16S sequencing and MALDI bio typing	Big defensins of 2 types Cg-BigDef1 and Cg-BigDef5	Antimicrobial activity	(De San Nicolas et al. 2022)
Pacific Ocean	*C. japonicus* Yellow (a yellow specimen of *Cnidopus japonicus*) and *C. japonicus* Red (a red specimen of *Cnidopus japonicus*)		NGS, LC-MS / MS	AnmTXCj 3a-1, AnmTXCj 3a-2AnmTX Cj 1a-1 etc	Neurotoxins, active biomolecule compounds	(Kalina et al. 2022)
Estuary Ria de Aveiro, Portugal	*Emericellopsiscladophorae*	MB & GN	UPLC- QToF–MS/MS, and genome sequencing (Illumina HiSeq)	9,12,13-Trihydroxyoctadec-10-enoic acid, hymeglusin Novo-biocin A, daunomycinone, isoreserpin, flavopiridol, 2'-O-Galloylhyperin	Antifungal, antibacterial, anti-cancer, Anti-inflammatory	(Gonçalves et al. 2022)
Parangipettai, located in southeast coast of India	Staphylococcus aureus (CASMTK1)	GN	Sequencing, Enzyme assays, FTIR assay	Hyaluronidase enzyme	Antioxidant and anti-breast cancer properties	(Thirumurthy et al. 2023)

MB – metabolomics; TC – transcriptomics; GN – genomics; PT – proteomics.

tunicates, also referred to as sea squirts, generate substances that show promise as anti-cancer agents (Zhu et al. 2022). Aplidine and didemnins are two examples of these bioactive compounds that have demonstrated inhibitory effects on angiogenesis and tumor cell proliferation. Didemnins have the potential to be used in cancer treatment, according to a study by (Costa-Lotufo et al. 2006), highlighting the importance of marine benthos as a source of anti-cancer pharmaceutical agents. Table 1 shows other isolated pharmaceutical compounds by using different omics technologies with their potential application and other details.

Marine bioprospection: challenges

Omics studies of marine resources present a range of challenges, from the initial sampling to the complex data analysis phase (Figure 1). First and foremost, obtaining representative and uncontaminated samples from the dynamic and often remote marine environments can be a logistical challenge, requiring specialized equipment and expertise. Once samples are collected, preserving their integrity during transportation and storage is crucial. Additionally, the vast diversity of marine species and ecosystems demands tailored experimental protocols and approaches, complicating standardization and comparability between studies. In terms of data analysis, the sheer volume and complexity of omics data, which often include genomics, proteomics, transcriptomics, and metabolomics, pose significant computational and bioinformatics challenges. Integrating and interpreting these multi-omics datasets can be intricate, requiring specialized tools and resources. Furthermore, issues related to data quality, reproducibility, and the need for well-annotated reference genomes and databases in marine science can hinder the robustness of omics studies. Lastly,

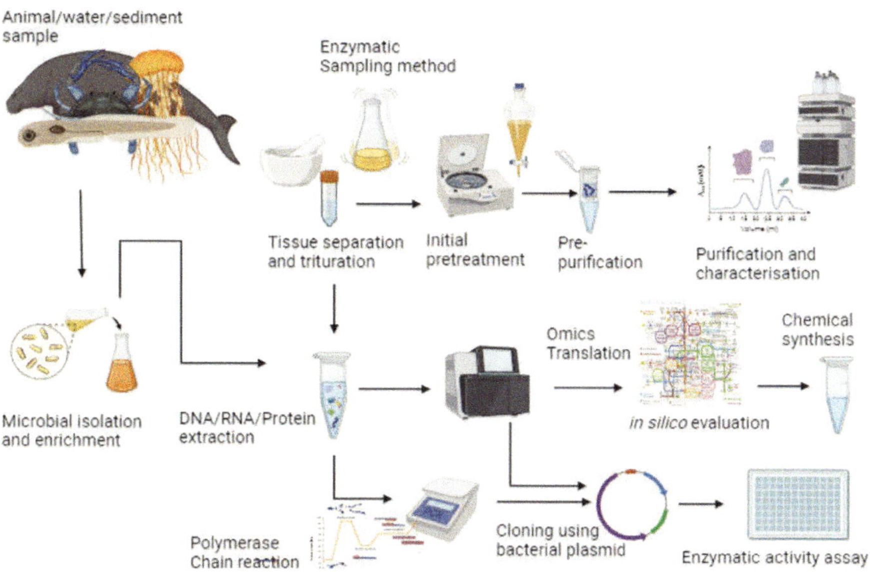

Figure 1. Standard steps of processing marine biodiversity for prospection.

ensuring the ethical collection of samples, especially from vulnerable ecosystems and endangered species, adds an ethical dimension to the challenges faced in these studies. Overcoming these hurdles is essential to harness the full potential of omics to understand and conserve marine resources.

Sampling

The immense diversity of marine life, which ranges from microscopic organisms to massive creatures with unique habitats, behaviors, and life cycles, is one of the main challenges. Gathering specimens from particular species or habitats necessitates specialized tools and in-depth knowledge of the intended organisms. Additionally, sampling operations may be hazardous and logistically challenging due to the harsh and unpredictable marine environment, which includes choppy seas, deep waters, and bad weather (Cario et al. 2019, Firestone et al. 2018, Schlacher et al. 2008). Another difficulty is maintaining sample integrity and avoiding contamination because the marine environment can introduce undesirable elements or deteriorate biological materials. The advancement of marine biotechnology necessitates the development of novel sampling techniques and technologies to tackle these challenges, in order to fully utilize the oceans for scientific research and biotechnological applications.

Cultivation methods

Cultivating marine microbes, plankton, seaweed, and other biodiversity for marine biotechnology purposes is a complex task that comes with its own set of challenges. It is challenging to sustain ideal growing conditions in the maritime environment due to its high degree of dynamism, in contrast to controlled laboratory settings. In the open ocean, a wide range of environmental factors, including temperature, salinity, nutrient availability, and light conditions, can affect microbes, plankton, and seaweed (Barra et al. 2014, Loureiro et al. 2015). Cultivation attempts may also be hampered by contamination from undesirable species and competition from other organisms. Transitioning from small-scale laboratory experiments to larger-scale, industrial production also presents scale-up challenges. The key to utilizing marine biodiversity for biotechnological applications is the development of economical and sustainable cultivation techniques. Overcoming these obstacles in marine agriculture is essential to the development of marine biotechnology, which has enormous potential for breakthroughs in industries like food production, biofuels, and pharmaceuticals.

Downstream processing

Downstream processing in marine biotechnology, which involves the extraction, purification, and processing of valuable compounds derived from marine biodiversity, presents its own set of challenges. Because marine organisms and their environments are unique, the targeted bioactive compounds are frequently present in low concentrations, necessitating a labor-intensive and expensive extraction process. To make matters more complicated, the variety of molecules originating from various marine species necessitates specialized purification methods. Downstream processing is made more difficult by the possibility of contaminants and impurities as well as the requirement for sustainable and ecologically friendly extraction techniques. Another challenge is scaling up production from laboratory-scale experiments to industrial levels, which necessitates efficient and economical technologies (Mores et al. 2021). To address these issues, researchers in marine biotechnology are continually developing innovative extraction and purification methods, as well as exploring sustainable practices to ensure that valuable compounds from marine biodiversity can be effectively harnessed for various applications while respecting the delicate marine ecosystems they are derived from.

Lack of comparative databases

The marine environment frequently supports a wide variety of species, each with a distinct genetic and biochemical make-up. It is difficult to recognize, evaluate, and comprehend the possible biotechnological uses of these species without strong, centralized databases that aggregate genetic, genomic, and metagenomic data from diverse marine organisms. Lack of these tools makes it more difficult to find new genes, enzymes, or bioactive substances—all of which are necessary for the creation of cutting-edge biotechnological goods and solutions. Moreover, standardizing data is a challenge due to the diversity of marine ecosystems, which impedes comparative research between various species and regions (Mineta and Gojobori 2016). Addressing this challenge calls for collaborative efforts to establish comprehensive and accessible marine biotechnology databases, which can foster scientific discoveries and support the sustainable utilization of marine resources for various applications, from drug discovery to renewable energy sources.

Conclusion

In conclusion, this chapter highlights the transformative potential of omics technologies in harnessing the untapped resources of our oceans for pharmaceutical applications. We explored how transcriptomics, proteomics, metabolomics, and genomics play pivotal roles in deciphering the genetic and biochemical secrets of diverse marine organisms, paving the way for groundbreaking discoveries. The exciting pharmaceutical findings from marine algae, seagrass, marine venoms, animals, and plankton underscore the vast treasure trove of bioactive compounds and therapeutic agents awaiting exploration. However, the path to realizing the blue bio-economy's full potential is not without its challenges, as we discussed the obstacles in marine sampling, cultivation, downstream processing, and the need for comparative databases. Overcoming these challenges is paramount, as marine biotechnology promises to revolutionize the pharmaceutical industry and contribute significantly to the sustainable development of the blue economy, ultimately benefiting both human health and the preservation of our marine ecosystems. This chapter serves as a testament to the incredible promise that lies beneath the waves, waiting to be unlocked through the power of omics and human ingenuity.

References

Abida, H., Ruchaud, S., Rios, L., Humeau, A., Probert, I., De Vargas, C. et al. (2013). Bioprospecting marine plankton. Mar. Drugs. https://doi.org/10.3390/md11114594.

Alencar, P.O.C., Lima, G.C., Barros, F.C.N., Costa, L.E.C., Ribeiro, C.V.P.E., Sousa, W.M. et al. (2019). A novel antioxidant sulfated polysaccharide from the algae Gracilariacaudata: *In vitro* and *in vivo* activities. Food Hydrocoll. https://doi.org/10.1016/j.foodhyd.2018.12.007.

Alseekh, S., Aharoni, A., Brotman, Y., Contrepois, K., D'Auria, J., Ewald, J. et al. (2021). Mass spectrometry-based metabolomics: a guide for annotation, quantification and best reporting practices. Nat. Methods. https://doi.org/10.1038/s41592-021-01197-1.

Ameen, F., AlNAdhari, S. and Al-Homaidan, A.A. (2022). Marine fungi showing multifunctional activity against human pathogenic microbes and cancer. PLoS One. https://doi.org/10.1371/journal.pone.0276926.

Assunão, J., Catarina Guedes, A. and Xavier Malcata, F. (2017). Biotechnological and pharmacological applications of biotoxins and other bioactive molecules from dinoflagellates. Mar. Drugs. https://doi.org/10.3390/md15120393.

Balch, C., Matei, D.E., Huang, T.H.M. and Nephew, K.P. (2010). Role of epigenomics in ovarian and endometrial cancers. Epigenomics. https://doi.org/10.2217/epi.10.19.

Barra, L., Chandrasekaran, R., Corato, F. and Brunet, C. (2014). The challenge of ecophysiological biodiversity for biotechnological applications of marine microalgae. Mar. Drugs. https://doi.org/10.3390/md12031641.

Bayona, L.M., de Voogd, N.J. and Choi, Y.H. (2022). Metabolomics on the study of marine organisms. Metabolomics. https://doi.org/10.1007/s11306-022-01874-y.

Bhatia, S., Rathee, P., Sharma, K., Chaugule, B.B., Kar, N. and Bera, T. (2013). Immuno-modulation effect of sulphated polysaccharide (porphyran) from Porphyravietnamensis. Int. J. Biol. Macromol. https://doi.org/10.1016/j.ijbiomac.2013.03.012.

Bhatnagar, I. and Kim, S.K. (2010). Immense essence of excellence: Marine microbial bioactive compounds. Mar. Drugs. https://doi.org/10.3390/md8102673.

Bovee, T.F.H., Hendriksen, P.J.M., Portier, L., Wang, S., Elliott, C.T., Van Egmond, H.P. et al. (2011). Tailored microarray platform for the detection of marine toxins. Environ. Sci. Technol. https://doi.org/10.1021/es2011385.

Calado, R., Leal, M.C., Gaspar, H., Santos, S., Marques, A., Nunes, M.L. et al. (2018). How to succeed in marketing marine natural products for nutraceutical, pharmaceutical and cosmeceutical markets. *In*: Grand Challenges in Biology and Biotechnology. https://doi.org/10.1007/978-3-319-69075-9_9.

Campos, A., Danielsson, G., Farinha, A.P., Kuruvilla, J., Warholm, P. and Cristobal, S. (2016). Shotgun proteomics to unravel marine mussel (Mytilus edulis) response to long-term exposure to low salinity and propranolol in a Baltic Sea microcosm. J. Proteomics. https://doi.org/10.1016/j.jprot.2016.01.010.

Campos, A.M.O. and de Almeida, A.M. (2016). Top-down proteomics and farm animal and aquatic sciences. Proteomes. https://doi.org/10.3390/proteomes4040038.

Cardoso, A.M., Coutinho, F.H., Silveira, C.B., Ignacio, B.L., Vieira, R.P. and Salloto, G. (2012). Metagenomics in polluted aquatic environments. Water Pollution 89–104. https://doi.org/10.5772/30489.

Cario, A., Oliver, G.C. and Rogers, K.L. (2019). Exploring the deep marine biosphere: challenges, innovations, and opportunities. Front. Earth Sci. https://doi.org/10.3389/feart.2019.00225.

Chang, S.C., Huq, R., Chhabra, S., Beeton, C., Pennington, M.W., Smith, B.J. et al. (2015). N-terminally extended analogues of the K+ channel toxin from Stichodactyla helianthus as potent and selective blockers of the voltage-gated potassium channel Kv1.3. FEBS J. https://doi.org/10.1111/febs.13294.

Coelho, F.J.R.C., Santos, A.L., Coimbra, J., Almeida, A., Cunha, Â., Cleary, D.F.R. et al. (2013). Interactive effects of global climate change and pollution on marine microbes: The way ahead. Ecol. Evol. https://doi.org/10.1002/ece3.565.

Collins, F.S. and Varmus, H. (2015). A new initiative on precision medicine. N. Engl. J. Med. https://doi.org/10.1056/nejmp1500523.

Costa-Lotufo, L.V., Pessoa, C., Moraes, M.E.A., Paixão Almeida, A.M., de Moraes, M.O. and da Cruz Lotufo, T.M. (2006). Marine organisms from Brazil as source of potential anticancer agents. Adv. Phytomedicine. https://doi.org/10.1016/S1572-557X(05)02011-8.

Cumashi, A., Ushakova, N.A., Preobrazhenskaya, M.E., D'Incecco, A., Piccoli, A., Totani, L. et al. (2007). A comparative study of the anti-inflammatory, anticoagulant, antiangiogenic, and antiadhesive activities of nine different fucoidans from brown seaweeds. Glycobiology. https://doi.org/10.1093/glycob/cwm014.

D'Ayala, G.G., Malinconico, M. and Laurienzo, P. (2008). Marine derived polysaccharides for biomedical applications: Chemical modification approaches. Molecules. https://doi.org/10.3390/molecules13092069.

Dalmaso, G.Z.L., Ferreira, D. and Vermelho, A.B. (2015). Marine extremophiles a source of hydrolases for biotechnological applications. Mar. Drugs. https://doi.org/10.3390/md13041925.

Danaraj, J., Yosuva, M. and Ayyappan, S. (2020). Comparative metabolomics analysis of wild and suspension cultured cells (SCC) of seagrass Halodulepinifolia (Miki) hartog of cymodoceaceae family. Aquat. Bot. https://doi.org/10.1016/j.aquabot.2020.103278.

Das, S. and Patel, B. (2020). Marine resources and animals in modern biotechnology. *In*: Animal Biotechnology: Models in Discovery and Translation. https://doi.org/10.1016/B978-0-12-811710-1.00027-6.

de la Calle, F. (2017). Marine microbiome as source of natural products. Microb. Biotechnol. https://doi.org/10.1111/1751-7915.12882.

De San Nicolas, N., Asokan, A., Rosa, R.D., Voisin, S.N., Travers, M.A., Rocha, G. et al. (2022). Functional diversification of oyster big defensins generates antimicrobial specificity and synergy against members of the microbiota. Mar. Drugs. https://doi.org/10.3390/md20120745.

Donia, M.S., Ruffner, D.E., Cao, S. and Schmidt, E.W. (2011). Accessing the hidden majority of marine natural products through metagenomics. ChemBioChem. 12(8): 1230–1236.

Dunstan, G.A., Volkman, J.K., Barrett, S.M., Leroi, J.M. and Jeffrey, S.W. (1993). Essential polyunsaturated fatty acids from 14 species of diatom (Bacillariophyceae). Phytochemistry. https://doi.org/10.1016/S0031-9422(00)90525-9.

Dyshlovoy, S.A., Otte, K., Venz, S., Hauschild, J., Junker, H., Makarieva, T.N. et al. (2017). Proteomic-based investigations on the mode of action of the marine anticancer compound rhizochalinin. Proteomics. https://doi.org/10.1002/pmic.201700048.

El-Rafie, H.M., El-Rafie, M.H., Abd Elsalam, H.M. and El-Sayed, W.A. (2016). Antibacterial and anti-inflammatory finishing of cotton by microencapsulation using three marine organisms. Int. J. Biol. Macromol. https://doi.org/10.1016/j.ijbiomac.2016.01.039.

Emwas, A.H., Roy, R., McKay, R.T., Tenori, L., Saccenti, E., Nagana Gowda, G.A. et al. (2019). Nmr spectroscopy for metabolomics research. Metabolites. https://doi.org/10.3390/metabo9070123.

Erpel, F., Mateos, R., Pérez-Jiménez, J. and Pérez-Correa, J.R. (2020). Phlorotannins: From isolation and structural characterization, to the evaluation of their antidiabetic and anticancer potential. Food Res. Int. https://doi.org/10.1016/j.foodres.2020.109589.

Fedorov, S., Dyshlovoy, S., Monastyrnaya, M., Shubina, L., Leychenko, E., Kozlovskaya, E. et al. (2010). The anticancer effects of actinoporin RTX-A from the sea anemone Heteractis crispa (=Radianthusmacrodactylus). Toxicon. https://doi.org/10.1016/j.toxicon.2009.11.016.

Fernández-Acero, F.J., Amil-Ruiz, F., Durán-Peña, M.J., Carrasco, R., Fajardo, C., Guarnizo, P. et al. (2019). Valorisation of the microalgae Nannochloropsisgaditana biomass by proteomic approach in the context of circular economy. J. Proteomics. https://doi.org/10.1016/j.jprot.2018.10.015.

Firestone, J., Bidwell, D., Gardner, M. and Knapp, L. (2018). Wind in the sails or choppy seas?: People-place relations, aesthetics and public support for the United States' first offshore wind project. Energy Res. Soc. Sci. https://doi.org/10.1016/j.erss.2018.02.017.

Fitton, J.H., Stringer, D.N., Karpiniec, S.S. and Park, A.Y. (2019). The manufacture, characterization, and uses of fucoidans from macroalgae. *In*: Enzymatic Technologies for Marine Polysaccharides. https://doi.org/10.1201/9780429058653-3.

Fonseca, S., Amaral, M.N., Reis, C.P. and Custódio, L. (2023). Marine natural products as innovative cosmetic ingredients. Mar. Drugs. https://doi.org/10.3390/md21030170.

Gleason, L.U. (2019). Applications and future directions for population transcriptomics in marine invertebrates. Curr. Mol. Biol. Reports. https://doi.org/10.1007/s40610-019-00121-z.

Gonçalves, M.F.M., Hilário, S., Van de Peer, Y., Esteves, A.C. and Alves, A. (2022). Genomic and metabolomic analyses of the marine fungus Emericellopsiscladophorae: insights into saltwater adaptability mechanisms and its biosynthetic potential. J. Fungi. https://doi.org/10.3390/jof8010031.

Grindberg, R.V., Ishoey, T., Brinza, D., Esquenazi, E., Coates, R.C., Liu, W.-t. et al. (2011). Single cell genome amplification accelerates identification of the apratoxin biosynthetic pathway from a complex microbial assemblage. PloS One 6(4): e18565.

Han, X. (2016). Lipidomics for studying metabolism. Nature Reviews Endocrinology 12(11): 668–679. doi:10.1038/nrendo.2016.98.

Han, X., Yang, K. and Gross, R.W. (2012). Multi-dimensional mass spectrometry-based shotgun lipidomics +and novel strategies for lipidomic analyses. Mass Spectrometry Reviews 31(1): 134–178.

He, R., Wakimoto, T., Takeshige, Y., Egami, Y., Kenmoku, H., Ito, T. et al. (2012). Porphyrins from a metagenomic library of the marine sponge Discodermia calyx. Molecular BioSystems 8(9): 2334–2338.

Holdt, S.L. and Kraan, S. (2011). Bioactive compounds in seaweed: Functional food applications and legislation. J. Appl. Phycol. https://doi.org/10.1007/s10811-010-9632-5.

Hu, S., Zhu, H., Chen, S., Wan, X., Liu, Y., Ren, Z. et al. (2023). Structural characterization and effects on insulin resistance of a novel chondroitin sulfate from Halaelurusburgeri skin. Marine Drugs 21(4): 221.

Huang, R.H., Lin, W., Zhang, P., Liu, J.Y., Wang, D., Li, Y.Q. et al. (2020). Anti-phytopathogenic bacterial metabolites from the seaweed-derived fungus Aspergillus sp. D40. Front. Mar. Sci. https://doi.org/10.3389/fmars.2020.00313.

Hughes, A.H., Magot, F., Tawfike, A.F., Rad-Menéndez, C., Thomas, N., Young, L.C. et al. (2021). Exploring the chemical space of macro-and micro-algae using comparative metabolomics. Microorganisms. https://doi.org/10.3390/microorganisms9020311.

Hwang, H.J., Patnaik, B.B., Chung, J.M., Sang, M.K., Park, J.E., Kang, S.W. et al. (2021). *De novo* transcriptome sequencing of triton shell Charonialampassauliae: Identification of genes related to neurotoxins and discovery of genetic markers. Mar. Genomics. https://doi.org/10.1016/j.margen.2021.100862.

Inoue, K., Waite, J.H., Matsuoka, M., Odo, S. and Harayama, S. (1995). Interspecific variations in adhesive protein sequences of *Mytilus edulis*, *M. galloprovincialis*, and *M. trossulus*. Biol. Bull. https://doi.org/10.2307/1542155.

Ji, C., Cao, X., Yao, C., Xue, S. and Xiu, Z. (2014). Protein-protein interaction network of the marine microalga Tetraselmissubcordiformis: Prediction and application for starch metabolism analysis. J. Ind. Microbiol. Biotechnol. https://doi.org/10.1007/s10295-014-1462-z.

Johansson, O.N., Töpel, M., Egardt, J., Pinder, M.I., Andersson, M.X., Godhe, A. et al. (2019). Phenomics reveals a novel putative chloroplast fatty acid transporter in the marine diatom Skeletonemamarinoi involved in temperature acclimation. Scientific Reports 9(1): 15143.

Kalina, R.S., Kasheverov, I.E., Koshelev, S.G., Sintsova, O.V., Peigneur, S., Pinheiro-Junior, E.L. et al. (2022). Nicotinic acetylcholine receptors are novel targets of APETx-like toxins from the Sea Anemone Heteractis magnifica. Toxins (Basel). https://doi.org/10.3390/toxins14100697.

Kendel, M., Wielgosz-Collin, G., Bertrand, S., Roussakis, C., Bourgougnon, N. and Bedoux, G. (2015). Lipid composition, fatty acids and sterols in the seaweeds Ulva armoricana, and Solieriachordalis from Brittany (France): An analysis from nutritional, chemotaxonomic, and antiproliferative activity perspectives. Marine Drugs 13(9): 5606–5628.

Kesraoui, O., Marzouki, M.N., Maugard, T. and Limam, F. (2011). *In vitro* evaluation of antioxidant activities of free and bound phenolic compounds from Posidonia oceanica (L.) Delile leaves. African J. Biotechnol. https://doi.org/10.5897/ajb10.847.

Kong, S., Zhao, Y.G., Guo, L., Gao, M., Jin, C. and She, Z. (2020). Transcriptomics of Planococcuskocurii O516 reveals the degrading metabolism of sulfamethoxazole in marine aquaculture wastewater. Environ. Pollut. https://doi.org/10.1016/j.envpol.2020.114939.

Kwon, D., Park, J.M., Duong, V.A., Hong, S.J., Cho, B.K., Lee, C.G. et al. (2020). Comparative proteomic profiling of marine and freshwater synechocystis strains using liquid chromatography-tandem mass spectrometry. J. Mar. Sci. Eng. https://doi.org/10.3390/jmse8100790.

Leprêtre, M., Palos-Ladeiro, M., Faugere, J., Almunia, C., Lemoine, J., Armengaud, J. et al. (2020). From shotgun to targeted proteomics: rapid Scout-MRM assay development for monitoring potential immunomarkers in Dreissena polymorpha. Anal. Bioanal. Chem. https://doi.org/10.1007/s00216-020-02868-2.

Li, C.Q., Ma, Q.Y., Gao, X.Z., Wang, X. and Zhang, B.L. (2021). Research progress in anti-inflammatory bioactive substances derived from marine microorganisms, sponges, algae, and corals. Mar. Drugs. https://doi.org/10.3390/md19100572.

Li, H., Monteiro, C., Heinrich, S., Bartsch, I., Valentin, K., Harms, L. et al. (2020). Responses of the kelp Saccharina latissima (Phaeophyceae) to the warming Arctic: from physiology to transcriptomics. Physiol. Plant. https://doi.org/10.1111/ppl.13009.

Li, X., Chen, W., Zhangsun, D. and Luo, S. (2020). Diversity of conopeptides and their precursor genes of conus litteratus. Mar. Drugs. https://doi.org/10.3390/md18090464.

Lindquist, N. (2002). Chemical defense of early life stages of benthic marine invertebrates. J. Chem. Ecol. https://doi.org/10.1023/A:1020745810968.

Liu, J., Yang, S., Li, X., Yan, Q., Reaney, M.J.T. and Jiang, Z. (2019). Alginate oligosaccharides: production, biological activities, and potential applications. Compr. Rev. Food Sci. Food Saf. https://doi.org/10.1111/1541-4337.12494.

Loureiro, R., Gachon, C.M.M. and Rebours, C. (2015). Seaweed cultivation: Potential and challenges of crop domestication at an unprecedented pace. New Phytol. https://doi.org/10.1111/nph.13278.

Mazurais, D., Darias, M., Zambonino-Infante, J.L. and Cahu, C.L. (2011). Transcriptomics for understanding marine fish larval development. Can. J. Zool. https://doi.org/10.1139/z11-036.

McEneff, G., Barron, L., Kelleher, B., Paull, B. and Quinn, B. (2013). The determination of pharmaceutical residues in cooked and uncooked marine bivalves using pressurised liquid extraction, solid-phase extraction and liquid chromatography-tandem mass spectrometry. Anal. Bioanal. Chem. https://doi.org/10.1007/s00216-013-7371-6.

Mineta, K. and Gojobori, T. (2016). Databases of the marine metagenomics. Gene. https://doi.org/10.1016/j.gene.2015.10.035.

Mores, S., Vandenberghe, L.P. de S., Magalhães Júnior, A.I., de Carvalho, J.C., de Mello, A.F.M., Pandey, A. et al. (2021). Citric acid bioproduction and downstream processing: Status, opportunities, and challenges. Bioresour. Technol. https://doi.org/10.1016/j.biortech.2020.124426.

Nalbantsoy, A., Hempel, B.F., Petras, D., Heiss, P., Göçmen, B., Iğci, N. et al. (2017). Combined venom profiling and cytotoxicity screening of the Radde's mountain viper (Montivipera raddei) and Mount Bulgar Viper (Montivipera bulgardaghica) with potent cytotoxicity against human A549 lung carcinoma cells. Toxicon. https://doi.org/10.1016/j.toxicon.2017.06.008.

Penesyan, A., Kjelleberg, S. and Egan, S. (2010). Development of novel drugs from marine surface associated microorganisms. Mar. Drugs. https://doi.org/10.3390/md8030438.

Poli, A., Finore, I., Romano, I., Gioiello, A., Lama, L. and Nicolaus, B. (2017). Microbial diversity in extreme marine habitats and their biomolecules. Microorganisms. https://doi.org/10.3390/microorganisms5020025.

Powell, M.J., Sutton, J.N., Del Castillo, C.E. and Timperman, A.T. (2005). Marine proteomics: Generation of sequence tags for dissolved proteins in seawater using tandem mass spectrometry. Mar. Chem. https://doi.org/10.1016/j.marchem.2004.09.004.

Pradhan, B. and Ki, J.S. (2022). Phytoplankton toxins and their potential therapeutic applications: a journey toward the quest for potent pharmaceuticals. Mar. Drugs. https://doi.org/10.3390/md20040271.

Purves, K., Macintyre, L., Brennan, D., Hreggviðsson, G., Kuttner, E., Ásgeirsdóttir, M.E. et al. (2016). Using molecular networking for microbial secondary metabolite bioprospecting. Metabolites. https://doi.org/10.3390/metabo6010002.

Raimundo, I., Silva, R., Meunier, L., Valente, S.M., Lago-Lestón, A., Keller-Costa, T. et al. (2021). Functional metagenomics reveals differential chitin degradation and utilization features across free-living and host-associated marine microbiomes. Microbiome. https://doi.org/10.1186/s40168-020-00970-2.

Réalis-Doyelle, E., Schwartz, J., Cabau, C., Franc, L. Le, Bernay, B., Rivière, G. et al. (2021). Transcriptome profiling of the pacific oyster crassostrea gigas visceral ganglia over a reproduction cycle identifies novel regulatory peptides. Mar. Drugs. https://doi.org/10.3390/md19080452.

Reid, N.M. and Whitehead, A. (2016). Functional genomics to assess biological responses to marine pollution at physiological and evolutionary timescales: Toward a vision of predictive ecotoxicology. Brief. Funct. Genomics. https://doi.org/10.1093/bfgp/elv060.

Ren, C.-Z., Gao, H.-M., Dai, J., Zhu, W.-Z., Xu, F.-F., Ye, Y. et al. (2022). Taxonomic and bioactivity characterizations of Mameliella alba strain LZ-28 isolated from highly toxic marine dinoflagellate Alexandriumcatenella LZT09. Marine Drugs 20(5): 321.

Reverter, M., Rohde, S., Parchemin, C., Tapissier-Bontemps, N. and Schupp, P.J. (2020). Metabolomics and marine biotechnology: coupling metabolite profiling and organism biology for the discovery of new compounds. Front. Mar. Sci. https://doi.org/10.3389/fmars.2020.613471.

Rivera-Pérez, C. and Hernández-Saavedra, N.Y. (2021). Review: Post-translational modifications of marine shell matrix proteins. Comp. Biochem. Physiol. Part - B Biochem. Mol. Biol. https://doi.org/10.1016/j.cbpb.2021.110641.

Rodrigues, P.M., Silva, T.S., Dias, J. and Jessen, F. (2012). PROTEOMICS in aquaculture: Applications and trends. J. Proteomics. https://doi.org/10.1016/j.jprot.2012.03.042.

Romero, M.R., Pérez-Figueroa, A., Carrera, M., Swanson, W.J., Skibinski, D.O.F. and Diz, A.P. (2019). RNA-seq coupled to proteomic analysis reveals high sperm proteome variation between two closely related marine mussel species. J. Proteomics. https://doi.org/10.1016/j.jprot.2018.08.020.

Romeu, M.J., Domínguez-Pérez, D., Almeida, D., Morais, J., Araújo, M.J., Osório, H. et al. (2021). Quantitative proteomic analysis of marine biofilms formed by filamentous cyanobacterium. Environ. Res. https://doi.org/10.1016/j.envres.2021.111566.

Rotter, A., Barbier, M., Bertoni, F., Bones, A.M., Cancela, M.L., Carlsson, J. et al. (2021). The essentials of marine biotechnology. Front. Mar. Sci. https://doi.org/10.3389/fmars.2021.629629.

Safavi-Hemami, H., Brogan, S.E. and Olivera, B.M. (2019). Pain therapeutics from cone snail venoms: From Ziconotide to novel non-opioid pathways. J. Proteomics. https://doi.org/10.1016/j.jprot.2018.05.009.

Sainis, I., Fokas, D., Vareli, K., Tzakos, A.G., Kounnis, V. and Briasoulis, E. (2010). Cyanobacterial cyclopeptides as lead compounds to novel targeted cancer drugs. Mar. Drugs. https://doi.org/10.3390/md8030629.

Schlacher, T.A., Schoeman, D.S., Dugan, J., Lastra, M., Jones, A., Scapini, F. et al. (2008). Sandy beach ecosystems: Key features, sampling issues, management challenges and climate change impacts. Mar. Ecol. https://doi.org/10.1111/j.1439-0485.2007.00204.x.

Seneca, F.O. and Palumbi, S.R. (2015). The role of transcriptome resilience in resistance of corals to bleaching. Mol. Ecol. https://doi.org/10.1111/mec.13125.

Seppänen-Laakso, T. and Orešič, M. (2009). How to study lipidomes. Journal of Molecular Endocrinology 42(3): 185–190.

Simon, C. and Daniel, R. (2011). Metagenomic analyses: past and future trends. Applied and Environmental Microbiology 77(4): 1153–1161.

Stuart, K.A., Welsh, K., Walker, M.C. and Edrada-Ebel, R.A. (2020). Metabolomic tools used in marine natural product drug discovery. Expert Opin. Drug Discov. https://doi.org/10.1080/17460441.2020.1722636.

Suleria, H.A.R., Gobe, G., Masci, P. and Osborne, S.A. (2016). Marine bioactive compounds and health promoting perspectives; innovation pathways for drug discovery. Trends Food Sci. Technol. https://doi.org/10.1016/j.tifs.2016.01.019.

Suleria, H.A.R., Osborne, S., Masci, P. and Gobe, G. (2015). Marine-based nutraceuticals: An innovative trend in the food and supplement industries. Mar. Drugs. https://doi.org/10.3390/md13106336.

Sunagawa, S., Coelho, L.P., Chaffron, S., Kultima, J.R., Labadie, K., Salazar, G. et al. (2015). Structure and function of the global ocean microbiome. Science 348(6237): 1261359.

Tang, B., He, W.L., Zheng, C., Cheang, T.Y., Zhang, X.I.F., Wu, H. et al. (2012). Marine fungal metabolite 1386A alters the microRNA profile in MCF-7 breast cancer cells. Mol. Med. Rep. https://doi.org/10.3892/mmr.2011.697.

Tang, J. (2011). Microbial metabolomics. Curr. Genomics. https://doi.org/10.2174/138920211797248619.

Tangerina, M.M.P., Furtado, L.C., Leite, V.M.B., Bauermeister, A., Velasco-Alzate, K., Jimenez, P.C. et al. (2021). Metabolomic study of marine Streptomyces sp.: Secondary metabolites and the production of potential anticancer compounds. PLoS One. https://doi.org/10.1371/journal.pone.0244385.

Techtmann, S.M. and Hazen, T.C. (2016). Metagenomic applications in environmental monitoring and bioremediation. J. Ind. Microbiol. Biotechnol. https://doi.org/10.1007/s10295-016-1809-8.

Terracciano, M., De Stefano, L. and Rea, I. (2018). Diatoms green nanotechnology for biosilica-based drug delivery systems. Pharmaceutics. https://doi.org/10.3390/pharmaceutics10040242.

Thirumurthy, K., Kaliyamoorthy, K., Kandasamy, K., Ponnuvel, M., Viyakarn, V., Chavanich, S. et al. (2023). Antioxidant and anti-breast cancer properties of hyaluronidase from marine Staphylococcus aureus (CASMTK1). J. Mar. Sci. Eng. https://doi.org/10.3390/jmse11040778.

Thiyagarajan, V., Wong, T. and Qian, P.Y. (2009). 2D gel-based proteome and phosphoproteome analysis during larval metamorphosis in two major marine biofouling invertebrates. J. Proteome Res. https://doi.org/10.1021/pr800976u.

Tomanek, L. (2014). Proteomics to study adaptations in marine organisms to environmental stress. J. Proteomics. https://doi.org/10.1016/j.jprot.2014.04.009.

Trindade, M., van Zyl, L.J., Navarro-Fernández, J. and Elrazak, A.A. (2015). Targeted metagenomics as a tool to tap into marine natural product diversity for the discovery and production of drug candidates. Front. Microbiol. https://doi.org/10.3389/fmicb.2015.00890.

Vilasi, A., Monti, M.C., Tosco, A., De Marino, S., Margarucci, L., Riccio, R. et al. (2013). Differential in gel electrophoresis (DIGE) comparative proteomic analysis of macrophages cell cultures in response to perthamide C treatment. Mar. Drugs. https://doi.org/10.3390/md11041288.

Wagner, V., Gessner, G. and Mittag, M. (2005). Functional proteomics: A promising approach to find novel components of the circadian system. Chronobiol. Int. https://doi.org/10.1081/CBI-200062348.

Wen, Z. and Johnson, M.B. (2009). Microalgae as a Feedstock for Biofuel Production. Virginia Coop. Ext.

Wyche, T.P., Alvarenga, R.F.R., Piotrowski, J.S., Duster, M.N., Warrack, S.R., Cornilescu, G. et al. (2017). Chemical genomics, structure elucidation, and *in vivo* studies of the marine-derived Anticlostridial Ecteinamycin. ACS Chem. Biol. https://doi.org/10.1021/acschembio.7b00388.

Xie, B., Huang, Y., Baumann, K., Fry, B.G. and Shi, Q. (2017). From marine venoms to drugs: Efficiently supported by a combination of transcriptomics and proteomics. Mar. Drugs. https://doi.org/10.3390/md15040103.

Yang, K. and Han, X. (2016). Lipidomics: techniques, applications, and outcomes related to biomedical sciences. Trends in Biochemical Sciences 41(11): 954–969.

Zhang, S., Song, W., Nothias, L.F., Couvillion, S.P., Webster, N. and Thomas, T. (2022). Comparative metabolomic analysis reveals shared and unique chemical interactions in sponge holobionts. Microbiome. https://doi.org/10.1186/s40168-021-01220-9.

Zhu, Y., Han, S., Li, J., Gao, H. and Dong, B. (2022). Aqueous extract of Sea Squirt (Halocynthiaroretzi) with potent activity against human cancer cells acts synergistically with Doxorubicin. Mar. Drugs. https://doi.org/10.3390/md20050284.

Chapter 6

Industrially Important Red Seaweed *Kappaphycus alvarezii*

Emerging Applications in the Domain Marine Biotechnology

Digvijay Singh Yadav,[1] *Doniya Elze Mathew,*[1] *Sumit Sudhir Phakatkar,*[2] *Cathrine Sumathi Manohar*[2,3,*] and *Vaibhav A. Mantri*[1,3,*]

Introduction

Eucheumatoids, comprising two related red seaweeds, namely, Kappaphycus and Eucheuma, are the primary and dominant source of carrageenan globally (Campbell and Hotchkiss 2017). Carrageenan—a sulphated polysaccharide—is sought-after by industries as diverse as food, dairy, confectionary, cosmetics, and pharmaceuticals, along with emerging applications in the biotechnological domain. The industry estimates in 2022 reported a turnover of USD 871.7 million, which is assumed to reach USD 1.3 billion by 2030. This is essentially due to the burgeoning demand for commodity products by consumers worldwide (Hydrocolloids Market Size, Share & Trends Analysis Report 2023). Currently, China plays a dominant role in sectors like processing, trading, etc. of *K. alvarezii* and has established itself as the primary biomass source, contributing to about 70% of the global carrageenan feedstock, followed by Indonesia, which earned USD 329 million from export in 2019 (Sultana et al. 2023). It is worth reporting

[1] Applied Phycology and Biotechnology Department, CSIR- Central Salt and Marine Chemicals Research Institute, GijubhaiBadheka Marg, Bhavnagar, Gujarat, India, 364002.
[2] Biological Oceanography Division, CSIR-National Institute of Oceanography, Dona Paula, Goa, India 403004.
[3] Academy of Scientific and Innovative Research (AcSIR), Ghaziabad, Uttar Pradesh, India 201002.
* Corresponding author: cathrine@nio.res.in; vaibhav@csmcri.res.in

that about 11.6 million tons of fresh weight, which contributes to about 33.6% of all seaweeds of feedstock, has been cultivated in 2019 (Cai et al. 2021). Further, large-scale commercial cultivation of *K. alvarezii* is taking place in Indonesia, the Philippines, Malaysia, India, and Tanzania (FAO 2022). Indonesia, with 9.79 million tons of fresh weight (84.28%), tops the feedstock production, followed by the Philippines with 1.49 million tons of fresh weight (12.90%) and Malaysia with 0.18 million tons of fresh weight (1.62%) (Cai et al. 2021).

In trade, carrageenan is generally termed as 'Carrageenan' or 'Processed Eucheuma Seaweed' (PES)—however, carrageen comes in two forms: either 'refined carrageenan' (RC) or 'semi-refined carrageenan' (SRC). It has been reported that about 80% of the global carrageenan production is utilized in three major application sectors: (i) processed meats; (ii) dairy and (iii) desserts and jellies, with growth registered at 1.5–3% year^{-1} (Langford 2023). The meat and dairy sector utilize SRC, especially in ice cream and chocolate milk, while the jellies and confectionery sector utilizes RC. The smaller market segment utilizing carrageenan includes beer clarifying agents, toothpastes, and pet foods. The high value and niche market applications are also registered in the pharmaceutical segment. This chapter highlights emerging carrageenan applications in the marine biotechnological domain. To address this, we have included recent research on biomedical applications, cosmetics, nutritional supplements, biofuels, bioplastics, bioremediation, biorefineries, and bio-composites. Additionally, we discuss farming techniques, as feedstock availability plays a crucial role for all application domains.

Methods developed for cultivation for sustainable feedstock production

The cultivation of *K. alvarezii* has been successful in the tropical waters of the Southeast Asian region, primarily the Brunei–Indonesia–Malaysia–Philippines (East Association of Southeast Asian Nations (ASEAN) Growth Area – BIMP-EAGA – integrated countries) (Hurtado et al. 2013, Msuya and Hurtado 2017). The diverse methods of farming that have been developed suit the local environment. The success of farming has been attributed to simple, non-cumbersome methods, utilizing infrastructure from the locally available material. The farming of this seaweed has been practiced by employing vegetative propagules, obtained from wild stock or from the previous harvest. The selected healthy propagules are tied to a soft plastic rope or a flat binder, by adopting a method which is popularly known as 'tie-tie'. The propagules are inserted into the loops that have been wound around a cultivation rope. However, over time a decline in seaweed productivity, vigor, growth rate, and tolerance to stress and diseases has been seen in cultivated *Kappaphycus* species. The production of elite seedlings through micropropagation of tissue cultured plants is helpful in selection of seaweed variety with enhanced phyconomic traits, providing a consistent supply of seeding material (Hurtado and Critchley 2020). The selected variety is first pre-acclimatized in a liquid medium for 3 days at 23 ± 2°C and then cut into small fragments of 8–10 mm. The fragments are then transferred into sufficient seawater supplemented with nutrient medium and growth stimulants like AMPEP,

IAA, and kinetin. The fragments of *Kappaphycus* thallus are cultured in liquid medium under continuous aeration (Sethuraman et al. 2024). Earlier studies have claimed a higher survival (up to 80%), early and fast shooting, and callus induction rate (up to 54%) using micropropagation of *K. alvarezii* (Mo et al. 2020, Luhan and Mateo 2017). Due to limitations of sexual reproduction and spore formation in *Kappaphycus* species micropropagation is a method of choice. However, a series of systematic experiments would be required to establish the standard procedure for the same, which may require more time and expertise. Several Asian countries like India, Indonesia, Malaysia, and Philippines have employed these techniques for increased seaweed productions (Rama et al. 2018, Reddy et al. 2017, Ali et al. 2020).

The methods enlisted below are most popular and widely used for commercial farming; (1) fixed-off bottom or off-bottom mono-line technique, (2) floating bamboo raft method, (3) the floating tubular net method, and (4) net method. These methods are used extensively in Indonesia, the Philippines, Malaysia, India, and Tanzania. The methods like fixed-off bottom or floating bamboo raft method are being used for waters where wave action is less or moderate, whereas techniques like floating tubular net method or net method are being practiced where water movement is high. More recently innovative methods like free swing, hanging long line, hanging basket, multiple raft long line, single raft long line, spider web and triangular method have been developed for deeper waters and are being practiced in in the Philippines and Malaysian waters (Hurtado et al. 2013). Nevertheless, more innovative techniques like triangular rafts, octagonal rafts and vertical raft alignment have been practiced in India (Mantri et al. 2017). Figure 1 demonstrates the representative image of bamboo rafts used in cultivation of *K. alvarezii* worldwide.

Figure 1. *Kappaphycus alvarezii* cultivation. (A) Series of floating rafts along Santa Elena province, Ecuador. (B) *K. alvarezii* growth on raft, Tamil Nadu, India (Image source (A), open access: Montúfar-Romero et al. 2023).

Marine biotechnological applications

Kappaphycus alvarezii has been studied for its use in food, cosmetic, printing, textile, pharmaceutical and biotechnological industries due to the presence of sulphated polysaccharides, hormones, amino acids, and pigments (Rupert et al. 2022). *K. alvarezii* is commercially cultivated for hydrocolloids. The unique biochemical profile of the algae makes it a promising candidate for novel interventions in healthcare due to known anti-inflammatory, anti-diabetic, anti-bacterial, anti-viral and antioxidant capabilities (Vaghela et al. 2022) and is driving the need for increased production (Kanatt 2023). Efforts to enhance cultivation performance through quality germplasm selection, tissue culture, and application of biostimulants would further help us to combat the reduced productivity, loss of vigor, and low crop quality (Hurtado and Critchley 2020, Yeong et al. 2014).

Biomedical applications

The abundance of vitamins and minerals, proteins, lipids, polyphenols, and bioactive substances in seaweeds, highlighting antimicrobial properties (against bacteria, virus, fungi) underscores their potential as food supplements or sources for extracting bioactive compounds (Holdt and Kraan 2011). The bioeconomic significance of *K. alvarezii* is rooted in the presence of κ carrageenan, utilized in the production of diverse products such as gels, dairy items and meat items, pet foods, and as immobilized biocatalysts. It is also employed across pharmaceuticals, cosmetics, fertilizers, and textile industries (Rajaram et al. 2021). In India, its powder due to the protein, crude fiber, ash, Vitamin E, trace amounts of Vitamin B2 and niacin content has been proposed as an ingredient in spice preparation to enhance the nutritional quality of the spices (Senthil et al. 2011). As per reports, *K. alvarezii* chloroform extracts displayed potent antimicrobial effects against *B. subtilis*, *K. pneumoniae*, *Vibrio harveyi*, *V. parahaemolyticus*, *V. alginolyticus*. A GC-MS based analysis revealed ninety-four bioactive compounds, including phenol, decane, vanillin, diphenylamine, benzophenone, dotriaconate, benzene, phytol, butanoic acid, etc. contributing significantly to the antioxidant and antibacterial properties (Das et al. 2023). It has also demonstrated bioefficiency in producing anti-inflammatory, antibacterial, and anticancer silver nanoparticles, as studied by Deepa et al. (2023). The nanoparticles exhibited anti-inflammatory properties, characterized through FTIR, UV-visible spectroscopy, EDX analysis, SEM and TEM. This bioefficiency is attributed to the flavonoid-based chemicals. Additionally, Kiruba et al. (2015) reported that the presence of bioactive quinones, sterols, and isoflavonols in *K. alvarezii* extract actively inhibits microbial growth in a dose-dependent manner. Carrageenan, derived from *K. alvarezii*, is a polysaccharide of considerable medical significance, with antimicrobial, antioxidant, anti-inflammatory and antitumor effects. Its biological activities are intricately influenced by factors such as molecular weight, sulfation pattern and yield. With enhanced stability, biodegradability, and biocompatibility, carrageenan holds promise for applications in drug delivery, wound healing, and tissue engineering (James et al. 2024). It is viewed that the antioxidant activity of carrageenan is directly influenced by various components like polysaccharides, proteins and small peptides, pigments, etc. (Siriwardhana et al. 2003). Methanolic extracts of *K. alvarezii* demonstrate varying levels of antioxidant activities, suggesting potential health benefits for humans. Owing to its diverse phycocolloid contents, fatty acid, and amino acid profiles, chemical composition, coupled with bioactive potentials and low lipid content, *K. alvarezii* is recommended in diets (Rajaram et al. 2021).

Considering the detrimental effects of lipid-filled diets on coronary health, incorporating seaweeds like *K. alvarezii* into human diets may contribute to longevity by mitigating heart diseases. Studies with rats on a high-cholesterol/high-fat diet supplemented with 5% *K. alvarezii* powder revealed lower body weight gain, decreased plasma total cholesterol and triacylglyceride levels, reduced plasma low-density lipoprotein cholesterol, increased high-density lipoprotein cholesterol, and a decreased atherogenic index. These findings suggest a protective effect of seaweed, potentially reducing the risk of cardiovascular diseases (Mantajun et al. 2010). Studies have shown that *K. alvarezii* extracts possess diverse bioactive compounds with hypoglycemic effects, as demonstrated in alloxan-induced diabetic rats. The aqueous extract of *K. alvarezii* significantly lowered blood glucose levels, restored glycosylated hemoglobin, and increased plasma insulin (Cyriac and Eswaran 2016). Diyana et al. (2015) further illustrated the antioxidant activity of *K. alvarezii*. Additionally, along with its sulphated polysaccharides, it exhibited antiviral properties against enveloped viruses like HIV, HSV, and HCMV (Fitton 2011). *K. alvarezii* is currently used in the carrageenan industry for gelling and thickening food additives, but its versatility extends to potential pharmacological applications, leveraging bioactive secondary metabolites. Its role as a functional food in human diets, offering richness in vitamins, protein, minerals, dietary fiber, and trace elements for enhanced health is also gaining attention.

Cosmetics and nutritional supplements

Seaweeds are a sustainable source of biomolecules, showing health benefits and application as nutraceutical compounds (Suleria et al. 2015). Their unique polysaccharide, pigments and polyphenolics are known to be responsible for their bioactivity (Ganesan et al. 2019, Nunes et al. 2020). In recent times, lifestyle changes have greatly affected the health and well-being of humans across all age groups. This is attributed to increased consumption of sugar-rich food and a sedentary lifestyle which leads to glucose accumulation in the blood, its autoxidation, oxidative phosphorylation, and the formation of advanced glycation end products, leading to the production of reactive oxygen species (ROS). Healthy individuals effectively eliminate ROS, using the body's defense systems (Patlevič et al. 2016). However, excessive ROS production overwhelms the natural detoxification system and leads to oxidative stress, and the induction of cellular pathways leading to the onset of many metabolic disorders (Mancini et al. 2018, Jagtap and Manohar 2021). The prevention of oxidative stress using algal biomolecules with antioxidant and validated anti-glycemic activities along with prebiotic activity could be an effective strategy for management of the metabolic syndrome which is a global health issue. Hence, there is a huge demand for natural sources of bioactive compounds exhibiting

ROS scavenging, anti-glycemic, and prebiotic activity for managing metabolic syndrome. *Kappaphycus* species is a much-studied seaweed species and its metabolic capabilities are greatly analyzed and commercially available. The biochemical activities of the seaweeds are attributed to their phenolic, flavonoids and protein content along with their polysaccharide composition which is efficiently tested in animal models and has also proved efficient during human trials (Wanyonyi et al. 2017, Mahadi et al. 2021, Salma et al. 2021). The cosmetic application of the various biomolecules of *Kappaphycus* has also been tested widely and compared with various seaweed species. *Kappaphycus* potential for cosmetic application varies from skin applications due to its unique polysaccharide and lipid content. The pigments are also known to provide protection against UV radiation and prevent damage due to the harmful radiation. Hence the *Kappaphycus* extracts and biomolecules have wide applications for protecting against increasing reactive oxygen stress and preventing various cosmetic damage (Kasanah et al. 2022).

Bioplastics and biofuels

The resource-efficient nature of plastics is overshadowed by their slow degradability and potential health risks, that pose a threat to soil, marine life, and human well-being (Kumar et al. 2023, Thakur et al. 2018). A shift towards bio-based and biodegradable plastics has been made to address this issue, with bioplastics being a promising alternative. Bioplastics are made from renewable resources like, but not limited to, corn, sugars, potatoes, and seaweeds which can be degraded by using microbes, ultraviolet light, or sunlight. This kind of biodegradation does not leave any toxic residue. Seaweeds provide a source of high-biomass, durable, and microwave-resistant biopolymers, offering a potential alternative to conventional plastics (Bezirhan and Bilgen 2015). Seaweeds are abundantly available, easy to cultivate, and have been known for their various pharmaceutical and industrial applications. Biopolymers extracted from seaweeds (hydrocolloids/polysaccharides) are edible and have excellent gelling, emulsifying, and thickening properties for food and medical uses (Simona et al. 2021). Commonly used seaweed species for biopolymers extraction are *Gracilaria* sp., *Sargassum* sp., *Ulva* sp., *Kappaphycus* sp., *Fucus* sp., *Ascophyllum* sp., *Gelidium* sp., *Euchema* sp., *Laminaria* sp., and *Saccharina* sp. (Zaimis et al. 2021). *Kappaphycus alvarezii* shows potential as a bio-preservative due to its antibacterial, antiviral, and antifungal properties. Carrageenan extracted from *K. alvarezii* has been reported to be a non-digestible and non-carcinogenic compound with biological activities like antioxidant, antitumor, antimicrobial, anti-diabetic, and antihypertensive effects in addition to their prebiotic and growth-promoting effects on gut microbiota. Thus *K. alvarezii* serves as a remarkable resource which has a rapid growth rate and extensive environmental tolerance (Cheong et al. 2018, Júnior and Turan 2022). Purified carrageenan is economical and easier to use, with high polysaccharide content and concentrations reaching up to 37% of algal dry weight (Wan Yahaya et al. 2023). Biofilms from *Kappaphycus alvarezii* may be produced following the earlier methods (Hanry and Surugau 2020, Rahmawati et al. 2019). κ-carrageenan from *K. alvarezii* is isolated and used in crude, semi-refined, and refined forms to produce biofilms. Semi-refined κ-carrageenan is obtained after processing κ-carrageenan with alkali while removing cellulosic materials, resulting in refined κ-carrageenan (Ghosh et al. 2006). Carrageenan-based biofilms pose challenges in industrial use due to their inherent hydrophilic nature and high water-vapor permeability. However, grafting or blending the biopolymers with other natural polymers like lignin, sorbitol, or glycerol significantly improves their tensile strength, thickness, opacity, solubility, moisture content, and water resistance while reducing water vapor permeability and stability in water (Rizal et al. 2022, Sudhakar et al. 2021). Similarly, the addition of natural bioactive extracts also helps with the enrichment of bioplastic properties. For example, adding neem leaf extract, lemon leaf extract, and gum and their combinations enhanced the antimicrobial and antioxidant properties of the biofilm from *K. alvarezii*, which helped extend the shelf life of food items (Murugan et al. 2023).

The potential of biopolymers extends far beyond bioplastics, encompassing the sustainable production of diverse materials like biofuels, bio-implants, drug delivery systems and potential scaffold materials (Faizan et al. 2021, Rizal et al. 2022, Toskas et al. 2012). Biofuels like bioethanol may serve as an alternative fuel source for sustainability, climate change mitigation, and a healthier future beyond fossil fuels (Masami et al. 2007). However, currently, bioethanol production from plants like corn and sugarcane competes for valuable land and resources, potentially increasing the prices of these food commodities. On the other hand, seaweeds grow fast and contain high amounts of carbohydrates for biofuel production. Seaweed cultivation does not impinge on agricultural land and purifies the air and the oceans through carbon sequestration. The pioneering efforts for sustainable biofuel production started with producing ethanol from laminaran and mannitol from brown seaweeds may be fermented to ethanol as reported by Horn et al. 2000. *K. alvarezii* is amongst the best species of red algae for bioethanol production due to easily hydrolysable polysaccharides, low lignin content, abundance, ease in cultivation, and high growth rate (4–7%) (Khambhaty et al. 2012, Ramachandra and Hebbale 2020). The D-galactose present in the algae can be converted into ethanol using acid hydrolysis and yeast (*Saccharomyces cerevisiae*) fermentation, achieving 66% conversion compared to pure galactose (Meinita et al. 2012). Currently the production of this red algae is only for phycocolloid, κ-carrageenan. However, the process of making ethanol from *Kappaphycus alvarezii* sap, and the production of other high-value compounds like hydroxymethyl furfural, formic acid and levulinic acid was reported from India a decade ago and has been granted a European patent (Mody et al. 2015, Mondal et al. 2013). United Nations and European Union lead the world by producing 46% of ethanol and 32% of biodiesel demand, respectively, whereas India ranks fifth in ethanol and fifteenth in biodiesel production. However, the global need for ethanol and biodiesel production is projected to increase to more than 150 bn L and 67 bn L respectively and tap $129.93 billion worth of market by 2030, at 6.40% cumulative annual growth (OECD/FAO 2023, Precedence Research 2023). Cellulosic biofuels are expected to contribute to 70% to the global biofuel market (Carriquiry et al. 2011).

Bioremediation

South East countries are the major cultivators of seaweeds of various red macroalgae such as *Gelidium amansii*, *Gracilaria verrucosa* and *Eucheuma denticulatum* including *Kappaphycus alvarezii.* The major target of seaweed biomass cultivation was for bioethanol production to produce green fuel and studies have shown that the residual biomass could be used to remove toxic heavy metals such as Cd(II), Pb(II), Cu(II) and Cr(III) ions through biosorption (Kang et al. 2011, Sunwoo et al. 2016). Bioremediation of industrial effluent was also tested using cultivated *Kappaphycus alvarezii* species and wild-grown *Sargassum wightti*. The study showed that both the seaweed species showed efficiency to remediate the industrial effluents and *Kappaphycus* species tested could accumulate various metal ions with greater efficiency than *Sargassum wightti* (Mahesh et al. 2020). Phytoremediation of heavy metals was tested in all three cultivated varieties of *Kappaphycus alvarezii* from Port Okha from the Northwest coast of India. Analysis using a comparative study of three colour kappa was observed in which brown algae showed more metal absorbance of cadmium. This was confirmed through volumetric biosorption studies using inductively coupled plasma atomic emission spectroscopy (Kumar et al. 2008). Integrated farming of eucheumatoids like *Kappaphycus alvarezii* var. *tambalang*, *Kappaphycus striatus* var. *sacol* and *Eucheuma denticulatum* var. *spinosumcultivation* also proved to be efficient for bioremediation of effluents from aquaculture industry. Such integrated seaweed cultivation proves to be efficient for removing excess nutrients released from industries and to prevent eutrophication (Hayashi et al. 2008, Narvarte et al. 2023). However the uptake of nutrient (NH^{4+}) depends on phosphate (PO_4^{-3}) levels.

Biorefinery

Biorefinery converts biomass into valuable products like energy, fuels, chemicals, food, and feed using physical, chemical, thermochemical, and biochemical technologies to maximize resource utilization, minimize waste, and increase profitability (Uzoejinwa et al. 2024). *Kappaphycus* stands fifth in the world's most cultivated macroalgae which contributes to 90% of the total carrageenan production in the world and has a mature and well-established technology (Rudke et al. 2020, Rupert et al. 2022). A biorefinery for *Kappaphycus* should primarily be focused on the production of carrageenan followed by the use of as many compounds of the harvested algal biomass as possible for the production of valuable products. Effluents and seaweed sap may be used as fertilizers, plant growth stimulators, and soil conditioners due to the presence of high minerals (Na, K, Cl, and S) and growth hormones (indole 3-acetic acid, gibberellin GA_3, kinetin and zeatin) (Singh et al. 2017, Bhuyar et al. 2020, Trivedi et al. 2021, Cokrowati et al. 2021). Also, the use of whole extracts of *K. alvarezii* as functional foods due to their reported nutraceutical activities would reduce the need of individual isolation of bioactive compounds and expedite the transition from research to practical applications. The sap and liquid waste may be subjected to the recovery of bioactive compounds and pigments by precipitation of aqueous extract fractions with $(NH_4)_2SO_4$ (Reddy et al. 2018).

The insoluble solid matter may be used for extraction of mannose, xylose, galactose, anhydro-galactose, and glucose after the extraction of carrageenan which can further serve as a carbon source for the production of bioethanol (Masarin et al. 2016). Furthermore, the extraction of proteins that make 5–15% of *K. alvarezii* may be very useful. The conventional carrageenan extraction process uses alkali treatment (6% KOH) which yields only 10% of the total protein content of the biomass. Alternatively, extraction of native proteins or individual amino acids from protein hydrolysates via electrodialysis, fractional precipitation, and chromatography for production of fuels and chemicals using biotransformation approach may be employed (Li et al. 2018). Implementation of enzymatic (e.g., sulphurylases, cellulase) extractions may also eliminate the need for alkaline treatment and might offer new and beneficial by-products (Ortiz-Tena et al. 2017, Tarman et al. 2020). Therefore, an ideal process of biorefinery for *K. alvarezii* would additionally produce galactans, glucans, mannose, xylose, sulphates, lipids, and proteins, individual amino acids, pigments, sap, and bioactive compounds. Figure 2 summarizes the integrated processes reported for the biorefinery of *K. alvarezii*. A standalone self-sustainable process of total utilization of *K. alvarezii* for the production of carrageenan, hydroxymethyl furfural, formic acid, levulinic acid, $K_3Na(SO_4)_2$, K_2SO_4 fertilizers, leaving spent aqueous stream for foliar spray demonstrated the economic viability of the process (Mondal et al. 2013). Different techniques may be employed for the extraction of specific high value compounds: for example, superficial fluid extraction (nonpolar molecules), pressurized liquid extraction (pigments, phenolic compounds, carrageenan), ultrasound assisted extraction (carrageenan), microwave assisted extraction (carrageenan) and subcritical water extractions (sugars) (Rudke et al. 2020).

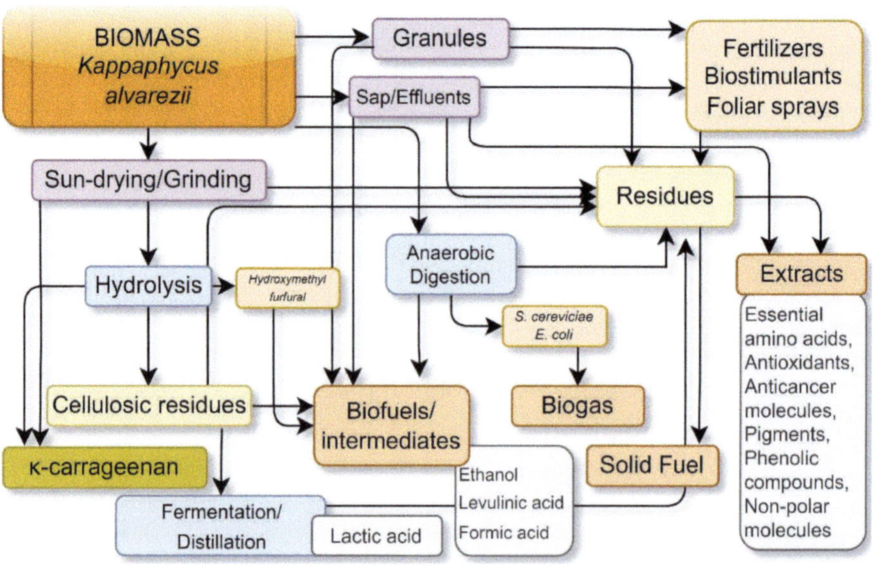

Figure 2. Integrated processes design for a biorefinery for *Kappaphycus alvarezii*.

Biocomposites

Seaweeds, particularly *Kappaphycus*, have emerged as a crucial source for biocomposites in polymer applications. The ability to blend polymeric biomaterials derived from *Kappaphycus* makes it a distinctive and sustainable material with diverse applications. Globally and in India, the increasing biomass availability of *Kappaphycus*, a well-studied seaweed with commercial cultivation, has attracted the attention of researchers and inventors due to its unique physical and chemical characteristics. Carrageenan, the predominant polysaccharide in *Kappaphycus* species, is renowned for its gel strength and high yield in cultivable species. This has led to its exploration in various applications in the healthcare, medical and food industry. Emphasis has been placed on bioplastic-related applications, especially in the formation of unique film properties. Additionally, *Kappaphycus* biocomposites have been studied for use in healthcare, including biocompatible films and bandages, as well as drug delivery applications. The film-forming properties of *Kappaphycus* polysaccharide have been extensively studied for all these applications. Limited attention has been given to the application of *Kappaphycus* biocomposites in chemistry-related applications, presenting an area with potential for further exploration. *Kappaphycus* biocomposites have been explored for their biomaterial capabilities in food packaging due to their nontoxic, biodegradable, and non-reactive properties. Also, carrageenan, widely used in food products, does not affect the taste of food. Studies have involved blending raw, refined, and semi-refined carrageenan with various materials, exploring their effects on biofilms, and reporting unique material properties and applications. The high gelling strength and versatility of carrageenan polymers make *Kappaphycus*-derived materials suitable for diverse applications. Carrageenan-based nanocomposites have been studied for applications such as drug delivery and antimicrobial purposes. Among the notable studies, *K. alvarezii* is studied and reported as a bio-stabilizer for superparamagnetic nanocomposites (Fe_3O_4-NCs) which comprised of montmorillonite (MMT) as matrix support, *Kappaphycus alvarezii* (SW) as bio-stabilizer and Fe_3O_4 as filler in the composites to form MMT/SW/Fe_3O_4-NCs, demonstrating promising drug delivery properties (Yew et al. 2019). Furthermore, *K. alvarezii* has proven effective as a modifier for degradable edible films, enhancing mechanical and anti-microbial characteristics (Salim et al. 2021). Applications of seaweed biocomposites of late has also diversified into their incorporation as construction materials. Seaweed-based particleboard has shown promise, with higher density values compared to starch-based alternatives (Liew et al. 2022). Current research is pursued in the realm of chemical modification and purification of *K. alvarezii* powdered biocomposites with Polylactic acid (PLA) to enhance strength and other essential properties for wider applications (Rosdi et al. 2023).

Conclusions

Kappaphycus alvarezii is a versatile resource for a sustainable future with an immense potential across various sectors due to the presence of unique bioactive compounds. The current carrageenan market is driven by rising consumer demand

and projected to reach $1.3 billion by 2030. Simple and adaptable farming methods like tie-tie and net methods are largely used, with variations for different water movement levels but innovative techniques like hanging long lines and triangular rafts show promise for deeper waters. *K. alvarezii* is a rich source of vitamins, minerals, dietary fiber, and protein, making it a valuable addition to diets to cure trending lifestyle disorders like diabetes and heart disease due to its antioxidant and anti-inflammatory properties. Extracts from *K. alvarezii* demonstrate antimicrobial, antiviral, and anti-glycemic activities. It also shows potential for use in cosmetics due to its unique polysaccharide and lipid content and offers a sustainable alternative to conventional plastics due to its biodegradability. This seaweed is also a source for bioethanol production, a sustainable alternative fuel source, and has been known to effectively remove heavy metals and excess nutrients from water bodies, contributing to environmental cleanup as demonstrated by the integrated cultivation of *K. alvarezii* with aquaculture to manage effluents and prevent eutrophication. Further it also serves as a valuable feedstock for biorefinery offering a multitude of value-added products like sugars, proteins, and biofuels. This warrants for optimizing cultivation techniques and biorefinery processes to enhance the economic viability and environmental sustainability of utilizing this valuable marine resource.

Acknowledgement

The authors of the manuscript acknowledge the Director, CSIR-Central Salt and Marine Chemicals Research Institute, Bhavnagar and The Director, CSIR-National Institute of Oceanography, Dona Paula, Goa for the facilities provided.

References

Ali, M.K.M., Critchley, A.T. and Hurtado, A.Q. (2020). Micropropagation and sea-based nursery growth of selected commercial *Kappaphycus* species in Penang, Malaysia. Journal of Applied Phycology 32: 1301–1309. https://doi.org/10.1007/s10811-019-02003-4.

Bezirhan, E. and Bilgen, H.D. (2015). A review: investigation of bioplastics. Journal of Civil Engineering and Architecture 9: 188–192. https://doi.org/10.17265/1934-7359/2015.02.007.

Bhuyar, P., Rahim, M.H., Sundararaju, S., Maniam, G.P. and Govindan, N. (2020). Antioxidant and antibacterial activity of red seaweed *Kappaphycus alvarezii* against pathogenic bacteria. Global Journal of Environmental Science and Management 6: 47–58. https://doi.org/10.22034/GJESM.2020.01.04.

Cai, J., Lovatelli, A., Aguilar-Manjarrez, J., Cornish, L., Dabbadie, L., Desrochers, A. et al. (2021). Seaweeds and microalgae: an overview for unlocking their potential in global aquaculture development. FAO Fisheries and Aquaculture Circular 1229. Rome, FAO. https://doi.org/10.4060/cb5670en.

Campbell, R. and Hotchkiss, S. (2017). Carrageenan industry market overview. pp. 193–205. *In*: Hurtado, A.Q., Critchley, A.T. and Neish, I.C. (eds.). Tropical Seaweed Farming Trends, Problems and Opportunities. Springer International Publishing. https://doi.org/10.1007/978-3-319-63498-2_13.

Carriquiry, M.A., Du, X. and Timilsina, G.R. (2011). Second generation biofuels: Economics and policies. Energy Policy 39(7): 4222–4234. https://doi.org/10.1016/j.enpol.2011.04.036.

Cheong, K.-L., Qiu, H.-M., Du, H., Liu, Y. and Khan, B.M. (2018). Oligosaccharides derived from red seaweed: production, properties, and potential health and cosmetic applications. Molecules 23(10): Article 10. https://doi.org/10.3390/molecules23102451.

Cokrowati, N., Risjani, Y., Andayani, S., Firdaus, M. and Honiar, R. (2021). Identification and determination of growth hormones of *Kappaphycus alvarezii* cultivated in Ekas Bay East

Lombok. In IOP Conference Series: Earth and Environmental Science 890: 012040. https://doi.org/10.1088/1755-1315/890/1/012040.

Cyriac, B. and Eswaran, K. (2016). Anti-hyperglycemic effect of aqueous extract of *Kappaphycus alvarezii* (Doty) Doty ex. P. Silva in alloxan-induced diabetic rats. Journal of Applied Phycology Aug; 28: 2507–13. https://doi.org/10.1007/s10811-015-0762-7.

Das, D., Arulkumar, A., Paramasivam, S., Lopez-Santamarina, A., del Carmen Mondragon, A. and Miranda Lopez, J.M. (2023). Phytochemical constituents, antimicrobial properties, and bioactivity of marine red seaweed (*Kappaphycus alvarezii*) and seagrass (*Cymodocea serrulata*). Foods 24; 12(14): 2811. https://doi.org/10.3390/foods12142811.

Deepa, V.H., Sivakami, R. and Rajan, S. (2023). Characterization of *Kappaphycus alvarezii* derived siver nanoparticles and its bioefficiency as an anti-inflammatory agent. Biochemical & Cellular Archives 2: 23. 10.51470/bca.2023.23.S1.1499.

Diyana, A.F., Abdullah, A., Hisham, Z.S. and Chan, K.M. (2015). Antioxidant activity of red algae *Kappaphycus alvarezii* and *Kappaphycus striatum*. International Food Research Journal 1; 22(5): 1977.

Faizan, M., Nadeem, H., Arif, A. and Zaheer, W. (2021). Bioplastics from biopolymers: an eco-friendly and sustainable solution of plastic pollution. Polymer Science, Series C 63(1): 47–63. https://doi.org/10.1134/S1811238221010057.

FAO. (2022). The State of World Fisheries and Aquaculture 2022: Towards Blue Transformation. Food and Agriculture Organization, https://doi.org/10.4060/cc0461en.

Fitton, J.H. (2011). Therapies from fucoidan; multifunctional marine polymers. Marine Drugs 9: 1731–1760. https://doi.org/10.3390/md9101731.

Ganesan, M., Trivedi, N., Gupta, V., Madhav, S.V., Radhakrishna Reddy, C. and Levine, I.A. (2019). Seaweed resources in India–current status of diversity and cultivation: prospects and challenges. Botanica Marina 62(5): 463–482.

Ghosh, P.K., Siddhanta, A.K., Prasad, K., Meena, R. and Bhattacharya, A. (2006). Process of preparation of biodegradable films from semi refined kappa carrageenan (United States Patent US7067568B1). https://patents.google.com/patent/US7067568B1/en.

Hanry, E.L. and Surugau, N. (2020). Characteristics and properties of biofilms made from pure carrageenan powder and whole seaweed (*Kappaphycus* sp.). Journal of Advanced Research in Fluid Mechanics and Thermal Sciences 76(2): Article 2. https://doi.org/10.37934/arfmts.76.2.99110.

Hayashi, L., Yokoya, N.S., Ostini, S., Pereira, R.T., Braga, E.S. and Oliveira, E.C. (2008). Nutrients removed by *Kappaphycus alvarezii* (Rhodophyta, Solieriaceae) in integrated cultivation with fishes in re-circulating water. Aquaculture 277(3-4): 185–191.

Holdt, S.L. and Kraan, S. (2011). Bioactive compounds in seaweed: functional food applications and legislation. Journal of Applied Phycology 2011 Jun; 23: 543–97. https://doi.org/10.1007/s10811-010-9632-5.

Horn, S.J., Aasen, I.M. and Østgaard, K. (2000). Production of ethanol from mannitol by Zymobacterpalmae. Journal of Industrial Microbiology and Biotechnology 24(1): 51–57. https://doi.org/10.1038/sj.jim.2900771.

Hurtado, A., Gerung, G., Yasir, S. and Critchley, A. (2013). Cultivation of tropical red seaweeds in the BIMP-EAGA region. Journal of Applied Phycology 26: 707–718. https://doi.org/10.1007/s10811-013-0116-2.

Hurtado, A.Q. and Critchley, A.T. (2020). Chapter 4 - Time for applications of biostimulants in phyconomy: seaweed extracts for enhanced cultivation of seaweeds (SEECS). pp. 103–127. *In*: Torres, M.D., Kraan, S. and Dominguez, H. (eds.). Sustainable Seaweed Technologies, Advances in Green and Sustainable Chemistry. Elsevier. https://doi.org/10.1016/B978-0-12-817943-7.00024-X.

Hydrocolloids Market Size, Share & Trends Analysis Report by product (Gelatin, xanthan gum, carrageenan, alginates, pectin, guar gum, carboxy methyl cellulose), by function, by application, by region, and segment Forecasts, 2023–2030. (2023, March 24). https://www.grandviewresearch.com/industry-analysis/hydrocolloids-market.

Jagtap, A.S. and Manohar, C.S. (2021). Overview on microbial enzymatic production of algal oligosaccharides for nutraceutical applications. Marine Biotechnology 23: 159–176.

James, J., Verma, M. and Sharma, N. (2024). Nanotechnology-driven improvisation of red algae-derived carrageenan for industrial and bio-medical applications. World Journal of Microbiology and Biotechnology 2024 Jan; 40(1): 4. https://doi.org/10.1007/s11274-023-03787-x.

Júnior, J.S.A.R. and Turan, G. (2022). Biodegradable plastic and film production from seaweeds. Bulletin of Biotechnology 3(1): Article 1. https://doi.org/10.51539/biotech.1033959.

Kanatt, S.R. (2023). Antioxidants from the red algae *Kappaphycus alvarezii*. pp. 457–472. *In*: Kim, S.-K., Shin, K.-H. and Venkatesan, J. (eds.). Marine Antioxidants. Academic Press. https://doi.org/10.1016/B978-0-323-95086-2.00001-1.

Kang, O.L., Ramli, N., Said, M., Ahmad, M., Yasir, S.M. and Ariff, A. (2011). *Kappaphycus alvarezii* waste biomass: a potential biosorbent for chromium ions removal. Journal of Environmental Sciences 23(6): 918–922.

Kasanah, N., Ulfah, M., Imania, O., Hanifah, A.N. and Marjan, M.I.D. (2022). Rhodophyta as potential sources of photoprotectants, antiphotoaging compounds, and hydrogels for cosmeceutical application. Molecules 27(22): 7788.

Khambhaty, Y., Mody, K., Gandhi, M.R., Thampy, S., Maiti, P., Brahmbhatt, H. et al. (2012). *Kappaphycus alvarezii* as a source of bioethanol. Bioresource Technology 103(1): 180–185. https://doi.org/10.1016/j.biortech.2011.10.015.

Khanjan, T., Kumar, R., Vijayanand, K.G., Bhojani, G., Kubavat, D. and Ghosh, A. (2021) Structural and functional changes in soil microbes by foliar drift spray of seaweed extract as revealed by metagenomics. https://doi.org/10.21203/rs.3.rs-753007/v1.

Kiruba, N.J., Pradeep, M.A. and Juliana, S.J. (2015). Study of phytoconstituents and antibacterial activity of *Kappaphycus alvarezii*. International Journal of Current Microbiology and Applied Sciences 4: 1209–17.

Kumar, S., Ganesan, K. and Subba Rao, P. (2008). Heavy metal chelation by non-living biomass of three-color forms of *Kappaphycus alvarezii* (Doty). Journal of Applied Phycology 20: 63–66.

Kumar, M., Chaudhary, D.R. and Jha, B. (2023). Surface-associated bacterial assemblages on marine anthropogenic litter in the intertidal zone of the Arabian Sea, India. Marine Pollution Bulletin 193: 115211. https://doi.org/10.1016/j.marpolbul.2023.115211.

Langford, Z. (ed.). (2023). Globalisation and Livelihood Transformations in the Indonesian Seaweed Industry. Routledge, London. https://doi.org/10.4324/9781003183860.

Li, SY., Ng, IS., Chen, P.T., Chiang, C.J. and Chao, Y.P. (2018). Biorefining of protein waste for production of sustainable fuels and chemicals. Biotechnology for Biofuels 11: 256. https://doi.org/10.1186/s13068-018-1234-5.

Liew, K., Tan, Y., Albert, C.M., Raman, V. and Boyou, M. (2022). Potential of using natural and synthetic binder in wood composites. Forests 13(6): 844.

Luhan, M.R.J. and Mateo, J.P. (2017). Clonal production of *Kappaphycus alvarezii* (Doty) Doty *in vitro*. J. Appl. Phycol. 29: 2339–2344. https://doi.org/10.1007/s10811-017-1105-7.

Mahadi, M., Azizul, M.S., Eshak, Z., Ibrahim, A.H., Akim, A.M. and Kamal, N. (2021). *Kappaphycus alvarezii* supplementation attenuates blood pressure, blood cholesterol, ACE and antioxidant activities in hypertensive and hypercholesteroleamic rats. Malaysian Journal of Medicine & Health Sciences 17.

Mahesh, V.B., Jayasankar, R. and Madhavan, A. (2020). Bioremediation of industrial effluent at source using aquatic macrophytes. Journal of the Marine Biological Association of India 62(2): 63–71.

Mancini, F.R., Affret, A., Dow, C., Balkau, B., Bonnet, F., Boutron-Ruault, M.C. et al. (2018). Dietary antioxidant capacity and risk of type 2 diabetes in the large prospective E3N-EPIC cohort. Diabetologia 61: 308–316.

Mantajun, P., Mohamed, S., Kharidah, M. and Noordin, M.M. (2010). Comparison of cardiovascular protective effects of tropical Seaweeds, *Eucheuma cottonii*, *Caulerpa lentillifera*, and *Sargassum polycystum*, on high-cholesterol/high-fat diet in rats. J. Med. Food 13: 792–800. https://doi.org/10.1089/jmf.2008.1212.

Mantri, V.A., Eswaran, K., Shanmugam, M., Ganesan, M., Veeragurunathan, V., Thiruppathi, S. et al. (2017). An appraisal on commercial farming of *Kappaphycus alvarezii* in India: Success in diversification of livelihood and prospects. Journal of Applied Phycology 29(1): Article 1.

Masami, K., Donald, M. and William, W. (2007). Considering trade policies for liquid biofuels. The International Bank for Reconstruction and Development (The World Bank), Washington. Renewable Energy 4: 7.

Masarin, F., Cedeno, F.R.P., Chavez, E.G.S. Oliveira, L.E., Gelli, V.C. and Monti, R. (2016). Chemical analysis and biorefinery of red algae Kappaphycus alvarezii for efficient production of glucose from residue of carrageenan extraction process. Biotechnology for Biofuels 9: 122. https://doi.org/10.1186/s13068-016-0535-9.

Meinita, M.D.N., Meinita, M.D.N., Ji-Young Kang, Kang, J.-Y., Gwi-Taek Jeong, Jeong, G.-T. et al. (2012). Bioethanol production from the acid hydrolysate of the carrageenophyte *Kappaphycus alvarezii* (cottonii). Journal of Applied Phycology 24(4): 857–862. https://doi.org/10.1007/s10811-011-9705-0.

Mo, V.T., Cuong, L.K., Tung, H.T., Van Huynh, T., Nghia, L.T., Khanh, C.M. et al. (2020). Somatic embryogenesis and plantlet regeneration from the seaweed *Kappaphycus striatus*. Acta Physiol Plant 42: 104. https://doi.org/10.1007/s11738-020-03102-3.

Mody, K.H., Ghosh, P.K., Sana, B., Gnanasekaran, G., Shukla, A.D., Eswaran, K. et al. (2015). Process for integrated production of ethanol and seaweed sap from *Kappaphycus alverezii* (United States Patent US8969056B2). https://patents.google.com/patent/US8969056B2/en.

Mondal, D., Sharma, M., Maiti, P., Prasad, K., Meena, R., Siddhanta, A.K. et al. (2013). Fuel intermediates, agricultural nutrients, and pure water from *Kappaphycus alvarezii* seaweed. RSC Advances 3(39): 17989–17997. https://doi.org/10.1039/C3RA42919A.

Montúfar-Romero, M., Rincones-León, R.E., Cáceres-Farias, L.B., Espinoza-Vera, M.M., Avendaño, U., Cruz-Jaime, T. et al. (2023). Feasibility of aquaculture cultivation of elkhorn sea moss (*Kappaphycus alvarezii*) in a horizontal long line in the Tropical Eastern Pacific. Scientific Reports 13(1): Article 1. https://doi.org/10.1038/s41598-023-41795-x.

Msuya, F. and Hurtado, A. (2017). The role of women in seaweed aquaculture in the Western Indian Ocean and South-East Asia. European Journal of Phycology 52: 482–494. https://doi.org/10.1080/09670262.2017.1357084.

Murugan, A., Banu, T., Duraikkannu, S.L., Munisamy, S., Vannamuthu, K. and Sivakumar, S. (2023). Fabrication and characterization of *Kappaphycus alvarezii* biomass based thin-film and its applications. Food and Humanity 1: 589–598. https://doi.org/10.1016/j.foohum.2023.06.029.

Narvarte, B.C.V., Hinaloc, L.A.R., Gonzaga, S.S.C. and Roleda, M.Y. (2023). Impacts of aquaculture nutrient sources: ammonium uptake of commercially important eucheumatoids depends on phosphate levels. Journal of Applied Phycology. https://doi.org/10.1007/s10811-023-03073-1.

Nunes, N., Valente, S., Ferraz, S., Barreto, M.C. and Pinheiro de Carvalho, M.A. (2020). Biochemical study of attached macroalgae from the Madeira Archipelago and beach-cast macroalgae from the Canary Islands: multivariate analysis to determine bioresource potential. Botanica Marina 63(3): 283–298.

OECD/FAO. (2023). OECD-FAO Agricultural Outlook 2023–2032. OECD Publishing. https://doi.org/10.1787/08801ab7-en.

Ortiz-Tena, J.G., Schieder, D. and Sieber, V. (2017). Carrageenan and more: biorefinery approaches with special reference to the processing of *Kappaphycus*. pp. 155–164. *In*: Hurtado, A.Q., Critchley, A.T. and Neish, I.C. (eds.). Tropical Seaweed Farming Trends, Problems and Opportunities: Focus on *Kappaphycus* and *Eucheuma* of Commerce. Springer International Publishing, Cham.

Patlevič, P., Vašková, J., Švorc Jr, P., Vaško, L. and Švorc, P. (2016). Reactive oxygen species and antioxidant defense in human gastrointestinal diseases. Integrative Medicine Research 5(4): 250–258.

Precedence Research. (2023, October). Liquid Biofuels Market Size to Hit US$ 148.06 Billion by 2032. Liquid Biofuels Market Size to Hit. https://www.precedenceresearch.com/liquid-biofuels-market. Accessed 1 February, 2024.

Rahmawati, M., Arief, M. and Satyantini, W.H. (2019). The effect of sorbitol addition on the characteristic of carrageenan edible film. IOP Conference Series: Earth and Environmental Science 236(1): 012129. https://doi.org/10.1088/1755-1315/236/1/012129.

Rajaram, R., Muralisankar, T., Paray, B.A. and Al-Sadoon, M.K. (2021). Phytochemical profiling and antioxidant capacity of *Kappaphycus alvarezii* (Doty) Doty collected from seaweed farming sites of tropical coastal environment. Aquaculture Research 2021 Jul; 52(7): 3438–48. https://doi.org/10.1111/are.15188.

Rama, R., Aslan, L.O.M., Iba, W., Nurdin, A.R., Armin, A. and Yusnaeni, Y. (2018). Seaweed cultivation of micropropagated seaweed (*Kappaphycus alvarezii*) in BunginPermai Coastal Waters, Tinanggea Sub-District, South Konawe Regency, South East Sulawesi. IOP Conf. Ser.: Earth Environ. Sci. 175: 012219. https://doi.org/10.1088/1755-1315/175/1/012219.

Ramachandra, T.V. and Hebbale, D. (2020). Bioethanol from macroalgae: Prospects and challenges. Renewable and Sustainable Energy Reviews 117: 109479. https://doi.org/10.1016/j.rser.2019.109479.

Reddy, C.R.K., Yokoya, N.S., Yong, W.T.L., Luhan, M.R.J. and Hurtado, A.Q. (2017). Micro-propagation of *Kappaphycus* and *Eucheuma*: Trends and prospects. pp. 91–110. *In*: Hurtado, A.Q., Critchley, A.T. and Neish, I.C. (eds.). Tropical Seaweed Farming Trends, Problems and Opportunities: Focus on *Kappaphycus* and *Eucheuma* of Commerce. Springer International Publishing, Cham. https://doi.org/10.1007/978-3-319-63498-2_5.

Reddy, C.R., Baghel, R.S., Trivedi, N., Kumari, P., Gupta, V., Prasad, K. et al. (2018). U.S. Patent No. 10,000,579. Washington, DC: U.S. Patent 10000579, June 19, 2018.

Rizal, S., Alfatah, T., Abdul Khalil, H.P.S., Yahya, E.B., Abdullah, C.K., Mistar, E.M. et al. (2022). Enhanced functional properties of bioplastic films using lignin nanoparticles from oil palm-processing residue. Polymers 14(23): Article 23. https://doi.org/10.3390/polym14235126.

Rosdi, F.N.M., Salim, N., Roslan, R., Bakar, N.H.A. and Sarmin, S.N. (2023). Potential red algae fibre waste as a raw material for biocomposite. Journal of Advanced Research in Applied Sciences and Engineering Technology 30(1): 303–310.

Rudke, A.R., de Andrade, C.J. and Ferreira, S.R.S. (2020). *Kappaphycus alvarezii* macroalgae: An unexplored and valuable biomass for green biorefinery conversion. Trends in Food Science & Technology 103: 214–224. https://doi.org/10.1016/j.tifs.2020.07.018.

Rupert, R., Rodrigues, K.F., Thien, V.Y. and Yong, W.T.L. (2022). Carrageenan from *Kappaphycus alvarezii* (Rhodophyta, Solieriaceae): Metabolism, structure, production, and application. Front Plant Sci. 13: 859635. https://doi.org/10.3389/fpls.2022.859635.

Salim, N., Abdullah, Y. and Hashim, R. (2021). Potential of seaweed (*Kappaphycus alvarezii*) as a particleboard. pp. 82–87. *In*: Materials Science Forum (Vol. 1025). Trans Tech Publications Ltd.

Salma, S., Hadju, V., Jompa, J., Stang, S., Sundari, S. and Usman, A.N. (2021). The effect of red seaweed (*Kappaphycus alvarezii*) biscuits on hemoglobin levels and body weight among the first trimester pregnant women. Open Access Macedonian Journal of Medical Sciences 9(A): 1019–1023.

Senthil, A., Mamatha, B.S., Vishwanath, P., Bhat, K.K. and Ravishankar, G.A. (2011). Studies on development and storage stability of instant spice adjunct mix from seaweed (*Eucheuma*). Journal of Food Science and Technology 2011 Dec; 48: 712–7. https://doi.org/10.1007/s13197-010-0165-3.

Sethuraman, T., Selvakumar, M., Munisamy, S. and Ganesh, D. (2024). Evaluation of a low-cost prototype for micropropagation of *Kappaphycus alvarezii* and its application. pp. 65–72. *In*: Critchley, A.T., Hurtado, A.Q. and Neish, I.C. (eds.). Tropical Phyconomy Coalition Development: Focus on Eucheumatoid Seaweeds. Springer International Publishing, Cham. https://doi.org/10.1007/978-3-031-47806-2_6.

Simona, J., Dani, D., Petr, S., Marcela, N., Jakub, T. and Bosuslava, T. (2021). Edile films from carrageenan/orange essential oil/trehalose-Structure, optical properties, and antimicrobial activity. Polymers 13: 332. https://doi.org.org/10.3390/polym13030332.

Singh, S.K., Lal, S.S., Singh, R.K. and Zodape, S.T. (2017). Fertilizer potential of sea weed (*Kappaphycus* and *Gracilaria*) saps in potato crop. JAS 4. https://doi.org/10.21921/jas.v4i1.7416.

Siriwardhana, N., Lee, K.W., Kim, S.H., Ha, J.W. and Jeon, Y.J. (2003). Antioxidant activity of *Hizikia fusiformis* on reactive oxygen species scavenging and lipid peroxidation inhibition. Food Science Technology International 9: 339–347. https://doi.org/10.1177/1082013203039014.

Sudhakar, M.P., Magesh Peter, D. and Dharani, G. (2021). Studies on the development and characterization of bioplastic film from the red seaweed (*Kappaphycus alvarezii*). Environmental Science and Pollution Research 28(26): 33899–33913. https://doi.org/10.1007/s11356-020-10010-z.

Suleria, H.A.R., Osborne, S., Masci, P. and Gobe, G. (2015). Marine-based nutraceuticals: An innovative trend in the food and supplement industries. Marine Drugs 13(10): 6336–6351.

Sultana, F., Wahab, M.A., Nahiduzzaman, M., Mohiuddin, M., Iqbal, M.Z., Shakil, A. et al. (2023). Seaweed farming for food and nutritional security, climate change mitigation and adaptation, and women empowerment: A review. Aquaculture and Fisheries 8: 463–480. https://doi.org/10.1016/j.aaf.2022.09.001.

Sunwoo, I.Y., Ra, C.H., Jeong, G.T. and Kim, S.K. (2016). Evaluation of ethanol production and bioadsorption of heavy metals by various red seaweeds. Bioprocess and Biosystems Engineering 39: 915–923.

Tarman, K., Sadi, U., Santoso, J. and Hardjito, L. (2020). Carrageenan and its enzymatic extraction. pp. 147–159. *In*: Encyclopedia of Marine Biotechnology. John Wiley & Sons, Ltd.

Thakur, S., Chaudhary, J., Sharma, B., Verma, A., Tamulevicius, S. and Thakur, V.K. (2018). Sustainability of bioplastics: Opportunities and challenges. Current Opinion in Green and Sustainable Chemistry 13: 68–75. https://doi.org/10.1016/j.cogsc.2018.04.013.

Toskas, G., Heinemann, S., Heinemann, C., Cherif, C., Hund, R.-D., Roussis, V. et al. (2012). Ulvan and ulvan/chitosan polyelectrolyte nanofibrous membranes as a potential substrate material for the cultivation of osteoblasts. Carbohydrate Polymers 89(3): 997–1002. https://doi.org/10.1016/j.carbpol.2012.04.045.

Uzoejinwa, B.B., Nwoke, O.A., Asoiro, F.U. and Ohagwu, C.J. (2024). Overview of biorefinery technology. pp. 157–190. *In*: Abomohra, A. and Ende, S. (eds.). Value-added Products from Algae: Phycochemical Production and Applications. Springer International Publishing, Cham.

Vaghela, P., Das, A.K., Trivedi, K., Anand, K.G.V., Shinde, P. and Ghosh, A. (2022). Characterization and metabolomics profiling of *Kappaphycus alvarezii* seaweed extract. Algal Research 66: 102774. https://doi.org/10.1016/j.algal.2022.102774.

Wan Yahaya, W.A., Azman, N.A.M., Adam, F., Subramaniam, S.D., Abd Hamid, K.H. and Almajano, M.P. (2023). Exploring the potential of seaweed derivatives for the development of biodegradable plastics: a comparative study. Polymers 15(13). https://doi.org/10.3390/polym15132884.

Wanyonyi, S., Du Preez, R., Brown, L., Paul, N.A. and Panchal, S.K. (2017). *Kappaphycus alvarezii* as a food supplement prevents diet-induced metabolic syndrome in rats. Nutrients 9(11): 1261.

Yeong, H.Y., Phang, S.M., Reddy, C.R.K. and Khalid, N. (2014). Production of clona planting materials from *Gracilaria changii* and *Kappaphycus alvarezii* through tissue culture and culture of *G. changii* explants in airlift photobioreactors. Journal of Applied Phycology 26: 729–746. https://doi.org/10.1007/s10811-013-0122-4.

Yew, Y.P., Shameli, K., Mohamad, S.E.B., Nagao, Y., Teow, S.Y., Lee, K.X. et al. (2019). Potential anticancer activity of protocatechuic acid loaded in montmorillonite/Fe_3O_4 nanocomposites stabilized by seaweed *Kappaphycus alvarezii*. International Journal of Pharmaceutics 572: 118743.

Zaimis, U., Ozolina, S. and Jurmalietis, R. (2021). Production of seaweed derived bioplastics. 20th International Scientific Conference Engineering for Rural Development. https://doi.org/10.22616/ERDev.2021.20.TF370.

Chapter 7

Ulvan from Green Seaweed as a Sustainable Source of Sulfated Polysaccharide for Biomedical Applications

Dina Fransiska,[1,#] *Bakti Berlyanto Sedayu,*[2,#] *Ellya Sinurat,*[1,#] *Endar Marraskuranto*[1,#] and *Hari Eko Irianto*[1,3,#,*]

Introduction

Natural compounds have served as the predominant reservoir of promising prospects for innovative medications. Sulfated polysaccharides, which constitute the primary component of seaweed cell wall matrices, are hydrocolloids that possess pharmacological and functional properties. They are devoid of toxicity and side effects, making them suitable for use in the food, cosmetic, and pharmaceutical industries (Peasura et al. 2016). Ulvan, a polysaccharide derived from green seaweeds, has potential applications in various fields. Green seaweed species belonging to the genus Ulva exhibit rapid growth rates and high productivity under various geo-climatic conditions, making them valuable due to their easily exploitable biochemical profiles (Angell et al. 2014). Figure 1 shows *Ulva lactuca* from Lombok island, Nusatenggara Barat, Indonesia. Ulvan is a water-soluble sulfated polysaccharide found in the cell wall of Ulva algae (Ning et al. 2022, Rosidah et al.

[1] Research Center for Marine and Land Bioindustry, National Research and Innovation Agency of the Republic of Indonesia, Lombok, Indonesia.
[2] Research Center for Food Technology and Processing, National Research and Innovation Agency of the Republic of Indonesia, Yogyakarta, Indonesia.
[3] Faculty of Food Technology and Health, Sahid University, Jakarta, Indonesia.
* Corresponding author: hariekoirianto1960@gmail.com
[#] Contributorship statement: Dina Fransiska, Bakti Berlyanto Sedayu, Ellya Sinurat, Endar Marraskuranto and Hari Eko Irianto have an equal contribution in the preparation of the manuscript, while Hari Eko Irianto is the corresponding author.

2020, Sun et al. 2020). It was found that the composition of ulvan mainly consists of rhamnose, uronic acid (glucuronic acid and iduronic acid), xylose, and a small amount of galactose. The monosaccharide components and degree of sulfation vary depending on ulva species, harvesting season, cultivation technique, growth conditions, and extraction method. Ulvan, which constitutes 9 to 36% of the dry weight of ulva biomass, is a polysaccharide present in the cell wall (Kidgell et al. 2019, Olasehinde et al. 2019).

Ulvan has been studied for its anticancer properties, showing cytotoxic effects on human cancer cell lines. Ulvan has also been found to have antioxidant and anti-inflammatory properties, making it a candidate for cancer therapy and immunomodulation (Pradhan et al. 2023). Ulvan-based silver nanoparticles have shown antimicrobial activity, making them potential agents for oral hygiene products (García-Márquez et al. 2023).

Ulvan can be extracted using various methods such as hot water, hot acidic and hot alkaline solvent extraction, microwave-assisted hot water extraction, autoclave extraction, and ultrasonic-assisted hot water extraction (Kazemi et al. 2023). Among these methods, ultrasonic-assisted hot acidic solvent extraction has been found to be the most efficient, with optimal conditions including a temperature of 86°C, a solvent-to-sample ratio of 34 ml g^{-1}, a pH of 4.5, and a power of 200 W (Barakat et al. 2022). Another method that has been used is microwave-assisted hydrothermal processing, which has shown promising results in the extraction of ulvan from *Ulva* spp. green seaweed (André et al. 2023). Additionally, a green strategy using citric acid as both an extractant and a cross-linking agent has been developed to

Figure 1. Cultivated *Ulva* from Lombok Island, Indonesia.

extract ulvan from foliose and filamentous *Ulva* sp. (Guidara et al. 2020). It has shown to have good hydrogel film dressing characteristics and can be incorporated with silver nanoparticles to enhance antibacterial properties and potential burn treatment (Sulastri et al. 2023). Ulvan has also been found to possess antioxidant and anticancer properties, making it a potential candidate for cancer therapy and immune system-boosting (Pradhan et al. 2023). Ulvan is composed of sulphated rhamnose, glucuronic acid, iduronic acid, and xylose, and its breakdown into oligomers or monomers can be facilitated by ulvan saccharifying enzymes. Ulvan has versatile biological actions, including antioxidant, immunomodulatory, antiviral, and antibacterial effects making it useful in the development of biomaterials for various applications (Anisha et al. 2023).

One study evaluated the antibacterial activity of an ulvan-based silver nanoparticle system, demonstrating its efficacy as an oral antimicrobial agent (Mohandoss et al. 2023). Another study extracted ulvan from *Ulva lactuca* and found it to have antimicrobial activity against fish and human pathogenic microbes, as well as to demonstrate anti-fouling properties (Ibrahim et al. 2022), anticancer effects, immunomodulatory activity, and pharmacokinetic properties (Pradhan et al. 2023). Another study of ulvan showed its inhibitory effect on α-synuclein fibrillogenesis and its potential as a novel candidate drug for the treatment of Parkinson's disease (Wang et al. 2022).

Ulvan has also shown inhibitory effects on the aggregation of the amyloid-β protein, which is associated with Alzheimer's disease, and has the potential as a functional food ingredient for treating AD (Liu et al. 2018). Additionally, research has focused on ulvan saccharifying enzymes, such as ulvan lyases, which are involved in breaking down ulvan into oligomers or monomers for biobased economy applications. Furthermore, ulvan-enriched diets have been studied for their effects on growth, antioxidant systems, and immune defense systems in fish, showing positive results against *Flavobacterium columnaris* (Harikrishnan et al. 2021).

The potential of the sea is an important asset that requires wise and sustainable management to support the sustainability of marine ecosystems and human development. By utilizing the potential of the sea responsibly, humans can obtain long-term benefits from these rich and diverse natural resources.

Research bibliometrics of Ulvan

The 1000 samples taken from the Google Scholar website using the keyword "Ulvan" and then processed using the VOSviewer software can be seen in Figure 2. It is seen that during the last ten years, the development of research on ulvan in the world focuses on polysaccharide, biorefinery and the aspects of the applications. The application stated in this bibliometric analysis is for antioxidant, tissue engineering, antimicrobial, and drug delivery.

There are 7 clusters of ulvan research.

Cluster 1 is antimicrobial, antioxidant, biomaterial, biopolymers, chitosan, enzyme, macroalga, marine alga, ulvan, seaweed, tissue engineering and wound dressing.

Cluster 2 is anticoagulant activity, antimicrobial activity, cytotoxic activity, oligosaccharides, polysaccharide lyase, sulfated polysaccharide, and ulvan lyase.

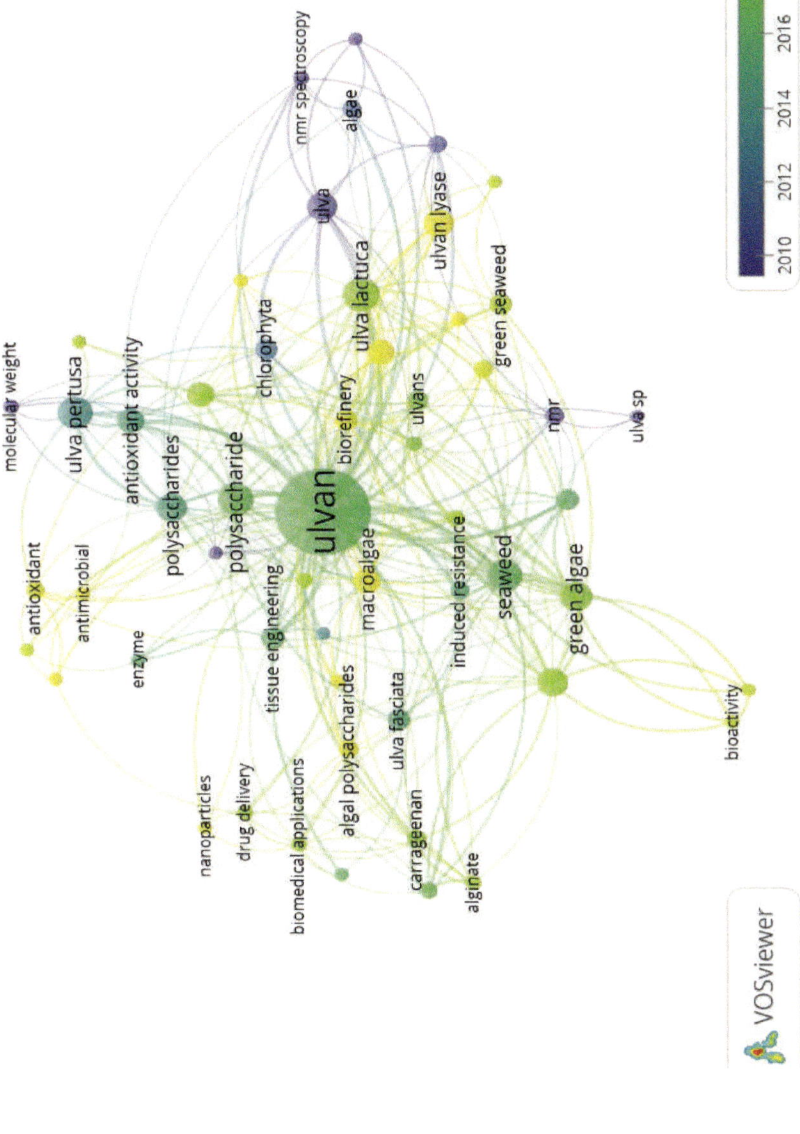

Figure 2. The 'ulvan' as a keyword in bibliometric analysis in VOSviewer.

Cluster 3 is algal polysaccharides, alginate, biomedical application, drug delivery, induced reference, nanoparticles, and sulfated polysaccharides.

Cluster 4 is biorefinery, NMR spectroscopy.

Cluster 5 is antioxidant activity, molecular weight, oxidative stress, and polysaccharide.

Cluster 6 is bioactivity, biotechnology, green algae, and sulfated polysaccharide.

Cluster 7 is extraction, NMR and *Ulva* sp.

However, the extensive use and development of ulvan resources face several obstacles due to factors such as its large molecular weight, limited solubility, and low bioavailability. However, ulvan oligosaccharides, derived from ulvan degradation, not only retain the bioactivity of the original polysaccharides effectively but also overcome the drawbacks of ulvan insolubility in water and low availability. This has led to increasing interest in the preparation and exploration of the biological activity of ulvan oligosaccharides in the field of marine biological resources development and research (Ning et al. 2022, Paulert et al. 2021).

Extraction and purification of ulvan

Extraction

The extraction process involves various factors such as pH, temperature and mechanical agitation, which affect the amount of material extracted and its metal content. Acid extraction leads to almost complete desulfation of the isolated ulvan, whereas purification by means of enzymatic methods retains significant SO_3 substituents (Tsubaki et al. 2016).

Before and after polysaccharide extraction, it is important to consider the presence of interfering substances that may be co-extracted with the target molecule, as well as other contaminants that enter during the extraction process. These substances and contaminants could potentially affect the purity of the extract or interfere with the biological response to the polysaccharide (Alves et al. 2013). The chemical makeup of ulvans is influenced by certain factors such as the *Ulva* sp. used as the extraction source, the harvesting season, and the methods used for extraction and purification. As a result, its composition becomes more complicated.

The main extraction method used is solvent extraction, along with two additional extraction techniques derived from the solvent extraction method, namely physical-assisted extraction and enzyme-assisted extraction. Solution extraction can be categorized into various subtypes, including hot water extraction, pressurized hot water extraction, acid extraction, and weak base extraction.

The quantitative yield and quality of ulvan are subject to considerable variation, mainly influenced by the extraction and purification methods applied, the biomass source (including species, whether derived from natural or cultivated feedstock, and location), storage conditions of the collected biomass, and pre-extraction processing. As a result, there is a wide range of extraction conditions documented in literature, leading to variations in the yield and quality of the resulting ulvans (Li et al. 2023).

Typically, there are two extraction methods: acid and enzyme. The amount and quality of extracted ulvan is shaped by the extraction process and the choice of solvent for extraction. The extraction procedure involves various parameters such as pH, temperature, and mechanical agitation, which determine the amount of extract obtained and its metal content (Tsubaki et al. 2016). Acid extraction leads to almost complete desulfation of the isolated ulvan, whereas purification via enzymatic methods preserves important levels of SO_3 substituents (Wahlström et al. 2020).

In recent years, various extraction methods have been investigated for several ulva species. For example, to improve the purity of ulvan, *U. ohnoi* was extracted at 90°C with pH 2.2 (Glasson et al. 2019). In addition, ulvan from *U. fasciata* underwent two extractions using distilled water, followed by a 72-hour oxidation process (de Carvalho et al. 2020). To increase the quantity of ulvan, *U. pertusa* underwent hot water extraction, followed by alcohol precipitation. Recent techniques involve a one-step freeze-dried extraction method (Li et al. 2020, Ponce et al. 2020).

For enzyme-based extraction of ulvan from *U. prolifera*, ulvan lyase obtained from the fermentation of *Catenovulum* sp. was used to catalyze the initial depolymerization step of ulvan. The choice of extraction method generally depends on the physicochemical properties of ulvan, which can interact with the cell wall components upon contact with the solvent (Fabrowska et al. 2016, Gajaria et al. 2017).

The extraction process yields various active components from ulvan, including polysaccharides (such as starch, cellulose, xyloglucan, and glucuronan). Co-extraction allows the retrieval of other components such as proteins, lipids, minerals, and secondary metabolites. This review specifically focuses on ulvan extraction from *U. lactuca*, emphasizing on methods that offer high yields, selectivity, and minimal degradation (Kidgell et al. 2019).

The compilation of ulvan extraction techniques includes: hot water extraction (at temperatures below 100°C), pressurized hot water extraction, acid extraction, weak alkaline extraction, extraction using enzymes, ultrasonic-assisted extraction, microwave-assisted extraction, enzymatic and chemical extraction, and ultrasonic and enzyme-assisted extraction (Li et al. 2023).

Enzyme extraction involves suspending onozula, macerozymes pectinase, α-amylase, and proteinase K in a buffer. Enzyme extraction yields are in the range of 15% to 47% (Guidara et al. 2020, Wahlström et al. 2020). There is still room to improve these yields as no optimization has been done. Besides the three commonly used methods, alternative approaches include multistage extraction using water, followed by $NaCO_3$ and NaOH solvents, obtaining yields in the range of 3.14% and 6.50%. Another method is a sequential extraction involving acid and ammonium oxalate solutions, which yielded average ulvan yields (Mhatre et al. 2019).

The most commonly used method is extraction using pure distilled water (He et al. 2016, Robic et al. 2009). This process yields ulvan consisting of glucuronic acid, glucose, arabinose, and xylose, with average extraction yields ranging from 25% to 40% (Li et al. 2023).

A higher extraction temperature facilitates enhanced dissolution of ulvan, while a lower pH value increases the selectivity of ulvan. In addition, extending the extraction duration contributes to higher ulvan yields. However, caution is essential to maintain the integrity of the ulvan structure. Factors such as high temperature, low pH, and long extraction time can compromise stability. Time duration and lower pH levels result in significant depolymerization of ulvan. Therefore, based on available literature, the optimal extraction conditions to achieve high yields, selectivity, and minimal degradation involve temperatures between 80°C and 90°C, pH levels ranging from 2 to 4.5, and durations of 1 to 3 hours (Mo'o et al. 2020).

In another investigation, it was reported that films using *U. lactuca* polymers remained stable in the temperature range of –50 to 250°C (Guidara et al. 2020). The acid extraction yield decreased when maintaining a constant pH and temperature. In contrast, the yield tends to increase with higher extraction pH under constant temperature conditions (Yaich et al. 2015). The polysaccharides present in ulvan form strong ligand bonds, a characteristic that may contribute to potential therapeutic effects (El Azm et al. 2019). Table 1 summarizes various methods of ulvan extraction,

Table 1. Various methods for ulvan extraction.

Methods	Conditions	Yield of Ulvan	Reference
Microwave-assisted extraction	temperature 160°C	14.33%	André et al. 2023
	300 W temperature 80°C 20 ml g^{-1} 30 min	20.41%	Kazemi et al. 2023
Autoclave	1.2 bar temperature 121°C 20 ml g^{-1} 30 min	17.89%	Kazemi et al. 2023
Hot water	temperature 85–90°C in a closed reflux system for 20 h	43.66%	Barakat et al. 2022
	pH 2.5 temperature 90°C 90 min	73.2%	Glasson et al. 2019
	temperature at 80°C 20 ml g^{-1}, 3 h	17.76%	Kazemi et al. 2023
Enzyme-based extraction	hot water solution and cellulose	17.95%	Guidara et al. 2020
Acid extraction	temperature 80°C and 90°C pH 2 and pH 1.5	3.4–14.22%	Guidara et al. 2020
	temperature 80°C pH 3 20 ml g^{-1} 3 h	23.21%	Kazemi et al. 2023
Alkaline extraction	pH 10 temperature 80°C 20 ml g^{-1} 3 h	16.11%	Kazemi et al. 2023
Ultrasonic-assisted hot acidic solvent extraction	150 W temperature 80°C 20 ml g^{-1} 30 min	22.73	Kazemi et al. 2023

highlighting the techniques employed, extraction conditions, and key findings, providing a comprehensive comparison of approaches to optimize yield.

The optimal conditions for ulvan extraction are determined by three criteria: (1) achieving high yields (2) ensuring high selectivity and (3) minimizing degradation; meaning ulvan is extracted without extensive hydrolysis. Based on these criteria and available literature data, the recommended extraction conditions are as follows to achieve high yield, selectivity, and low degradation such as temperature: 80–90°C, pH range: 2–4.5 and duration: 1–3 hours.

The methods to isolate and purify ulvan, aim to achieve two desired outcomes: (1) high recovery and (2) a pure product. Chemical and enzymatic methods for polysaccharide isolation and purification have been extensively explored and reviewed (Kidgell et al. 2019).

After ulvan extraction, the extract is separated from residual biomass through filtration and/or centrifugation. Subsequently, ulvan isolation is performed using methods such as precipitation, evaporation, or concentration via ultrafiltration, followed by lyophilization. Spray drying is another efficient process used in industrial settings (Barbosa et al. 2015). In a laboratory setting, ethanol precipitation is the primary method for isolating ulvan, typically, adding 70–96% ethanol content to concentrated extracts to precipitate the ulvan, leaving ethanol-soluble compounds such as pigments and metabolites in the solution (Dürig et al. 1997). Although ethanol precipitation is easy to perform, this method is constrained by the low solubility of salts and the higher solubility of low molecular weight ulvans in ethanol-water mixtures. The precipitation method can be challenging to replicate, and in the absence of dialysis or ultrafiltration, this method can lead to the increased salt content in the ulvan extract.

Fundamentally, the choice of extraction technique should ideally aim to limit the co-extraction of macromolecular impurities and minimize the need for intensive purification techniques. However, due to the salt content carried by the extractor and attached to marine species, as well as the similar physical properties of ulvans and contaminating polysaccharides and proteins, purification becomes a necessary step (Yaich et al. 2013).

Purification

Initial purification steps may include the removal of photosynthetic pigments (decolorization) and lipids (degreasing). Various methods, such as supercritical extraction, soxhlet extraction, or simple immersion in organic solvents can be used to achieve decolorization and lipid removal (Alves et al. 2013).

Effective desalting of ulvan extracts can be done through dialysis, ultrafiltration, or chromatography. Dialysis and ultrafiltration facilitate the reduction or removal of excess salts and small molecules that may contaminate the ulvan extract. Dialysis is very efficient in removing excess salts from analytical samples and other applications that require very low residual ash. Ultrafiltration is usually a two-step process involving concentration (volume reduction) followed by diafiltration (washing or buffer exchange). Both dialysis tubes and membrane filters for ultrafiltration are available in a wide range of pore sizes or molecular weight cut-offs (MWCO). The

MWCO range for ulvans is generally between 3.6 and 12 kDa (August et al. 2005, Siddhanta et al. 2001). In dialysis, pore size selection is based on ulvan retention, whereas for ultrafiltration, pore size selection considers ulvan retention and permeate flow rate, which decreases with decreasing pore size. The optimal balance between ulvan retention and time efficiency is usually achieved at around 10 kDa MWCO (Glasson et al. 2019, Kidgell et al. 2019, Robic et al. 2009).

Considering the substantial impurities present in many ulvan extracts, it is somewhat unexpected that chromatographic techniques are not more commonly used for purification prior to investigating the physicochemical properties and biological activities of ulvan. The poly-anionic and polydisperse characteristics of ulvan make it particularly suitable for anion exchange chromatography (AEC) and size exclusion chromatography (SEC). AEC is effective in removing protein and neutral polysaccharide impurities from ulvan samples. Both weak anion exchangers (such as diethyl aminoethyl (DEAE) or media functionalized with tertiary amines) (Siddhanta et al. 2001) and strong anion exchangers (such as media functionalized with quaternary amines (Q)) (Robic et al. 2009) have shown to successfully purify ulvan. SEC is commonly used as an analytical technique to determine molecular weight and molecular weight distribution.

Nevertheless, SEC has wider applications. For example, SEC can be used to fractionate oligosaccharides and polysaccharides, assisting in elucidating structures more clearly (Robic and Lahaye 2007) and generating size classes for structure-function relationships (Dumas et al. 2010, Jingjing et al. 2009). In the context of the former, SEC has been used to fractionate ulvan oligosaccharides for clearer structure analysis, in particular using NMR techniques (Lahaye 1998). Various size exclusion media are available, targeting different molecular size ranges. Commonly used media to purify ulvan samples include Sepharose CL-6B (Dumas et al. 2010), Sephacryl S-400/HR (Wang et al. 2008), (Ghosh et al. 2004) Sephacryl S-1000 and Sephadex G-200 (Raymond and Simard 2006), with purification often performed via QFF columns (Chi Li et al. 2020).

Ultrafiltration can be used to remove salts and further improve the ulvan purity of *U. ohnoi* (Glasson et al. 2019).

Typically, the main impurities in this context are salts, and their optimal removal involves the use of dialysis or ultrafiltration methods. Depending on the extraction technique chosen, protein removal may also be required, which can be achieved through chemical or enzymatic means or by using chromatographic methods (Shi 2016). The latter approach is more consistent and facilitates the acquisition of highly pure ulvan fractions. Practical decision-making in the selection of extraction, isolation, and purification procedures should be strategic and consider the available resources. With limited resources, one may choose to employ selective acid extraction, concentrate via evaporation, purify through dialysis, and isolate by precipitation. Conversely, in situations with sufficient resources, selective acid extraction is preferred, and less selective extraction processes may be acceptable, followed by concentration and diafiltration using ultrafiltration. Additional purification steps, such as anion exchange chromatography, together with subsequent lyophilization or spray drying, may then be used to isolate the purified ulvan.

Applications of ulvan

Within the seaweed Ulva genus, around fifty species have been identified. Extensive investigations have been carried out on ulvan compounds obtained through the extraction of various *Ulva* sp. species from diverse regions worldwide. The results convincingly demonstrate that these compounds possess numerous functional properties and offer a range of benefits. Subsequently, simpler derivative compounds, oligosaccharides, not only effectively preserve the biological activity of the original polysaccharide but also enhance the solubility of ulvan in water and its bioavailability. Overcoming the limitations that previously restricted broader applications, this development has made the investigation to attain ulvan oligosaccharides one of the most sought-after studies focusing on the present, particularly in the utilization of marine biodiversity.

Based on research reports, ulvan is primarily known for its biological activities such as anti-inflammatory, anticoagulant, anti-tumor, antioxidant, immune regulation, and others. Therefore, ulvan is widely applied in the biomedical and functional food fields. However, with the hydrogel nature of its polysaccharide structure, the potential utilization of ulvan extends even further, encompassing areas of polymer materials and energy (biofuel). Here are the potential applications of ulvan in various industrial sectors.

Ulvan in biomedical applications

Due to its diverse biological activities, ulvan is recognized as highly promising for therapeutic applications in human health (Table 2). Numerous studies have demonstrated the various biological activities of ulvan extracted from various green seaweeds, including *U. fasciata, U. lactuca, U. intestinalis, U. pertusa, U. ohnoi*, and *Ulothrix flacca*. It was revealed that the biological activity of ulvan depends on the molecule size of ulvan. Because of their small molecular size, low-molecular-weight polysaccharides are easier to dissolve, absorb, and make bioavailable than their high-molecular-weight counterparts (Chi, Zhang et al. 2020, Flórez-Fernández et al. 2023, Kidgell et al. 2020, Li et al. 2020, Li et al. 2018a, Tabarsa et al. 2018, Zhao et al. 2020). This allows for more effective biological activity. Furthermore, the efficacy of their biological functions is also determined by the degree of sulfation (Kidgell et al. 2019, Qi and Sun 2015). Other research also revealed that ulvan derivatives and/or fractions showed biological activities, such as anti-coagulant (Adrien et al. 2019, de Carvalho et al. 2018) and immunostimulatory effect (Tabarsa et al. 2018).

Recently it has been seen that the mechanisms underlying ulvan's ability to inhibit or impede virus replication involve interactions with viral envelope glycoproteins or binding to cell surface receptors. Additionally, it has been observed that the antiviral effectiveness of ulvan is influenced by the molecular weight of the ulvan; a higher molecular weight correlates with greater antiviral activity. Notably, ulvan has shown promise in protecting cells from infection from the deadly SARS-CoV-2 virus. This protective effect is attributed to the interaction of the negatively charged groups of ulvan with proteins on the virus's surface, as reported by Shefer et al. (2021) and Li

Table 2. Biological activities of ulvan from various green seaweeds.

No.	Species	Bioactivity	Reference
1	*Ulva fasciata* (now *U. lactuca*)	Anti-coagulant	de Carvalho et al. 2018
		Antimicrobial; antioxidant; anti-arthritic	Barakat et al. 2022
2	*Ulva lactuca*	Antifungal; antioxidant; anticancer; anti-coagulant	Adrien et al. 2019, El Fayoumy et al. 2022, Pradhan et al. 2023, Yaich et al. 2017
		Anti-coagulant; Anti-viral; Anti-hyperlipidemic; Anti-peroxidative; anti-hyperlipidemic; anti-oxidant; immunostimulating; anti-proliferative	Alves et al. 2013
		Cytotoxic activity	Thanh et al. 2016
		Antioxidant; antimicrobial	Amin 2020
		Antitumor; immunoregulatory effect	Zhao et al. 2020
		Antitumor;anticoagulant	Guidara et al. 2021
		Antiviral; antimicrobial; antioxidant; anticancer	Maray et al. 2023
		Anticoagulant; antithrombotic	Reis et al. 2020
3	*Ulva* sp.	Anti-inflammatory	Flórez-Fernández et al. 2023
4	*Ulva intestinalis* (formerly *Enteromorpha intestinalis*)	Anticancer	Pradhan et al. 2023
		Antioxidant; anti-bacterial; anti-proliferative effect	Kazemi et al. 2023
		Immunostimulatory effect	Tabarsa et al. 2018
5	*Ulva ohnoi*	Inflammatory effect	Kidgell et al. 2020
6	*Ulva pertusa*	Antiviral	Chi et al. 2020
		Antioxidant; anti-hyperlipidemic	Li et al. (2020), Li et al. (2018b), Li et al. (2018c)
7	*Ulothrix flacca*	Anti-coagulant; immunomodulating activity	Li et al. (2018a)

et al. (2023). The effectiveness of ulvan in suppressing virus replication has been investigated across various virus types, particularly those linked to pandemics and those posing threats to both human and livestock health. The viruses tested include herpes simplex virus (HSV), dengue virus (DENV), yellow fever virus (YFV), West Nile virus (WNV), influenza (H1N1), avian influenza virus (AIV), and measles virus (MeV).

Ulvan, which is the main active constituent in *Ulva* sp. has bioactivities, including anti-inflammatory properties (Kidgell et al. 2020); anti-tumor, anticoagulant (Guidara et al. 2021); antimicrobial, antiviral, antioxidant and anticancer properties (Maray et al. 2023).

Ulvan has been recognized for its anti-inflammatory benefits, whereas the body physiologically responds to damage or the presence of external factors. This response is ultimately aimed at restoring the structure and function of body tissues. The anti-inflammatory properties are proven by *in vitro* tests on various experimental animals such as mice, chicken, and Wistar rats. In this context, inflammation typically relies on cytokine production and is primarily a consequence of the activation of NF-kB, a protein complex regulating transcription, through pathogen-associated molecular patterns (PAMPs) on toll-like receptors (TLRs). Immune cells, particularly macrophages, initiate a pro-inflammatory response by releasing cytokines, prostaglandin E2 (PGE2), and tumor necrosis factor-alpha. On the other hand, anti-inflammatory cytokines play a role in preventing damage from excessive macrophage activation by inhibiting the release of pro-inflammatory cytokines (Kidgell et al. 2019).

Ulvan also exhibits a substantial role in regulating the immune system, functioning as an immunomodulator alongside its anti-inflammatory properties. This modulation occurs through the activation of macrophages, as previously mentioned. The increased functions of macrophages, collectively referred to as macrophage activation, encompass improvements in morphology, metabolism, and functionality to efficiently eliminate infectious agents within the body. Seaweed polysaccharides or glycoproteins may activate macrophages by binding to specific receptors including TLR4, CD14, complement receptor 3 (CR3), scavenger receptor (SR), dectin-1, and mannose receptor (MR). In addition to the macrophage activation mechanism, the inflammatory function of ulvan may also involve the activation mechanism of lymphocytes (Erniati and Ezraneti 2020).

Moreover, anti-radical ability from the ulvan was also observed through inhibiting lipid peroxidation and boosting the activity of antioxidant enzymes (Pradhan et al. 2023). Antioxidants, responsible for scavenging free radicals, are renowned for their significant roles in safeguarding against diseases induced by reactive oxygen species (ROS). By neutralizing these harmful molecules, antioxidants contribute to maintaining cellular health and preventing oxidative stress-related disorders (Qi and Sun 2015). Previous findings suggest that the potent antioxidant activity of ulvan is associated to the degree of sulfate group substitution along its polymeric backbone (Pradhan et al. 2023). Several studies have also demonstrated the anticoagulant activity of ulvan. For example, investigations by de Carvalho et al. 2020 on the *U. fasciata* genus by using the activated partial thromboplastin time (APTT) assay, showed that the anticoagulant efficacy of polycarboxyl ulvans depends on a combination of high carboxyl and sulfate content with low molecular weight. Additionally, *U. lactuca* underwent *in vitro* testing, including observations on prothrombin time (PT) and thrombin time (TT), indicating a reduction in thrombus weight in mice associated with anti-inhibiting anti-factor Xa and IIa. These findings provide strong evidence regarding the anticoagulant potential of sulfated polysaccharides isolated from *U. lactuca* (Reis et al. 2020).

In addition to its diverse range of bioactive compounds, Ulva polysaccharide is rich in both soluble and insoluble dietary fiber, with a notable prevalence of the soluble one (SDF). The SDF contributes to the host's health by boosting gastrointestinal immunity, lowering blood glucose and lipids, and reducing the risks of colorectal cancer and cardiovascular issues. In contrast, insoluble dietary fiber (IDF) is associated with the excretion of bile acids, a reduction in the passage time through the gastrointestinal tract, and the enhancement of fecal bulk. The binding capacity of both SDF and IDF with organic and inorganic complexes, as well as their breakdown and fermentation by intestinal microflora, holds additional significance (Yu-Qing et al. 2016). This puts Ulva as a potential alternative healthy food for future development.

Ulvan as biomaterials in tissue engineering

Some *Ulva* species feature a three-dimensional porous structure based on ulvan, making them a promising option for medical applications, specifically in tissue engineering. Additionally, ulvan has the capacity to form a polymer hydrogel, providing valuable applications in the valorization for cartilage tissue engineering (Mo'o et al. 2020). The research conducted by Barros et al. 2013 involved the combination of ulvan and chitosan, resulting in the formation of carboxymethylated ulvan (CMU) and carboxymethyl chitosan polymers. These compounds exhibit non-cytotoxic properties, making them suitable candidates for thorough in vitro and *in vivo* evaluations in tissue engineering. Additionally, Alves et al. 2012 has also reported that the combination of ulvan with poly(D-lactic acid) has demonstrated promise in producing innovative scaffolds intended for applications in bone tissue engineering.

Limitations of ulvan technology

Ulvan has significant potential for a wide variety of applications, but there are some limitations and challenges. These limitations can impact the scalability, efficiency, and practical implementation of Ulvan-based solutions. Some of these limitations are:

1. *Availability of Ulva feedstock*: To date, commercial-scale cultivation of *Ulva* sp. grass does not exist. *Ulva* sp. cultivation is still laboratory-scale and comes from nature. This slows down the commercialization of Ulvan production to an industrial scale and the inconsistency of Ulvan products produced.
2. *Extraction Efficiency*: Ulvan extraction from macroalgae can be a labor-intensive and expensive process. Current extraction methods may not yet produce large amounts of Ulvan efficiently, limiting its widespread adoption.
3. *Purity and Quality Control*: Ensuring the purity and quality of the extracted Ulvan is critical for its application in various industries. However, maintaining consistent quality standards poses a challenge, mainly due to variations in ulva species, growth conditions, and extraction techniques.
4. *Structural Complexity*: The complex and heterogeneous structure of ulvans presents challenges in fully understanding their properties and functions. This complexity can hinder the development of standardized processing methods and customized formulations for specific applications.

5. *Biological activity and health benefits*: Knowledge of the biological activity and other health benefits of ulvans is still relatively limited, especially about the detailed mechanism of action using animal models. Research related to the mechanism of action using experimental animals may be a research priority in the future. Similarly, detailed clinical studies are also needed regarding pharmacokinetics and safety-related pharmacokinetics and safety along with health-beneficial activities.
6. *Stability and Shelf Life*: The susceptibility of ulvan to degradation under certain environmental conditions, such as pH variation and enzymatic activity raises concerns about its stability and shelf life in practical applications. Developing strategies to improve ulvan's stability without compromising its bioactivity is critical for long-term storage and commercial viability.
7. *Scale-up Challenges*: The transition from laboratory-scale production to industrial-scale production of ulvan-based products involves significant challenges, including optimizing the extraction process, ensuring consistent product quality, and managing production costs. Scalability issues may hinder the widespread adoption of Ulvan technology in various industries.
8. *Regulatory Hurdles*: The regulatory landscape governing the use of ulvan in food, pharmaceutical, and cosmetic products varies across regions. Navigating regulatory requirements and obtaining necessary approvals for ulvan-based products can be a time-consuming and complicated process, which impacts market access and commercialization efforts.

To overcome these limitations, collaborative efforts from researchers, industry stakeholders, and policymakers are needed to overcome technical, economic, and regulatory barriers. Future research efforts should focus on developing innovative extraction techniques, refining purification methods, elucidating structure-function relationships, and optimizing processing protocols to unlock the full potential of ulvan technology.

Challenges towards commercialization

Seaweed has great potential beyond its use as food, with increasing interest in its role in advancing a bio-based economy. However, merely producing seaweed-based products in small-scale laboratory settings and comparing them to existing standards is insufficient for understanding the barriers to commercialization. A comprehensive approach is needed to address scalability, cost-effectiveness, market demand, regulations, and sustainability, paving the way for the successful development of economically viable seaweed-derived bioproducts. Moreover, more companies are recently looking to invest in the seaweed economy, acknowledging the value of adopting biology-based management approaches. This involves utilizing eco-friendly techniques for collecting and processing seaweeds in alignment with their life cycles and bioavailability.

Numerous factors must be carefully considered to ensure the successful commercial development of seaweed for biomedical or cosmetic applications. This includes securing a sustainable supply of raw materials, particularly by enhancing seaweed cultivation methods to optimize ulvan production. By cultivating seaweed varieties resilient to natural threats like diseases or pests, a reliable raw material source can be established to meet industrial demands. Additionally, it's crucial to implement efficient post-harvest handling techniques to preserve the functional substances and bioactive properties of seaweed-derived compounds throughout the processing and extraction processes.

Another important consideration is the potential alteration of seaweed compounds during the extraction and purification phases. Thus, there's a critical need to prioritize biomedical safety and security. Understanding how these processes affect the chemical structure of seaweed components, including factors like molecular weight, bioactivity, and toxicity, is essential for developing new seaweed-based drugs (Cotas et al. 2024). Therefore, encouraging rigorous analysis of the final product and the establishment of a certification system to endorse seaweed compounds as viable drugs is an obligation. Developing new biomedical products is a highly regulated and complex process that relies on the expertise and advanced research skills of the medical community. Clinical trials involving human and animal participants require meticulous care and adherence to strict laws and safety measures. The ultimate goal of drug development is to minimize suffering and introduce innovative treatments to improve human health and well-being.

Currently, there are limited commercial seaweed products available worldwide, highlighting the importance of fostering collaboration between research and industry for their broader use. While utilizing seaweed compounds for biomedical purposes poses significant challenges, it remains an intriguing avenue of exploration. Despite these hurdles, research into seaweed compounds holds promise due to their potential therapeutic benefits and the opportunity for creating new products. However, fully realizing the medicinal potential of seaweed compounds requires further research efforts to overcome these challenges.

Acknowledgement

This review is one of the outputs of Group 3 activities - Satreps Project, a research collaboration between the University of Tsukuba, Japan and the National Research and Innovation Agency (BRIN) of the Republic of Indonesia. The authors express their gratitude for the support for the project.

References

Adrien, A., Bonnet, A., Dufour, D., Baudouin, S., Maugard, T. and Bridiau, N. (2019). Anticoagulant activity of sulfated ulvan isolated from the green macroalga ulva rigida. Marine Drugs 17(5). https://doi.org/10.3390/md17050291.

Alves, A., Duarte, A.R.C., Mano, J.F., Sousa, R.A. and Reis, R.L. (2012). PDLLA enriched with ulvan particles as a novel 3D porous scaffold targeted for bone engineering. The Journal of Supercritical Fluids 65: 32–38.

Alves, A., Sousa, R.A. and Reis, R.L. (2013). A practical perspective on ulvan extracted from green algae. Journal of Applied Phycology 25(2): 407–424. https://doi.org/10.1007/s10811-012-9875-4.

Amin, H.H. (2020). Ulvan as a new trend in agriculture, food processing and medicine. Asian Journal of Fisheries and Aquatic Research 6(4): 47–54. https://doi.org/10.9734/ajfar/2020/v6i430105.

André, J., Flórez-Fernández, N., Domínguez, H. and Torres, M.D. (2023). Microwave-assisted extraction of *Ulva* spp. including a stage of selective coagulation of ulvan stimulated by a bio-ionic liquid. International Journal of Biological Macromolecules 225: 952–963. https://doi.org/10.1016/J.IJBIOMAC.2022.11.158.

Angell, A.R., Mata, L., de Nys, R. and Paul, N.A. (2014). Variation in amino acid content and its relationship to nitrogen content and growth rate in Ulva ohnoi (Chlorophyta). Journal of Phycology 50(1): 216–226. https://doi.org/https://doi.org/10.1111/jpy.12154.

Anisha, G.S., Augustianath, T., Padmakumari, S., Singhania, R.R., Pandey, A. and Patel, A.K. (2023). Ulvan from green macroalgae: Bioactive properties advancing tissue engineering, drug delivery systems, food industry, agriculture and water treatment. *In*: Bioresource Technology Reports (Vol. 22). Elsevier Ltd. https://doi.org/10.1016/j.biteb.2023.101457.

August, P., Liang, Y., Huang, J., Song, X., Zhang, P. and Xu, H. (2005). Four viral proteins of white spot syndrome virus (WSSV) that attach to shrimp cell membranes. 66: 81–85.

Barakat, K.M., Ismail, M.M., Abou El Hassayeb, H.E., El Sersy, N.A. and Elshobary, M.E. (2022). Chemical characterization and biological activities of ulvan extracted from Ulva fasciata (Chlorophyta). RendicontiLincei 33(4): 829–841. https://doi.org/10.1007/s12210-022-01103-7.

Barbosa, J., Borges, S., Amorim, M., Pereira, M.J., Oliveira, A., Pintado, M.E. et al. (2015). Comparison of spray drying, freeze drying and convective hot air drying for the production of a probiotic orange powder. Journal of Functional Foods 17: 340–351. https://doi.org/10.1016/j.jff.2015.06.001.

Barros, A.A.A., Alves, A., Nunes, C., Coimbra, M.A., Pires, R.A. and Reis, R.L. (2013). Carboxymethylation of ulvan and chitosan and their use as polymeric components of bone cements. Acta Biomaterialia 9: 9086–9097.

Chi, Y., Li, H., Wang, P., Du, C., Ye, H., Zuo, S. et al. (2020). Structural characterization of ulvan extracted from Ulva clathrata assisted by an ulvan lyase. Carbohydrate Polymers 229. https://doi.org/10.1016/j.carbpol.2019.115497.

Cotas, J., Lomartire, S., Gonçalves, A.M.M. and Pereira, L. (2024). From ocean to medicine: harnessing seaweed's potential for drug development. International Journal of Molecular Sciences 25: 797.

Chi, Y., Zhang, M., Wang, X., Fu, X., Guan, H. and Wang, P. (2020). Ulvan lyase assisted structural characterization of ulvan from Ulva pertusa and its antiviral activity against vesicular stomatitis virus. International Journal of Biological Macromolecules 157: 75–82. https://doi.org/10.1016/j.ijbiomac.2020.04.187.

de Carvalho, M.M., de Freitas, R.A., Ducatti, D.R.B., Ferreira, L.G., Gonçalves, A.G., Colodi, F.G. et al. (2018). Modification of ulvans via periodate-chlorite oxidation: Chemical characterization and anticoagulant activity. Carbohydrate Polymers 197: 631–640. https://doi.org/10.1016/j.carbpol.2018.06.041.

de Carvalho, M.M., Noseda, M.D., Dallagnol, J.C.C., Ferreira, L.G., Ducatti, D.R.B., Gonçalves, A.G. et al. (2020). Conformational analysis of ulvans from Ulva fasciata and their anticoagulant polycarboxylic derivatives. International Journal of Biological Macromolecules 162: 599–608. https://doi.org/10.1016/j.ijbiomac.2020.06.146.

Dumas, B., Jaulneau, V., Lafitte, C., Jacquet, C., Fournier, S., Salamagne, S. et al. (2010). Ulvan, a sulfated polysaccharide from green algae, activates plant immunity through the jasmonic acid signaling pathway. Journal of Biomedicine and Biotechnology 2010. https://doi.org/10.1155/2010/525291.

Dürig, J., Bruhn, T., Zurborn, K.H., Gutensohn, K., Bruhn, H.D. and Béress, L. (1997). Anticoagulant fucoidan fractions from Fucus vesiculosus induce platelet activation *in vitro*. Thrombosis Research 85(6): 479–491. https://doi.org/10.1016/S0049-3848(97)00037-6.

El Azm, N.A., Fleita, D., Rifaat, D., Mpingirika, E.Z., Amleh, A. and El-Sayed, M.M.H. (2019). Production of bioactive compounds from the sulfated polysaccharides extracts of ulvalactuca: post-extraction enzymatic hydrolysis followed by ion-exchange chromatographic fractionation. Molecules 24(11). https://doi.org/10.3390/molecules24112132.

El Fayoumy, R.A., El-Sheekh, M.M. and Abu Ahmed, S.E. (2022). Potential of ulvan polysaccharide from *Ulva lactuca* as antifungal against some foodborne fungi isolated from spoiled tomato sauce cans.

Journal of Aquatic Food Product Technology 31(7): 658–671. https://doi.org/10.1080/10498850.2022.2093149.

Erniati, E. and Ezraneti, R. (2020). Aktivitasimunomodulatorekstrakrumputlaut. Acta Aquatica: Aquatic Sciences Journal 7(2): 79–86. https://doi.org/10.29103/aa.v7i2.2463.

Fabrowska, J., Ibañez, E., Łęska, B. and Herrero, M. (2016). Supercritical fluid extraction as a tool to valorize underexploited freshwater green algae. Algal Research 19: 237–245. https://doi.org/10.1016/j.algal.2016.09.008.

Flórez-Fernández, N., Rodríguez-Coello, A., Latire, T., Bourgougnon, N., Torres, M.D., Buján, M. et al. (2023). Anti-inflammatory potential of ulvan. International Journal of Biological Macromolecules 253. https://doi.org/10.1016/j.ijbiomac.2023.126936.

Gajaria, T.K., Suthar, P., Baghel, R.S., Balar, N.B., Sharnagat, P., Mantri, V.A. et al. (2017). Integration of protein extraction with a stream of byproducts from marine macroalgae: A model forms the basis for marine bioeconomy. Bioresource Technology 243: 867–873. https://doi.org/10.1016/j.biortech.2017.06.149.

García-Márquez, J., Moreira, B.R., Valverde-Guillén, P., Latorre-Redoli, S., Caneda-Santiago, C.T., Acién, G. et al. (2023). *In vitro* and *in vivo* effects of ulvan polysaccharides from Ulva rigida. Pharmaceuticals 16(5). https://doi.org/10.3390/ph16050660.

Ghosh, P., Adhikari, U., Ghosal, P.K., Pujol, C.A., Carlucci, M.J., Damonte, E.B. et al. (2004). *In vitro* anti-herpetic activity of sulfated polysaccharide fractions from Caulerpa racemosa. Phytochemistry, 65(23): 3151–3157. https://doi.org/10.1016/J.PHYTOCHEM.2004.07.025.

Glasson, C.R.K., Donnet, L., Angell, A., Vucko, M.J., Lorbeer, A.J., Vamvounis, G. et al. (2019). Multiple response optimisation of the aqueous extraction of high quality ulvan from Ulva ohnoi. Bioresource Technology Reports 7. https://doi.org/10.1016/j.biteb.2019.100262.

Guidara, M., Yaich, H., Amor, I. Ben, Fakhfakh, J., Gargouri, J., Lassoued, S. et al. (2021). Effect of extraction procedures on the chemical structure, antitumor and anticoagulant properties of ulvan from *Ulva lactuca* of Tunisia coast. Carbohydrate Polymers 253. https://doi.org/10.1016/j.carbpol.2020.117283.

Guidara, M., Yaich, H., Benelhadj, S., Adjouman, Y.D., Richel, A., Blecker, C. et al. (2020). Smart ulvan films responsive to stimuli of plasticizer and extraction condition in physico-chemical, optical, barrier and mechanical properties. International Journal of Biological Macromolecules 150: 714–726. https://doi.org/10.1016/J.IJBIOMAC.2020.02.111.

He, J., Xu, Y., Chen, H. and Sun, P. (2016). Extraction, structural characterization, and potential antioxidant activity of the polysaccharides from four seaweeds. International Journal of Molecular Sciences 17(12). https://doi.org/10.3390/ijms17121988.

Harikrishnan, R., Devi, G., Van Doan, H., Balasundaram, C., Arockiaraj, J. and Jagruthi, C. (2021). Efficacy of ulvan on immune response and immuno-antioxidant gene modulation in *Labeo rohita* against columnaris disease. Fish & Shellfish Immunology, doi: 10.1016/J.FSI.2021.08.004.

Ibrahim, M.I.A., Amer, M.S., Ibrahim, H.A.H. and Zaghloul, E.H. (2022). Considerable production of ulvan from *Ulva lactuca* with special emphasis on its antimicrobial and anti-fouling properties. Applied Biochemistry and Biotechnology 194(7): 3097–3118. https://doi.org/10.1007/s12010-022-03867-y.

Jingjing, Z., Quanbin, Z., Jing, W., Xuelian, S.H.I. and Zhongshan, Z. (2009). Analysis of the monosaccharide composition of fucoidan by precolumn derivation HPLC *. 27(3): 578–582. https://doi.org/10.1007/s00343-009-9205-0.

Kazemi, M., Fathi, M., Jahanbin, K., Taghdir, M. and Abbaszadeh, S. (2023). Optimization of ultrasonic-assisted hot acidic solvent extraction of ulvan from Ulva intestinalis of the Persian Gulf: Evaluation of structural, techno-functional, and bioactivity properties. Food Hydrocolloids 142: 108837. https://doi.org/10.1016/J.FOODHYD.2023.108837.

Kidgell, J.T., Glasson, C.R.K., Magnusson, M., Vamvounis, G., Sims, I.M., Carnachan, S.M. et al. (2020). The molecular weight of ulvan affects the *in vitro* inflammatory response of a murine macrophage. International Journal of Biological Macromolecules 150: 839–848. https://doi.org/10.1016/j.ijbiomac.2020.02.071.

Kidgell, J.T., Magnusson, M., de Nys, R. and Glasson, C.R.K. (2019). Ulvan: A systematic review of extraction, composition and function. *In*: Algal Research (Vol. 39). Elsevier B.V. https://doi.org/10.1016/j.algal.2019.101422.

Lahaye, M. (1998). NMR spectroscopic characterisation of oligosaccharides from two *Ulva rigida* ulvan samples (Ulvales, Chlorophyta) degraded by a lyase. Carbohydrate Research 314(1-2): 1–12.

Le, B., Golokhvast, K.S., Yang, S.H. and Sun, S. (2019). Optimization of microwave-assisted extraction of polysaccharides from ulva pertusa and evaluation of their antioxidant activity. Antioxidants 8(5). https://doi.org/10.3390/antiox8050129.

Li, B., Xu, H., Wang, X., Wan, Y., Jiang, N., Qi, H. et al. (2020). Antioxidant and antihyperlipidemic activities of high sulfate content purified polysaccharide from *Ulva pertusa*. International Journal of Biological Macromolecules 146: 756–762. https://doi.org/10.1016/j.ijbiomac.2019.11.061.

Li, C., Tang, T., Du, Y., Jiang, L., Yao, Z., Ning, L. et al. (2023). Ulvan and Ulva oligosaccharides: a systematic review of structure, preparation, biological activities and applications. Bioresources and Bioprocessing 10(1). https://doi.org/10.1186/s40643-023-00690-z.

Li, P., Wen, S., Sun, K., Zhao, Y. and Chen, Y. (2018a). Structure and bioactivity screening of a low molecular weight ulvan from the green alga Ulothrix flacca. Marine Drugs 16(8). https://doi.org/10.3390/md16080281.

Li, W., Jiang, N., Li, B., Wan, M., Chang, X., Liu, H. et al. (2018b). Antioxidant activity of purified ulvan in hyperlipidemic mice. International Journal of Biological Macromolecules 113: 971–975. https://doi.org/10.1016/j.ijbiomac.2018.02.104.

Li, W., Wang, K., Jiang, N., Liu, X., Wan, M., Chang, X. et al. (2018c). Antioxidant and antihyperlipidemic activities of purified polysaccharides from *Ulva pertusa*. Journal of Applied Phycology 30(4): 2619–2627. https://doi.org/10.1007/s10811-018-1475-5.

Liu, F., Zhao, W., Zhao, F., Dong, Q., Ying, W., Wei, W. et al. (2020). Dual effect of the acidic polysaccharose ulvan on the inhibition of amyloid-β protein fibrillation and disintegration of mature fibrils. ACS Applied Materials & Interfaces. doi: 10.1021/ACSAMI.0C14292.

Maray, S.O., Abdel-Kareem, M.S.M., Mabrouk, M.E.M., El-Halmouch, Y. and Makhlof, M.E.M. (2023). *In vitro* assessment of antiviral, antimicrobial, antioxidant and anticancer activities of ulvan extracted from the green seaweed *Ulva lactuca*. Thalassas 39(2): 779–790. https://doi.org/10.1007/s41208-023-00584-z.

Metwaly, H.R., El-Khair, A., El-Sayed, B. and Amin, H.F. (2023). Chemical and biochemical properties of marine algae *Ulva lactuca* and nannocholoropsisoculata. Egyptian Journal of Aquatic Biology & Fisheries 27(3): 19–34. www.ejabf.journals.ekb.eg.

Mhatre, A., Gore, S., Mhatre, A., Trivedi, N., Sharma, M., Pandit, R. et al. (2019). Effect of multiple product extractions on bio-methane potential of marine macrophytic green alga *Ulva lactuca*. Renewable Energy 132: 742–751. https://doi.org/10.1016/j.renene.2018.08.012.

Mohandoss, S., Murugaboopathy, V., Haricharan, P.B., Hebbal, M.I., Saadaldin, S., Soliman, M. et al. (2023). Ulvan as a reducing agent for the green synthesis of silver nanoparticles: a novel mouthwash. Inorganics 11(1). https://doi.org/10.3390/inorganics11010005.

Mo'o, F.R.C., Wilar, G., Devkota, H.P. and Wathoni, N. (2020). Ulvan, a polysaccharide from Macroalga *Ulva* sp.: A review of chemistry, biological activities and potential for food and biomedical applications. Applied Sciences (Switzerland) 10(16). https://doi.org/10.3390/app10165488.

Ning, L., Yao, Z. and Zhu, B. (2022). Ulva (Enteromorpha) polysaccharides and oligosaccharides: a potential functional food source from green-tide-forming macroalgae. *In*: Marine Drugs (Vol. 20, Issue 3). MDPI. https://doi.org/10.3390/md20030202.

Olasehinde, T.A., Mabinya, L.V., Olaniran, A.O. and Okoh, A.I. (2019). Chemical characterization of sulfated polysaccharides from Gracilariagracilis and *Ulva lactuca* and their radical scavenging, metal chelating, and cholinesterase inhibitory activities. International Journal of Food Properties 22(1): 100–110. https://doi.org/10.1080/10942912.2019.1573831.

Paulert, R., Brunel, F., Melcher, R.L.J., Cord-Landwehr, S., Niehues, A., Mormann, M. et al. (2021). The non-sulfated ulvanobiuronic acid of ulvans is the smallest active unit able to induce an oxidative burst in dicot cells. Carbohydrate Polymers 270. https://doi.org/10.1016/j.carbpol.2021.118338.

Peasura, N., Laohakunjit, N., Kerdchoechuen, O., Vongsawasdi, P. and Chao, L.K. (2016). Assessment of biochemical and immunomodulatory activity of sulphated polysaccharides from *Ulva intestinalis*.

International Journal of Biological Macromolecules 91: 269–277. https://doi.org/10.1016/J.IJBIOMAC.2016.05.062.

Ponce, M., Zuasti, E., Anguís, V. and Fernández-Díaz, C. (2020). Effects of the sulfated polysaccharide ulvan from Ulva ohnoi on the modulation of the immune response in Senegalese sole (Solea senegalensis). Fish and Shellfish Immunology 100: 27–40. https://doi.org/10.1016/j.fsi.2020.02.054.

Pradhan, B., Bhuyan, P. and Ki, J.-S. (2023). Immunomodulatory, antioxidant, anticancer, and pharmacokinetic activity of ulvan, a seaweed-derived sulfated polysaccharide: an updated comprehensive review. Marine Drugs 21(5): 300. https://doi.org/10.3390/md21050300.

Qi, H. and Sun, Y. (2015). Antioxidant activity of high sulfate content derivative of ulvan in hyperlipidemic rats. International Journal of Biological Macromolecules 76: 326–329. https://doi.org/https://doi.org/10.1016/j.ijbiomac.2015.03.006.

Raymond, P. and Simard, R. (2006). The occurrence of antibiotic resistance genes in Taq polymerases and a decontamination method applied to the detection of genetically modified crops. 3517: 321–325. https://doi.org/10.1007/s10529-005-5931-3.

Reis, S.E., Andrade, R.G.C., Accardo, C.M., Maia, L.F., Oliveira, L.F.C., Nader, H.B. et al. (2020). Influence of sulfated polysaccharides from *Ulva lactuca* L. upon Xa and IIa coagulation factors and on venous blood clot formation. Algal Research 45: 1017. https://doi.org/10.1016/j.algal.2019.101750.

Robic, A., Gaillard, C., Sassi, J.F., Leral, Y. and Lahaye, M. (2009). Ultrastructure of Ulvan: A polysaccharide from green seaweeds. Biopolymers 91(8): 652–664. https://doi.org/10.1002/bip.21195.

Robic, A. and Lahaye, M. (2007). Reviews structure and functional properties of ulvan, a polysaccharide from green seaweeds related papers. Biomacromolecules 8(6): 1765–1774.

Rosidah, I., Ningsih, S., Renggani, T.N., Agustini, K. and Efendi, J. (2020). ProfilHematologiTikus (Rattus norvegicus) GalurSparague-Dawley JantanUmur 7 dan 10 Minggu. JurnalBioteknologi&Biosains Indonesia Homepage 7(1): 142.

Shefer, S., Robin, A., Chemodanov, A., Lebendiker, M., Bostwick, R., Rasmussen, L. et al. (2021). Fighting SARS-CoV-2 with green seaweed *Ulva* sp. extract: Extraction protocol predetermines crude ulvan extract anti-SARS-CoV-2 inhibition properties in *in vitro* Vero-E6 cells assay. PeerJ 9. https://doi.org/10.7717/peerj.12398.

Shi, L. (2016). Bioactivities, isolation and purification methods of polysaccharides from natural products: A review. International Journal of Biological Macromolecules 92: 37–48. https://doi.org/10.1016/j.ijbiomac.2016.06.100.

Siddhanta, A.K., Goswami, A.M., Ramavat, B.K., Mody, K.H. and Mairh, O.P. (2001). Water soluble polysaccharides of marine algal species of Ulva (Ulvales, Chlorophyta) of Indian waters. In Indian Journal of Marine Sciences (Vol. 30).

Sulastri, E., Lesmana, R., Zubair, M.S., Abdelwahab Mohammed, A.F., Elamin, K.M. and Wathoni, N. (2023). Ulvan/Silver nanoparticle hydrogel films for burn wound dressing. Heliyon 9(7). https://doi.org/10.1016/j.heliyon.2023.e18044.

Sun, Y., Liu, Z., Song, S., Zhu, B., Zhao, L., Jiang, J. et al. (2020). Anti-inflammatory activity and structural identification of a sulfated polysaccharide CLGP4 from Caulerpa lentillifera. International Journal of Biological Macromolecules 146: 931–938. https://doi.org/10.1016/J.IJBIOMAC.2019.09.216.

Tabarsa, M., You, S.G., Dabaghian, E.H. and Surayot, U. (2018). Water-soluble polysaccharides from Ulva intestinalis: Molecular properties, structural elucidation and immunomodulatory activities. Journal of Food and Drug Analysis 26(2): 599–608. https://doi.org/10.1016/j.jfda.2017.07.016.

Thanh, T.T.T., Quach, T.M.T., Nguyen, T.N., Vu Luong, D., Bui, M.L. and Tran, T.T. Van. (2016). Structure and cytotoxic activity of ulvan extracted from green seaweed Ulva lactuca. International Journal of Biological Macromolecules 93: 695–702. https://doi.org/https://doi.org/10.1016/j.ijbiomac.2016.09.040.

Tsubaki, S., Oono, K., Hiraoka, M., Onda, A. and Mitani, T. (2016). Microwave-assisted hydrothermal extraction of sulfated polysaccharides from *Ulva* spp. and Monostromalatissimum. Food Chemistry 210: 311–316. https://doi.org/10.1016/j.foodchem.2016.04.121.

Wahlström, N., Nylander, F., Malmhäll-Bah, E., Sjövold, K., Edlund, U., Westman, G. et al. (2020). Composition and structure of cell wall ulvans recovered from *Ulva* spp. along the Swedish west coast. Carbohydrate Polymers 233. https://doi.org/10.1016/j.carbpol.2020.115852.

Wang, J., Zhang, Q., Zhang, Z. and Li, Z. (2008). Antioxidant activity of sulfated polysaccharide fractions extracted from Laminaria japonica. International Journal of Biological Macromolecules 42(2): 127–132. https://doi.org/10.1016/j.ijbiomac.2007.10.003.

Wang, W., Wang, X., Gao, W., Cui, Z., Zhang, H., Lu, F. et al. (2022). Ulvan inhibits α-synuclein fibrillation and disrupts the mature fibrils: *In vitro* and *in vivo* studies. International Journal of Biological Macromolecules 211: 580–591. https://doi.org/10.1016/J.IJBIOMAC.2022.05.045.

Yaich, H., Garna, H., Besbes, S., Paquot, M., Blecker, C. and Attia, H. (2013). Effect of extraction conditions on the yield and purity of ulvan extracted from *Ulva lactuca*. Food Hydrocolloids 31(2): 375–382. https://doi.org/10.1016/j.foodhyd.2012.11.013.

Yaich, H., Garna, H., Bchir, B., Besbes, S., Paquot, M., Richel, A. et al. (2015). Chemical composition and functional properties of dietary fibre extracted by Englyst and Prosky methods from the alga Ulva lactuca collected in Tunisia. Algal Research 9: 65–73. https://doi.org/10.1016/j.algal.2015.02.017.

Yaich, H., Amira, A. Ben, Abbes, F., Bouaziz, M., Besbes, S., Richel, A. et al. (2017). Effect of extraction procedures on structural, thermal and antioxidant properties of ulvan from *Ulva lactuca* collected in Monastir coast. International Journal of Biological Macromolecules 105: 1430–1439. https://doi.org/https://doi.org/10.1016/j.ijbiomac.2017.07.141.

Yu-Qing, T., Shehzadi, K.M.R. and Ashraf, M.F. (2016). *Ulva lactuca* and its polysaccharides: food and biomedical aspects. Journal of Biology, Agriculture and Healthcare 6 (1): 140–151.

Zhao, C., Lin, G., Wu, D., Liu, D., You, L., Högger, P. et al. (2020). The algal polysaccharide ulvan suppresses growth of hepatoma cells. Food Frontiers 1(1): 83–101. https://doi.org/10.1002/fft2.13.

Chapter 8

Exploring the Therapeutic Potential of Marine Microorganisms for Novel Anticancer Drugs

Bhumi Javia,[1] *Anjana K. Vala*[2,*] and *Dushyant R. Dudhagara*[1,*]

Introduction

Cancer is the main cause of mortality worldwide, and it is thought to be the second major cause of death among people worldwide. There are no effective cancer therapies because some of the treatments that are currently used have undesirable side effects. Around 10 million people worldwide died from cancer in 2020, and there were an estimated 25 million new cases, according to WHO estimates. Cancer is the main cause of death, with 2.26 million cases of breast cancer, 2.21 million cases of lung cancer, 1.93 million cases of colon and rectum, 1.41 million cases of prostate, 1.20 million cases of skin (non-melanoma) and 1.09 million cases of stomach cancer (World Health Organization 2020).

The marine environment accounts for 70% of the earth's biosphere and serves as the greatest ecosystem for diverse living organisms, particularly microorganisms. It has been found that marine organisms, including sponges, algae and corals, but especially marine bacteria and fungi, produce unique secondary metabolite having distinctive and varied chemical structures that can be crucial to the synthesis of drugs for cancer treatment (Alves et al. 2018). Marine habitats are rich sources of potential anticancer drugs, particularly marine bacterial species like *Lactobacilli*

[1] Department of Life Sciences, Bhakta Kavi Narsinh Mehta University, Khadiya-362263, Junagadh, India.

[2] Department of Life Sciences, Maharaja Krishnakumarsinhji Bhavnagar University, Bhavnagar-364001, India.

* Corresponding authors: akv@mkbhavuni.edu.in; dushyant373@gmail.com

Table 1. Classification and sources of potential anticancer compounds derived from marine bacteria.

Type	Compound	Source	Reference
Peptides	Haliamide	*Haliangium ochraceum* SMP-2	Lin et al. 2016
	Mixirin A, B, and C	*Bacillus* sp.	Zhang et al. 2004
	Halobacillin	*Bacillus* sp.	Trischman et al. 1994
	Mojavensin A	*B. mojavensis* B0621	Ma et al. 2012
	Arenamides (A to C)	*Salinispora Arenicola* CNT-088	Asolkar et al. 2009
	Ieodoglucomide A and B	*B. licheniformis*	Tareq et al. 2012
	Urukthapelstatin A	*Mechercharimyces asporophorigenens* YM11-542	Lin et al. 2013
	Ariakemicins (A and B)	*Rapidithrix* sp.	Oku et al. 2008
	Halolitoralins (A-C)	*Halobacillus litoralis* YS3106	Yang et al. 2002
Proteins/peptide	Bacteriocins	*Lactobacilli*	Kathiresan and Thiruneelakandan 2008
Antibiotics	Proximicins (A to C)	*Verrucosispora* MG-37	Fiedler et al. 2008
	Aminocoumacin	*Bacillus* sp	Liu et al. 2013
Enzymes	Asparaginase	*Bacillus licheniformis* 4T1C26E	Mathew et al. 2023
	Glutaminase	*Pseudomonas* spp. RAS123	Elborai et al. 2023
	Methioninase	*Alcaligenes aquatilis* BJ-1	Javia et al. 2023
	Arginase	*Bacillus licheniformis* OF2	Selim et al. 2024
Dicarboxylic acids	Hydroxyphenylpyrrole dicarboxylic acids	*Halomonas. Stenophila*	Ruiz-Ruiz et al. 2011
Polysaccharides	O-specific polysaccharide (OPS) and O-deacylated lipopolysaccharide (DPS)	*Poseidonocella sedimentorum* KMM 9023T	Kokoulin et al. 2018
Polyketide	18-O-demethylpederin	*Labrenzia* sp. PHM005	Schleissner et al. 2017
	Macrolactin-A	Marine bacteria	Carte 1996
Imidazole	Bulbimidazoles	*Microbulbifer* sp.	Karim et al. 2020
Alkaloid	Psyriacitrin	*Burkholderia* sp. NBF227	Xu et al. 2019b
	Tetra(indol-3-yl) ethenone	*Pseudovibrio denitrificans* BBCC725	Rodrigues et al. 2017
	Cycloprodigiosin hydrochloride	*Pseudoalteromonas denitricans*	Kamata et al. 2001
Quinolones	Aflaquinolone	*Bacillus subtilis*	Abdel-Wahab et al. 2019

and *Noctiluca scintillans* and include other prokaryotic organisms. Additionally, algae produce secondary anticancer metabolites, and mangroves harbor anticancer compounds. Overall, marine ecosystems contribute to slightly more than 70% of the world's potential anticancer resources. Anticancer drugs derived from marine habitats include prokaryotic organism especially marine bacterial species such as *Lactobacilli* and *Noctiluca scintillans*, algae that produce secondary anticancer metabolites, mangroves with anti-cancer metabolites along with various types of marine living beings that make up slightly more than 70% of the world. Similarly, the marine microflora and microalgae species are highly diverse and account for 90% of the sea biomass in the world's oceans (Kim and Kalimuthu 2015).

Extreme marine environments are generally characterized by higher salinity and pressure, low oxygen content for cellular respiration, high light availability, and extremely intense environments with a wide range of marine temperature (Gomes et al. 2016). These challenging conditions have led to the evolution of marine-derived biomolecules such as peptides, pigments, enzymes, lipids, and antibiotics. These molecules have been used as molecular models in drug research and have shown promise in cancer prevention and treatment. It has been demonstrated that purified peptides from marine microorganisms exhibit cytotoxic properties against a variety of human cancer cell line, including leukemia, breast, lung, colon, pancreatic, and liver. While certain FDA-approved medications have been obtained from terrestrial sources, a significant number of anticancer drugs and important metabolites have been derived from marine sources. There are almost 30,000 known combinations of marine-derived natural resources (Kiuru et al. 2014, Chen et al. 2019). Peptides from a variety of marine species; terpenes with a wide range of structural diversity; high molecular weight organic sugars; polyketides with unique biological properties; and many designed classes of alkaloids from marine algae are examples of marine natural products (Table 1). All of these products play a major role in the combination of anticancer drugs based on their vital natural defenses against persistently fatal diseases (Ruiz-Torres et al. 2017).

Therefore, the marine environment with its vast biodiversity and unique conditions serves as a promising reservoir for anticancer drugs and could lead to significant advancements in future cancer therapies. This chapter aims to provide sources of anticancer drugs from marine origin, mechanisms behind cancer treatment, factors involved in the identification of anticancer drugs, and approaches to cancer treatment that are emerging from the field of modern research on marine species.

Bacteria

Marine microorganisms have a rich source of unique genes, the exploration of which could lead to the discovery of novel drugs. However, only a few bacteria can be cultivated in laboratory settings, highlighting the urgent need for the development of new cultivation techniques to isolate slow-growing bacteria and those that produce unique natural products. The secondary metabolites produced by these microscopic marine organisms have led to the discovery of several pharmacologically significant compounds, including novel anti-inflammatory agents such as Pseudopterosins, Scytonemin, Manoalide and Topsentins; antibiotics such as Marinone; anticancer compounds such as Bryostatins, Sarcodictyin Discodermolide and Eleutherobin (Table 1) (Jensen et al. 1996).

It has long been known that bacteria have the potential to act as anticancer agents. The first microbes to be used as anticancer agents were *Streptococcus* and *Clostridium*. Numerous bacterial genera that have shown to have anticancer properties and are utilized in anticancer treatments include *Bifidobacterium, Lactobacillus, Salmonella, Pseudomonas, Escherichia, Streptococcus, Clostridium, Mycobacterium, Proteus, Caulobacter* and *Listeria.* Currently, genetically modified bacteria are primarily used in anticancer therapies, utilizing a variety of anticancer mechanisms and strategies (Mengesha et al. 2007, Patyar et al. 2010, Kramer et al. 2018). The strategies include pro-drug cleavage, gene transfer, RNA interference, live bacterial toxicity, expression of cancer-specific agent, and combination therapy (Heimann et al. 2003, Nallar et al. 2017, Song et al. 2018, Wang et al. 2018, Sedighi et al. 2019).

Recently, certain small chemical compounds, including antibiotics have been used in cancer treatment due to their potential anticancer properties. Antibiotics are chemical compounds with antimicrobial characteristics that work by suppressing bacterial growth (Bhattacharya and Mukherjee 2015). Aminofuran antibiotics Proximicins A – C are produced from the marine bacteria Verrucosispora MG-37, which was found from marine sediments in Norway's Raune Fjord. Proximicins contain a structural element known as 4-amino-furan-2-carboxylic acid, a previously unknown gamma-amino acid. Proximicins have shown cytotoxicity against various cancer cell lines, including AGS (gastric adenocarcinoma), HepG2 (Liver) and MCF-7 (Breast) cells (Fiedler et al. 2008). Proximicin C showed strong cytotoxicity against glioblastoma and breast cancer cells, while proximicin B was reported to cause apoptosis in Hodgkin's lymphoma and T-cell leukemia cells in different investigations (Brucoli et al. 2012). Additionally, it has been discovered that proximicins up-regulate p53 and p21, the cyclin kinase inhibitor, which contributes to their lethal effect on tumor cells (Schneider et al. 2008). The antibiotics in the aminocoumacin group are members of a small isocoumarin family that has the 3,4-dihydro-8-hydroxyisocoumarin chromophore (Liu et al. 2013). PM-94128 ($C_{22}H_{34}N_2O_6$, M.Wt. 423.24 Da) is an isocoumarin produced by *Bacillus* sp. It has shown anticancer efficacy against a number of cancer cell lines, such as A-549, P-388 (murine leukemia) cells, HT-29 (human colonic adenocarcinoma) and MEL-28 (human melanoma cells) (Baz 1997).

Bacteriocins and toxins are a type of ribosomally synthesized bacterial peptide or protein with a wide range of antimicrobial properties that are effective against drug-resistant and pathogenic microbes due to their multiple killing mechanisms. Remarkably, a few of these substances have also shown anticancer properties (Baindara et al. 2017, Weerakkody and Witharana 2019). Live bacteria called probiotics have the potential to both treat and prevent many diseases (Kamaliya et al. 2023). Probiotic bacteria, particularly *Lactobacilli* and *Bifidobacteria* play an important role in pathogen control by producing antibacterial proteins such as bacteriocins and anticancer substances (Wollowski et al. 2001, Kathiresan and Thiruneelakandan 2008). There has been evidence that lactobacilli-containing dietary supplements reduce the risk of experimental colon cancer (Goldin and Gorbach 1992). *In vivo* studies have also demonstrated that *Lactobacilli* and *Noctiluca scintillans* can prevent colon cancer and melanoma (Carte 1996). *Lactobacilli* are known for their potential in mitigating the risk of colon cancer due to their ability to decrease the activity of certain enzymes, including azoreductase, nitroreductase, and β-glucuronidase, in the diet of rats (Goldin and Gorbach 1977, Mitall et al. 1995).

It is known that many marine animal groups produce toxins, and research indicates that these toxins could be produced by marine bacteria that live in symbiosis with these animals (Kodama et al. 1990). An example of this is Kahalalide F (KF), a depsipeptide that is extracted from the mollusk *Elysia rubefescens* found in Hawaii. It is thought that this compound is actually synthesized by the microbes that live in association with the mollusk. KF has been observed to have cytotoxic effects and can halt the cell cycle in the G1 phase in a p53 independent way. Laboratory studies have shown that KF is active against solid tumors with a particular selective action pattern in prostate cancer cell lines. Furthermore, comprehensive *in-vivo* studies have demonstrated the effectiveness of KF in treating both breast and colon cancers (Kodama et al. 1990). Microbial toxins are valuable in neurophysiological and neuropharmacological research. It has been reported that bacteria associated with *Noctiluca scintillans* are known to contribute to red tides. Macrolactin-A has been identified for its inhibitory effects on B16-F10 murine melanoma cancer cells as well as herpes simplex viruses (HSV) types I and II. Moreover, macrolactin-A has shown protective properties against HIV replication (Carte 1996).

Enzymes from marine microorganisms, such as L-asparaginase (Vala et al. 2018), L-glutaminase (Elborai et al. 2023), L-methioninase (Javia et al. 2023), and L-arginase (Selim et al. 2024) show promising anticancer properties and are being explored as potential therapeutic agents for cancer treatment. Tumor cells are auxotroph for certain amino acids such as L-methionine, L-glutamine, L-asparagine, and L-arginine and lack the intrinsic enzyme systems that synthesize these amino acids. Therefore, for these cancer cells to grow and proliferate, they require an exogenous supply of these amino acids, usually through diet. As a result, by removing their respective amino acids from blood plasma, these enzymes are extensively utilized in the treatment of cancer (Javia et al. 2024).

Marine bacterial phyla are increasingly recognized as significant sources for discovering novel bioactive compounds. For instance, haliamide, isolated from the marine myxobacterium *Haliangium ochraceum* SMP-2, is a product of hybrid polyketide synthase and non-ribosomal peptide synthetase biosynthesis. It shows moderate cytotoxicity against the HeLa cell line (Sun et al. 2016). Zhang et al. (2004) reported three novel cyclic acyl peptides, Mixirin A, B and C from marine *Bacillus*. These cyclopeptides displayed cytotoxic activities against human colon cancer cells (HCT-116) with IC50 values of 0.68, 1.6, and 1.3 mg/ml, respectively. Halobacillin is an iturin family compound produced by *Bacillus* sp. CND-914 isolated from deep-sea sediment near the Guaymas Basin, Mexico. Halobacillin showed anticancer activities against HCT-116 with an IC50 value of 0.98 mg/ml (Trischman et al. 1994). Mojavensin A, another member of the iturin family produced by the marine bacterium

B. mojavensis B0621 and exhibited cytotoxicity against the human leukemia (HL-60) cell line (Ma et al. 2012).

Arenamides A, B, and C are cyclohexadepsipeptides produced through the fermentation process of the marine bacterium *Salinispora arenicola* CNT-088, which was found in the Great Astrolabe Reef in the Kandavu Island chain, Fiji (Asolkar et al. 2009). Arenamides A and B have shown to exhibit moderate anticancer activity against HCT-116 cells, while Arenamide C does not display any cytotoxic effects. Remarkably, Arenamides A and B have been found to inhibit TNF-induced activation of NFkb in stably transfected 293/NFkB-Luc human embryonic kidney cells. Additionally, they also inhibit the production of nitric oxide (NO) and prostaglandin E2 (PGE2) in lipopolysaccharide-induced macrophages (Asolkar et al. 2009). Ieodoglucomide A-B are glycolipopeptides synthesized by the marine bacterium *B. licheniformis* isolated from Leodo Reef sediments, South Korea. These compounds contain amino acids, a novel fatty acid, succinic acid, and a sugar moiety. Ieodoglucomide B has shown cytotoxicity against lung and stomach cancer cells (Tareq et al. 2012). Urukthapelstatin A is a cyclic thiopeptide produced by *Mechercharimyces asporophorigenens* YM11-542, which was isolated from the marine lake at Urukthapel Island, Palau. It has proven to be cytotoxic to a variety of cancer cell lines, such as those from the breast, colon, lung, and ovarian cancers (Matsuo et al. 2007, Lin et al. 2013). Ariakemicins A and B are compounds produced by a marine gliding bacterium *Rapidithrix* sp., isolated from Ariake Inland Sea, Japan. These compounds are synthesized by a hybrid polyketide-nonribosomal peptide synthetase and exhibit low cytotoxicity towards A549 and BHK baby hamster kidney cells (Oku et al. 2008). Halolitoralins A, B, and C are cyclic peptides produced by *Halobacillus litoralis* YS3106, which was isolated from marine sediment. Halolitoralin A is a cyclic hexapeptide, while halolitoralins B and C are cyclic tetrapeptides. These compounds displayed moderate anticancer activity towards human gastric tumor cells *in vitro* (Yang et al. 2002).

It has been demonstrated that *Halomonas* spp. GWS-BW-H8hM inhibited the growth of MCF7, HepG2, and HM02 (gastric adenocarcinoma) cell lines, causing death via cell cycle arrest (Bitzer et al. 2006, Sagar et al. 2013). Exopolysaccharides (EPSs) and sulfated EPSs isolated from *H. stenophila* are examples of heterogeneous polymers that have shown to have pro-apoptotic effects on T-leukemia cells; however, primary T cells exhibit resistance to these effects (Ruiz-Ruiz et al. 2011). Marine *Halomonas* sp. was used to isolate cytotoxic hydroxyphenylpyrrole dicarboxylic acids, namely 3-(4-hydroxyphenyl)-4-phenylpyrrole-2,5-dicarboxylic acid (HPPD-1) and 3,4-di-(4-hydroxyphenyl) pyrrole-2,5-dicarboxylic acid (HPPD-2), as well as indole derivatives. Both HPPD-1 and HPPD-2 demonstrated potent anticancer activity by blocking 12-O-tetradecanoylphorbol-13-acetate induced expression of Epstein-Barr virus early antigen, with HPPD-2 being more effective (Wang et al. 2006). Extracts from *Sulfitobacterpontiacus* and *Halomonas axialensis* isolates suppressed the development of HeLa cells and DU145 cells by 50–70%, indicating that these isolates had cytotoxic action due to particular cytotoxic chemicals (Sagar et al. 2013).

Pseudovibrio denitrificans BBCC725 produced tetra (indol-3-yl) ethenone, a new cytotoxic indole (Rodrigues et al. 2017). Tetra (indol-3-yl) ethenone is moderately cytotoxic to the human lung tumor cell line A549 and mouse fibroblast cell line L929. Analogous to pederin, a novel tetrahydropyran-core polyketide produced by *Labrenzia* sp. PHM005 (Schleissner et al. 2017). This novel pederin, 18-O-demethylpederin, demonstrated strong anti-proliferative properties towards all four cell lines: human lung carcinoma cell line, colon adenocarcinoma, breast adenocarcinoma and pancreas adenocarcinoma.

The anticancer potential of O-deacylated lipopolysaccharide (DPS) and sulfated O-specific polysaccharide (OPS) from *Poseidonocella sedimentorum* KMM 9023T was investigated. According to Kokulin et al. (2018) KMM 9023T was found to have no cytotoxic impact and to suppress the proliferation of several human cancer cells, such as HT-29, SK-MEL-5, and MCF-7.

Bulbimidazoles A, B, and C (novel alkanoylimidazoles) produced by *Microbulbifer* sp. DC3-6 γ-proteobacterium which was isolated from the stony coral of the genus *Tubastraea* suppressed the proliferation of P388 murine leukemia cells (Karim et al. 2020).

Pseudoalteromonas denitricans produces cycloprodigiosin hydrochloride, a red pigment. According to Kamata et al. (2001), the pigment inhibited the transcriptional activity of nuclear factor kB (NF-kB), which enhanced the apoptosis induced by the tumor necrosis factor in HeLa cells. This finding offers one potential explanation for the immunosuppressive and anti-tumor properties of this molecule.

Chinese *Burkholderia* sp. NBF227 is the source of psyriacitrin, an alkaloid of marine origin with a characteristic beta-carboline scaffold. The following human cancer cell lines were used to examine the anticancer properties of pyriacitrin compounds: SGC-7901, A875, HepG2, and MARC145. A few of these compounds of beta-carbolines have cytotoxic effects ranging from moderate to high. With IC50 values of 6.82, 8.43, 7.69 and 7.19 μM, the compound containing a sulfonyl group had the strongest inhibitory efficacy against all cell lines (Xu et al. 2019b).

The following chemicals were isolated from the co-culture of *Bacillus subtilis* and *Aspergillus versicolor*, a sponge-derived fungus, on rice-solid media: a new aflaquinolone (22-epi-aflaquinolone B), two new anthraquinones, and thirty known compounds. The new analogues were only found in the co-culture extract, but not in the axenic fungus cultures. Furthermore, compared to the axenic fungal culture, the co-culture extract enhanced the production of the known compounds versicolorin B, averufin, and sterigmatocyctin by factors of 1.5, 2.0, and 4.7, respectively. Only five of the identified compounds, O-demethylsterigmatocystin, sterigmatocystin, sterigmatin, AGI-B4, and stephacidin A, had an IC50 range of 2.0–21.2 μM and moderately reduced the development of the mouse lymphoma cell line L5178Y (Abdel-Wahab et al. 2019).

Actinomycetes

For over 50 years, actinomycetes isolated from terrestrial origin have served as a valuable pharmaceutical resource for the discovery of antibiotics and associated bioactive substances. On the other hand, research on marine actinomycetes remains relatively rare. Actinomycetes, particularly *Micromonosporaceae* and *Streptomyces* (Table 2), show promise in the isolation of significant growth-inhibiting agents and new anticancer drugs. Marine actinomycetes in the *Micromonosporaceae* family are among the most promising bacteria that produce antibiotics. These bacteria have been proven to be a potent source of anticancer drugs that target proteasome function, and multiple pharmaceuticals have tested their industrial potential.

Two novel glycosylated polyketides, arisostatin A and B, were identified from the culture extract of *Micromonospora* sp. during the search for bioactive compounds from actinomycetes collected from the deep-sea water in Toyama Bay (Igarashi et al. 2000, Furumai et al. 2000). Arisostatins are a new class of cytotoxic chemicals, similar to tetrocarcins. Arisostatin A has an effective cytotoxic impact on human tumor cells and stimulates caspase 3, a crucial effector protease that causes apoptosis (Table 2) (Kim et al. 2003).

Micromonospora sp. L-25-ES25-008, an actinomycete was isolated from a sponge collected from the Indian Ocean near the coast of Mozambique. This organism produced IB-96212, a 26-membered macrolide with a spiroketal lactone structure. Canedo et al. (2000) reported that the isolate had cytotoxic activity against mouse leukemia P-388, human lung nonsmall cell A-549, colon adenocarcinoma HT-29, and melanoma MEL-28 cell lines (Fernandez et al. 2000).

According to reports, the actinomycetes strains ACT01 and ACT02 that were isolated from maritime sediment were potentially toxic to the MCF-7 and MDA-MB-231 breast cancer cell lines (Ravikumar et al. 2012). *Streptomyces* spp. from the marine ascidian *Synoicum indicum* yielded 10 pure compounds (Pyrazine-1,4-dione substituted cyclic dipepetide) (Nalli et al. 2017). Five of these ten compounds demonstrated noteworthy effects against interleukin-6 (IL-6) and tumor necrosis factor-a (TNF-a) (Nalli et al. 2017).

According to Maskey et al. (2004b), gutingimycin is a highly polar derivative of trioxacarcin derived from *Streptomyces* species and has been isolated from Laguna de Terminos silt in the Gulf of Mexico. In addition to the previously identified trioxacarcins A–C (Maskey et al. 2004a), the same *Streptomyces* species also produces trioxacarcins D–F. When tested against a lung cell line, doxacarcins A–C showed strong anti-tumor activity, with IC50 values ranging from 0.1, 6.0, 0.003 to 0.26 ng/mL, respectively.

The study conducted by Cho et al. (2007) investigated the potential of marine-derived actinomycetes as sources of antitumor compounds. The results showed that lucentamycins A–D, which are peptides containing 3-methyl-4-ethylideneproline and isolated from *Nocardiopsis lucentensis* strain CNR-712, showed notable *in vitro* cytotoxicity against HCT-116. The IC50 values for Lucentamycins A and B were 0.20 and 11 μM, respectively.

The marine-derived Mei37 isolate of *Streptomyces* sp. produced mansouramycin A–D and 3-methyl-7-(methylamino)-5,8-isoquinolinedione, which showed significant cytotoxicity and a high level of the selectivity to non-small-cell lung cancer, breast cancer, melanoma, and prostate cancer cells (Hawas et al. 2009).

Macrodiolide tartrolon D, isolated from *Streptomyces* sp. MDG-04-17-069, showed high cytotoxic action towards 3 human cancer cell lines including lung

Table 2. Classification and sources of anticancer compounds isolated from marine actinomycetes.

Type	**Compound**	**Source**	**Reference**
Polyketides	Arisostatin A and B	*Micromonospora* sp.	Igarashi et al. 2000
	Gutingimycin	*Streptomyces* spp.	Maskey et al. 2004b
	Ansamycins, divergolides	*Streptomyces* sp. KFD18	Zhou et al. 2019
	Macrodiolide tartrolon D	*Streptomyces* sp. MDG-04-17-069	Pérez et al. 2009
	Macrolide	*Micromonospora* sp. L-25-ES25-008	Canedo et al. 2000
	Gancidin W	*Streptomyces paradoxus* VITALK03	Ravi et al. 2017
	Neaumycin B	*Micromonospora* sp. CNY-010	Kim et al. 2018
	Dibohemamines A-C	*Streptomyces spinoverrucosus*	Fu et al. 2016
	Lobophorins	*Streptomyces* sp.	Lin et al., 2014
	Akazamicin, actinofuranone, N-formylanthranilic acid	*Nonomuraea* sp. AKA32	Yang et al. 2019
	Julichrome	*Streptomyces* sp.	Shen et al. 2021
	Anthracyclinones, boshramycinones	*Streptomyces* sp.	Shaaban et al. 2021
	Sekgranaticin, granaticins A	*Streptomyces* sp.	Lv et al. 2019
	Shellmycins	*Streptomyces* sp. Shell-016	Han et al. 2020
Lipid	Dokdolipids A, B, and, C	*Actinoalloteichus hymeniacidonis*	Choi et al. 2019a
Polyether	Terrosamycins	*Streptomyces* sp. RKND004	Sproule et al. 2019
Ester	1, 2-benzene dicarboxylic acid, mono 2-Ethylhexyl ester (DMEHE)	*Streptomyces* sp. VITSJK8	Krishnan et al. 2014
Xanthone	Kigamicin	*Streptomyces bingchenggensis* ULS14	Davies-Bolorunduro et al. 2019
Quinone	Lactoquinomycin	*Streptomyces* sp. SS17A	Zhou et al. 2020a
	Lactoquinomycin	*Nocardiopsis* sp. OUCMDZ-4936	Wang et al. 2019
Peptide	Pyrazine-1,4- dione	*Streptomyces* spp.	Nalli et al. 2017
	Lucentamycins A-D, 3-methyl-4-ethylideneproline	*Nocardiopsis lucentensis*	Cho et al. 2007
	Thiocoraline	*Micromonospora marina*	Erba et al. 1999

Table 2 contd. ...

...Table 2 contd.

Type	Compound	Source	Reference
	Dentigerumycin E	*Streptomyces albogriseolus* JB5	Shin et al. 2018
	Neothioviridamide	*Streptomyces* sp. MSB090213SC12	Kawahara et al. 2018
	Rakicidins G-I	*Micromonospora chalcea* FIM 02–523	Nie et al. 2018
	Streptodepsipeptides	*Streptomyces* sp. P11-23B	Ye et al. 2017
	Neo-actinomycin	*Streptomyces* sp. IMB094	Pérez et al. 2017
Alkaloids	Geranylpyrrol A and piericidin F	*Streptomyces* sp. CHQ-64	Han et al. 2017
Quinolones	Mansouramycin A-D, 3-methyl-7-(methylamino)-5,8-isoquinolinedione	*Streptomyces* spp.	Hawas et al. 2009
Lactone	Salinosporamide A	*Salinospora* spp.	Chauhan et al. 2005

(A549), colon (HT29), and breast (MDA-MB-231) cancer with IC50 = 0.16, 0.31, and 0.79 μM compared to doxorubicin as a reference (Pérez et al. 2009).

Coraline, isolated from *Micromonospora marina*, inhibits RNA assembly and has cytotoxic potential for lung and colon cancer cell lines as well as advanced skin disease. Thiocoraline is a new bioactive depsipeptide derived from *Micromonospora marina*, a type of marine bacterium found in the Mozambique Strait that suppresses RNA production. In vitro, thiocoraline demonstrated cytotoxic action against human colon cancer cell lines LOVO and SW620, with IC50 values of 15 nM and 500 nM, respectively. It is interesting to note that the compound specifically inhibits the growth of colon cancer cell lines through a defective p53 system (Erba et al. 1995). The cytotoxic activity of thiocoraline was mediated by a cell cycle arrest in the GI phase and the result was a reduction in the rate of S phase progression towards the G2/M phases was detected by using bromodeoxyuridine/DNA biparametric flow cytometric analysis (Erba et al. 1999).

Dentigerumycin E is a pyran-bearing polyketide acyl chain and cyclic hexapeptide that was isolated from the marine actinobacterium *Streptomyces albogriseolus* JB5. It contains three piperazic acids. In contrast to not being cytotoxic to the normal human breast epithelial cell line MCF-10A, it demonstrated considerable cytotoxicity against the lung carcinoma A549, colorectal cancer HCT116, breast cancer MDA-MB-231, liver cancer SK-HEP-1, and stomach cancer SNU638 cell lines (Shin et al. 2018).

Similarly, neothioviridamide is a polythioamide cyclic peptide that exhibits high cytotoxicity against human ovarian cancer (SKOV-3), malignant pleural mesothelioma (Meso-1), and immortal human T lymphocytes (Jurkat) cell lines. It was identified following the discovery of a new biosynthetic cluster (thioviridamide-like biosynthetic gene) from *Streptomyces* sp. MSB090213SC12 using genome mining and heterologous production of a bacterial artificial chromosome in *Streptomyces* (Kawahara et al. 2018).

The marine actinobacterium *Micromonospora chalcea* FIM 02–523 yielded three cyclic depsipeptides called rakicidins G–I, which had strong cytotoxic effects on the human pancreatic cancer cell line PANC-1 and the human colon carcinoma cell line HCT-8 (Nie et al. 2018). The length of the hydroxy fatty acid moiety of rakicidins G–I. *Micromonospora* sp. FIM05328 yielded the 26-membered polyene macrolactam FW05328-1, which exhibited strong bioactivity against three human oesophageal squamous cell cancer cell lines: KYSE30, KYSE180, and EC109 (Nie et al. 2018).

Neumycin B is a cytovaricin-ossamycin-oligomycin macrolide which was isolated from *Micromonospora* sp. CNY-010. Its stereo structure was determined by combining genomic data with nuclear magnetic resonance (NMR) studies. Although it is unstable, neumycin B exhibits strong anti-human glioblastoma cell line U87 action (Kim et al. 2018). Three novel dimeric bohemamine analogs, designated Dibohemamines A–C were obtained from a marine-derived *Streptomyces spinoverrucosus* strain. These analogues were dimerized via a methylene group. It was discovered that dibohemamines B and C exhibited nanometer cytotoxicity towards the non-small cell lung cancer cell line A549 (Fu et al. 2016).

Pactamides A–F are six novel polyketides that were obtained from *Streptomyces pactum* SCSIO 02999 through manipulating genetic promoters. According to Saha et al. (2017) pactamide A exhibited strong cytotoxic activity against human liver cancer cells (HepG2), human large-cell lung carcinoma (NCI-H460), human breast cancer cell (MCF-7), and human glioblastoma cell line (SF-268). Pactamides B–F had low to moderate cytotoxic activity against these human cell lines. According to Ye et al. (2017) two cyclodepsipeptides, streptodepsipeptides P11A and P11B were obtained from *Streptomyces* sp. P11-23B and demonstrated strong anti-proliferative bioactivity in four human glioblastoma cell lines (U251, U87-MG, SHG-44, and C6).

Four novel antimycin alkaloids, antimycins E–H have been found through research using *Streptomyces* sp. strain THS-55. These antimycins demonstrated a strong cytotoxic effect on HPV-transformed human cervix adenocarcinoma (HeLa) cells, as well as an average anti-proliferative effect against the human cervical cancer cell SiHa, human myelogenous leukemia cell line K562, and human leukaemia cell line. However, cytotoxic effects were detected in normal embryonic kidney cells 293T of human (Zhang et al. 2017). Two novel alkaloids, geranylpyrrol A and piericidin F have been identified in a strain of *Streptomyces* sp. CHQ-64 knockout mutant. Among these, piericidin F has shown strong anti-proliferative effects on a number of cancer cell lines, such as HeLa, NB4, human acute promyelocytic leukemia, and A549 and H1975 human lung carcinoma (Han et al. 2017).

Two novel chromodepsipeptides, neo-actinomycin A and neo-actinomycin B, were identified from the marine *Streptomyces* sp. IMB094 (Wang et al. 2017). These demonstrated a high cytotoxic effect on the human colorectal carcinoma cell line HCT116 and the human lung carcinoma cell line A549. A marine *Streptomyces caniferus* GUA-06-05-006A was the source of two novel macrolides, PM100117 and PM100118 (Pérez et al. 2016). Human lung carcinoma (A549), coloractal cancer

(HT-29), and breast adenocarcinoma (MDA-MB-231) all showed strong cytotoxic effects in response to PM100117 and PM100118. Two lobophorins, H and I were isolated from a symbiotic strain of *Streptomyces* sp. (strain 1053U.I.1a.3b) of cone snails (Lin et al. 2014). Lobophorins are a broad class of spirotetronates that have antibacterial and cytotoxic properties. Among these, lobophorin I exhibited strong cytotoxic action against the CEM-TART human T-cell leukemia cell line.

Marine *Streptomyces* provide the majority of bioactive substances utilized in the pharmaceutical and medical sectors as effective treatments against multiple diseases. Pure compound 1, 2-benzene dicarboxylic acid, mono 2-Ethylhexyl ester (DMEHE), derived from marine-derived actinomycete *Streptomyces* sp. VITSJK8, was tested for its cytotoxic activity against mouse embryonic fibroblast (NIH 3T3), human keratinocyte (HaCaT) normal cell lines, human hepatocellular liver carcinoma (HepG 2), and human breast adenocarcinoma (MCF-7) cell lines. The compound showed IC50 values of 42, 100, 250, and 500 lg/ml, respectively (Krishnan et al. 2014).

Streptomyces paradoxus VITALK03 produced a secondary metabolite, Gancidin W also known as 3-(2-methyl propyl)-octahydropyrrolo (1,2-a) piperazine-1, 4-dione. It demonstrated cytotoxicity for MCF-7 breast cancer cells, with an IC50 value of 1.56 lg/mL, while the VERO cells (control) had an IC50 value of 12.5 lg/mL. Thus, gancidin W may be utilized in the development of a medication to treat breast cancer (Ravi et al. 2017). Varghese et al. (2015) found that *Nocardiopsis* sp. VITKSM06 had substantial cytotoxicity against HeLa cells, with an IC50 value of 5–9 μg/ml. The cells displayed compacted chromatin, which might be attributed to inhibition of Histone Deacetylase Inhibitors (HDAC) (Varghese et al. 2015).

Streptoglutarimide H and known streptovitacin A were among the new streptoglutarimides A–J, identified from the marine-derived actinomycete *Streptomyces* sp. ZZ741, which suppressed the growth of human glioma U87MG and U251 cell lines. Streptovitacin A exhibited greater potency compared to doxorubicin, the positive control (IC50 values: 1.6 and 6.8 μM, respectively) (Zhang et al. 2019a).

Two known analogues and the four new polyketide ansamycins, divergolides (macrolides) T, U, V, and W have been obtained from the fermentation broth of the actinomycete *Streptomyces* sp. KFD18 isolated from mangroves. The compounds have shown to be cytotoxic against human gastric cancer cell line SGC-7901, a human leukemic cell line K562, cervical cancer cell line HeLa, and the lung cancer cell line A549. Compounds T and V demonstrated strong and selective cytotoxic effects on SGC-7901 cells (IC50 values of 2.8 and 4.7 μM, respectively). The result indicated that these compounds were more cytotoxic to the cell line than the two positive controls, adriamycin (IC50 = 6.9 μM) and imatinib (IC50 = 86.8 μM) (Zhou et al. 2019).

A novel and two known aromatic polyketide cytotoxic chemicals that hindered the growth of a murine B16 melanoma cancer cell line were generated by the deep-sea actinomycete *Nonomuraea* sp. AKA32. Akazamicin, actinofuranone C, and N-formylanthranilic acid constitute the novel chemical. According to Yang et al. (2019), the compounds' cytotoxic IC50 values were 1.7, 1.2, and 25 μM, respectively. *Streptomyces* spp. produced a novel monomeric julichrome, julichrome derivatives, and a gliotoxin known as diketopiperazine. Cytotoxic activity of the compound was tested against human hepatocarcinoma cell lines HepG-2 and SMMC-7721, human breast cancer cell lines MCF-7 and MDA-MB-231, and a human not cancerous hepatic cell line (L-02) (Shen et al. 2021).

Three novel anthracyclinones, boshramycinones A–C, 2-acetyl-1, 8-dihydroxy-3-methyl-anthraquinone, bafilomycin B1, bafilomycin B2, and C1-amide were all produced in the culture broth of the marine-derived *Streptomyces* sp. Mei 16-1, 2. The cytotoxic activity of the crude extract and the separated compounds was tested against 36 human cancer cell lines. With a mean IC50 value of 38.30 _M/mL, boshramycinone B demonstrated a non-selective and relatively low cytotoxic effect on the 36 human cancer cell lines. With an IC50 range of 0.51–17.27, boshramycinone A was remarkably more potent (Shaaban et al. 2021). The culture broth of *Streptomyces* sp. was used to produce sekgranaticin, granaticins A–B, and methyl granaticinate. A unique hybrid polyketide, sekgranaticin has a complicated 7-ring system. These compounds inhibited A549, MCF-7, P6C, and HCT-116 cancer cell lines with the IC50 values of 0.02–6.77 μM (Lv et al. 2019).

Shellmycins A–D were four novel cytotoxic tetrahydroanthra pyrone compounds isolated from *Streptomyces* sp. shell-016 in a marine shell sediment sample collected from Binzhou Shell Dike Island and Wetland National Nature Reserve in China. The compounds showed anticancer activity against five cancer cell lines: A375, H1299, HepG2, HT29, and HCC1937, with IC50 values ranging from 0.69–26.3 μM (Han et al. 2020). The *Actinoalloteichus hymeniacidonis* marine sediment actinomycete was collected from the beaches of Dokdo Island in the Republic of Korea. Dokdolipids A, B, and C are three novel hydroxylated rhamnolipids produced by the actinomycetes. The cytotoxic activity of the three compounds was assessed against six human cancer cell lines: PC-3, MDA-MB-231, ACHN, NCI-H23, HCT-15, and NUGC-3. The compounds had an IC50 of moderately harmful to the cell lines (Choi et al. 2019a).

Gigamicidin (ULDF4) and staurosporine (ULDF5), two known compounds, were obtained from the extract of *Streptomyces bingchenggensis* ULS14 from the Lagos Lagoon in Nigeria. The IC50 values of ULDF4 and ULDF5 against HeLa cell growth were 0.11 and 0.24 μM/mL, respectively (Davies-Bolorunduro et al. 2019). Lactoquinomycin C and D, two novel medermycin derivatives, and six other known compounds were produced via rice-solid fermentation of marine *Streptomyces* sp. SS17A. Lactoquinomycin C and D were not toxic to PC-3 and HCT-116 human malignant tumor cell lines. However, 3-methyl-antibiotic G-15F significantly suppressed their development with IC50 values of 0.02 and 0.04 μM, respectively. According to research, Lactoquinomycin C and D require the -lactone unit in order to have considerable cytotoxicity (Zhou et al. 2020a).

Three novel p-terphenyl derivatives, nocarterphenyls A–C, were produced by the marine-derived actinobacterial strain *Nocardiopsis* sp. OUCMDZ-4936. Among the 26 human cancer cell lines used for screening, these compounds demonstrated strong cytotoxic activity against several cell lines, with IC50 values ranging from 0.10 to 17.0 μM (Wang et al. 2019).

Terrosamycins A and B, two novel polycyclic polyether compounds were obtained through purification of a fermented broth of *Streptomyces* sp. RKND004 isolated from sediment on Prince Edward Island. The OSMAC method and a metabolomics assay based on UPLC-HRMS were used to identify the chemical substances. The structure of the compound was inferred from data analyses of X-ray diffraction, NMR, and HRMS. Additionally, the screening revealed that these compounds were more selective for cancer cells than normal cells, such as the human foreskin fibroblast cell line (BJ) (IC50 = 79 and 267 μM, respectively) and healthy Cercopithecus aethiops kidney epithelial cells (Vero) (IC50 = 36 and 93 μM, respectively) (Sproule et al. 2019).

Padanamides (A and B) are significantly modified linear tetrapeptides synthesized by *Streptomyces* sp. isolated from sediments from the marine environment. The molecular weights of padanamide A ($C_{31}H_{47}N_7O_9$) and padanamide B ($C_{31}H_{46}N_6O_9$) are 669.31 and 634.33 Da, respectively. According to reports, padanamide B exhibits cytotoxic activity against human lymphocytes, or Jurkat cells, whereas padanamide A exhibits very mild cytotoxicity towards same cell (Williams et al. 2011).

Cyclic hexadepsipeptides known as piperazimycins A, B and C are produced by *Streptomyces* sp. that have been isolated from maritime sediments close to the island of Guam. Rare amino acids include hydroxyacetic acid, a-methylserine, g-hydroxypiperazic acid, gchloropiperazic acid, 2-amino-8-methyl-4,6-nonadienoic acid, and 2-amino-8-methyl-4,6-decadienoic acid are the building blocks of piperazimycins. Piperazimycin A exhibited strong cytotoxicity against a variety of cancer cell lines with varying origins (Miller et al. 2007). Furthermore, compared to the evaluated leukemia cell lines, peperazimycin A demonstrated three times greater activity towards solid tumors in a different investigation (Li et al. 2009).

Marine fungi

Approximately 1.5 million to 3 million fungal species occur worldwide (Hawksworth et al. 2012). *Aspergillus*, *Fusarium*, and *Penicillium* are the three species of filamentous fungi of terrestrial origin that have a rich profile of physiologically active metabolites that have been identified (Lange 1996). Yet the marine fungi are the least explored as compared to other terrestrial counterparts along with other ecological groups. Marine fungi are found in the ocean and colonize a different niche. Obligate marine fungi remain an unknown resource, although marine facultative fungi have been examined due to their ability to produce novel compounds that are not present in terrestrial fungi. Compared to their terrestrial counterparts, marine fungi offer a variety of distinct sources for novel anticancer drugs (Butler et al. 2014). Marine fungi are untapped source of novel antibacterial, antiviral, anticancer, antiplasmodial and anti-inflammatory compounds (Dudhagara et al. 2023). Since the majority of fungi can thrive in peculiar and harsh conditions, they can produce peculiar and distinctive secondary metabolites (Fenical and Jensen 1993). Anticancer chemicals have been found in marine fungi that originate from various sources such as deep-sea sediments, algae, sponges, endophytic mangroves, along with other marine fungi (Table 3) (Gomes et al. 2015).

Table 3. Classification and sources of potential anticancer compounds derived from marine fungi.

Type	Compound	Source	Reference
Peptides	Simplicilliumtides	*Simplicillium obclavatum* EIODSF 020	Zhou et al. 2016
	Scopularide A and B	*Scopulariopsis brevicaulis*	Yu et al. 2008
	Sansalvamide A	Marine fungi	Vasko et al. 2010
Indole	Varioloid A	*Paecilomyces variotii* EN-291	Zhang et al. 2016b
Quinone	Leptosphaerodione	*Leptosphaeria oraemaris*	Suja et al. 2014
	Auxarthrols	*Sporendonema casei*	Ge et al. 2019
	Peniquinone	*Penicillium* sp. L129	Zhang et al. 2019b
	Penizofuran	*Penicillium* sp. L129	Zhang et al. 2019b
	Acremonin	*Acremonium* sp.	Abdel-Lateff et al. 2002
	Pericosines	*Periconia byssoides* OUPS-N133	Yamada et al. 2007
	Toluquinol	Marine fungus	García-Caballero et al. 2014
	Aspergiolide	*A. glaucus*	Du et al. 2007
	Anthracenedione	*Halorosellinia* spp. and *Guignardia* spp.	Zhang et al. 2010
Polyketide	Azaphilones	*Chaetomium globosum* MP4-S01-7	Wang et al. 2020
Alkaloid	Meleagrin	*Emericella dentata* Nq45	Hamed et al. 2021
	Roquefortine H and I,	*Penicillium* sp.	Du et al. 2010
	N-glutarylchaetoviridins	*Chaetomium globosum* HDN151398	Sun et al. 2019
	Ascomylactams	*Didymella* sp. CYSK-4	Chen et al. 2019
	Phomapyrrolidone	*Didymella* sp. CYSK-4	Chen et al. 2019
	Phomapyrrolidone	Didymella sp. CYSK-4	Chen et al. 2019
	Deoxyapoaranotin, acetylaranotin, acetylapoaranotin	*Aspergillus versicolor* KMD 901	Choi et al. 2011
	Chaetoglobosins	*Chaetomium globosum* C2F17	Luo et al. 2020
	7'-demethoxyrostratin	*Epicoccum nigrum* SD-388	Chi et al. 2020
	Ochrazepine	*Aspergillus ochraceus*	Fan et al. 2019
Xanthone	Xanthone derivative	*Wardomyces nomalus*	Abdel-Lateff et al. 2002
	Cytarabine	*Halorosellinia* sp.	Molinski et al. 2009
	Secalonic acid D	Mangrove endophytic fungus	Zhang et al. 2009

Table 3 contd. ...

...Table 3 contd.

Type	Compound	Source	Reference
Terpenoids	Insulicolidea	*Aspergillus flocculosus*	Insulicolidea
	Eutypellacytosporins	*Eutypella* sp. D-1	Zhang et al. 2019c
	Sorbicillactone-B	*Penicillium chrysogenum*	Bringmann et al. 2007
	Andrastane A	*Penicillium*	Xie et al. 2019
Terpene	Versicolorin	*Aspergillus versicolor* ZZ761	Zhang et al. 2021
	Aphidicolin	*Botryotinia fuckeliana*	Vinh et al. 2019
	Penicindopene	*Penicillium* sp. YPCMAC1	Liu et al. 2019b
	Ophiobolins	*Aspergillus flocculosus*	Choi et al. 2019b
	Scopararane I	*Eutypella* sp. FS46	Liu et al. 2017
	Diorcinol	*Aspergillus versicolor* ZZ761	Zhang et al. 2021
Steroid	16-_-methylpregna-17_, 19-dihydroxy-(9, 11)-epoxy-4-ene-3,18-dione-20-acetoxy	*Penicillium citrinum* SCSIO 41017	Salendra et al. 2021
	Penicisteroids	*Penicillium granulatum* MCCC 3A00475	Xie et al. 2019
Lactone	Penicilactones	*Penicillium chrysogenum* MCCC 3A00292	Niu et al. 2019
Polyketal	Lithocarols	*Phomopsis lithocarpus* FS508	Xu et al. 2019a
Polyketone	2-hydroxyl-3-pyrenocine-thio propanoic acid	*Penicillum citreonigrum* XT20-134	Tang et al. 2019
Ester	2,3,4-trihydroxybutyl cinnamate	*Penicillum citreonigrum* XT20-134	Tang et al. 2019
Sterol	Eburicol	*Clonostachys rosea* MMS1090	Dos Santos Dias et al. 2019
Cyclohexenoid	Monarubin	*Monascus ruber* BB5	Ran et al. 2020
Aromatic compound	P-terphenyl derivatives	*Aspergillus candidus*	Silchenko et al. 2019
	Prenylterphenyllins	*Aspergillus candidus* LDJ-5	Zhou et al. 2020b
Antibiotic	Spiroxin A	Unidentified fungus (strain LH-37H248	McDonald et al. 1999

Toluquinol is obtained from marine fungus that inhibits the unrestricted ability for replication of tumor cells, which is one of the hallmarks of cancer. The growth of the colon adenocarcinoma HT29 cell line, fibrosarcoma HT1080 cell line, and promyelocytic leukemia HL60 cell line is inhibited by toluquinol. In addition to suppressing cancer angiogenesis, IC50 values of the toluquinol were less than 10 μM in all three cell lines (García-Caballero et al. 2014).

Sorbicillactone-B has been produced from a salt-water culture of *Penicillium chrysogenum* derived from the Mediterranean sponge species *Ircinia fasciculata* (Bringmann et al. 2007). Diketopiperazines (DKPs) found in marine resources, particularly the ones isolated from marine-derived fungus have received increased interest because to their chemical structure and bioactivity variety. *Aspergillus ustus* NSC-F038 is the source of halimide ((-)-phenylahistin), a fungus prenylated DKP that stopped P388's cancer cell cycle in the G2/M phase (Kanoh et al. 1997). Scopararane I, a pimarane-type diterpene, and its anologus have been isolated from the culture medium of a marine sediment-derived fungus *Eutypella* sp. FS46, which showed moderate cytotoxic activity against MCF-7, NCI-H460, and SF-268 tumor cell lines (Liu et al. 2017).

The marine alga-derived endophytic fungus *Paecilomyces variotii* EN-291 has been known to produce varioloid A, which was cytotoxic to A549, HCT116, and HepG2 cell lines (Zhang et al. 2016b). Eight novel linear pepetides simplicilliumtides A-H were found in a culture broth of the deep-sea derived fungus strain *Simplicilliumobclavatum* EIODSF 020. Simplicilliumtides A, E, G, and H have limited anticancer action against the HL-60 or K562 cell lines (Liang et al. 2016). Novel anti-cancer compounds penicilazaphilones B and C, which have shown anti-cancer properties against human gastric cancer cells SGC-7901 and melanoma cells B-16 were the derivatization analogues of azaphilonidal (Zhou et al. 2016).

Calanic acid D, a mycotoxin that was extracted from the mangrove endophytic fungus, is another chemical belonging to the ergochrome class that has good cytotoxic effect on HL60 and K562 cells by inducing apoptosis in leukemia cells (Zhang et al. 2009). Acremonin A from *Acremonium* sp. (Abdel-Lateff et al. 2002) and Xanthone derivative from *Wardomycesnomalus* (Abdel-Lateff et al. 2003) are two examples of marine-derived fungus that produce antioxidants. Free radical reactions have been linked to diseases such as atherosclerosis, dementia, and cancer. Antioxidants have the potential to be used as medicines or food additives since they delay or prevent oxidation. Trabectedin (a treatment for soft tissue sarcoma) and Cytarabine (an antileukemic medication) are the source of the anthracenedione derivatives that were screened from the mangrove endophytic fungus *Halorosellinia* sp. and *Guignardia* sp. (Molinski et al. 2009, Tao et al. 2010).

Interestingly, two novel cyclodepsipeptides called Scopularide A and B, have been identified as originating from the marine fungus *Scopulariopsis brevicaulis* (Yu et al. 2008). These peptides inhibited the growth of pancreatic and colon tumor cell lines. It is yet unclear what precise mechanism is causing this inhibition. Sansalvamide A, a structurally distinct cyclic depsipeptide that was extracted from a variety of marine fungi showed cytotoxic properties against a range of cell lines, including those from the pancreas, colon, breast, prostate sarcoma, and melanoma in recent research. Yet, the exact mechanism is uncertain and suggests that Sansalvamide A interacts with the heat shock protein (HSP90) N-middle domain, preventing the formation of protein complexes that are essential for the growth of tumors in mammalian cell lines (Vasko et al. 2010).

Leptosphaerin, leptosphaerolide (and its O-dihydroquinone derivative), and leptosphaerodione from the lignicolous fungus *Leptosphaeria oraemaris* (Pleosporaceae) were produced by higher fungi (basidiomycetes), endophytic fungi, and filamentous fungi from marine habitats (Guerriero et al. 1991). The topoisomerase I inhibitor (+)-3,3,7,7,8,8-hexahydroxy-5,5-dimethylbianthraquinone, isolated from both *Aspergillus*, showed in vitro cytotoxicity and anticancer activity (Suja et al. 2014). The study conducted by Du et al. (2007) revealed that Aspergiolide A, which was obtained from the marine filamentous fungus *A. glaucus*, exhibited cytotoxicity against A-549, HL-60, BEL-7402, and P388 cell lines. Meanwhile, alkaloids that were recovered from *Penicillium* spp. and generated from deep-ocean sediment shown anticancer activity.

The study conducted by Du et al. (2007) revealed that Aspergiolide A, which was obtained from the marine filamentous fungus *A. glaucus*, exhibited cytotoxicity against A-549, HL-60, BEL-7402, and P388 cell lines. Meanwhile, alkaloids that were obtained from *Penicillium* spp. from deep-ocean sediment showed anticancer activity. Two new alkaloids, meleagrin D and E, and two new diketopiperazines, roquefortine H and I, showed lethal action against A-549 and HL-60 cells by apoptosis, arresting the cell cycle at the G2/M phase (Du et al. 2010). The anticancer efficacy of 14 anthracenedione derivatives of secondary metabolites from the mangrove endophytic fungus *Halorosellinia* spp. and *Guignardia* spp. has also been reported (Zhang et al. 2010). The 14 anthracenedione derivatives were discovered to work through apoptosis induction.

A novel diketopiperazine disulfide called deoxyapoaranotin, along with acetylaranotin and acetylapoaranotin was produced by a bioassay-guided fractionation of an EtOAc extract from a culture of the marine fungus *Aspergillus versicolor* KMD 901. The fungal species was isolated from a deep-sea sediment from the Ulleung Basin in the East Sea of Korea (Choi et al. 2011). It was discovered that the compounds could cause the HCT116 colon cancer cell line to undergo apoptosis via both intrinsic and extrinsic routes. They were also discovered to cleave caspases-3, -9, and -8, indicating that caspase 8 initiates the apoptotic process. These findings also showed that the pro-apoptotic protein Bax was up-regulated while the anti-apoptotic proteins Bcl-2 and Bcl-xL were down-regulated. Apart from its *in vitro* apoptotic action, acetylapoaranotin also significantly inhibited tumor development in an *in vivo* xenograft model. This compound showed the strongest proliferation inhibition effects in HCT116, AGS, A549, and MCF-7 cancer cell lines (Choi et al. 2011).

N-glutarylchaetoviridins A, B, and C are three novel azaphilone alkaloids containing glutamine residues that were extracted from the fungal extract of *Chaetomium globosum* HDN151398. Glutamate residues were first seen in N-glutarylchaetoviridins A–C. Five azaphilone-loaded different amino acid residues were formed when the fungal strain was incubated with amino acids. Thus, by cultivating with amino acids, this approach can enhance the structural diversity of this strain. However, as compared to a standard, adriamycin has an IC50 range of 0.1–0.6 μM. None of these compounds showed a strong cytotoxic action (IC50 5.7–19.4 μM) against a number of eleven human cancer cell lines which were examined (Sun et al. 2019).

Ascomylactams A, B, and C, as well as the known compounds phomapyrrolidone C and phomapyrrolidone A, which were isolated from the mangrove endophytic fungus *Didymella* sp. CYSK-4, all moderately inhibited the growth of MDA-MB-435, MDA-MB-231, SNB19, HCT116, NCI-H460 and PC-3 human cancer cell lines, respectively, within the IC50 range of 4.2–7.8 μM. A single-crystal X-ray diffraction investigation was also performed on the ascomylactam A and B moiety, which consists of tetracyclic skeletons joined with either a 12- or 13-membered macrocylic ring (Chen et al. 2019).

Sixteen structurally distinct chaetoglobosins, 10-(indol-3-yl)-(13) cytochalasans, and a novel 6-O-methyl-chaetoglobosin Q have been extracted from the coral-associated fungus *Chaetomium globosum* C2F17 through HPLC-DAD-guided extraction procedure. Only chaetoglobosins E and F with an IC50 range of 1.4–9.2 μM showed considerable cytotoxic action against K562, A549, Huh7, H1975, MCF-7, U937, BGC823, HL60, HeLa, and MOLT-4 cancer cell lines (Luo et al. 2020).

The deep-sea fungus *Epicoccum nigrum* SD-388 extract yielded the following four new thiodiketopiperazine alkaloids: 5'-hydroxy-6'-ene-epicoccin G, 7-methoxy-7'-hydroxyepicoccin G, 8'-acetoxyepicoccin D, 7'-demethoxyrostratin C, and two new enantiomeric diketopiperazines: (+/–)-5-hydroxydiphenylalazine A. The 7'-demethoxyrostratin C compounds demonstrated strong antiproliferative action against Huh7.5, with IC50 values of 9.52 μM. It appears that the disulfide bridge located at C-2/C-2' is essential for the anticancer activity (Chi et al. 2020).

Four novel conjugates of ochrazepine A, B, C, and D were isolated through the fermentation of broth using coral associated *Aspergillus ochraceus* strain LCJ11-102. These conjugates dimerized from 2-hydroxycircumdatin C and aspyrone by the nucleophilic addition to the epoxide, while Ochrazepine A–D were produced by a semisynthesis comprising the nucleophilic addition of 2-hydroxycircumdatin C to aspyrone. Ochrazepine C was active against A673, U87, and Hep3B cell lines with IC50 values of 2.5–11.3 M among the 26 examined human cancer cell lines, while Ochrazepine A was cytotoxic to ten human cancer cell lines (Fan et al. 2019).

Eutypella sp. D-1, an arctic fungus known for producing cytotoxic cyclopropyl- and cyclobutyl-fused pimarane diterpenoids on conventional media, was stimulated to produce extracts on rice medium. This led to the finding of four new meroterpenoids, eutypellacytosporins A-D, and the known compound cytosporin D. These novel compounds exhibited antiproliferative activity against DU145, SW1990, Huh7, and Panc-1 cell lines, with IC50 values ranging from 4.9 to 17.1 μM (Zhang et al. 2019c).

The isolation of marine compounds from *Aspergillus versicolor* ZZ761 yielded one novel chemical, indoloditerpene, as well as fifteen known compounds. Only versicolorin B and diorcinol showed antiproliferative activity against human glioma U87MG and U251 cell lines, with IC50 values of 11.3 and 30.5 μM and 4.4 and 6.2 μM, respectively (Zhang et al. 2021).

Two drimane derivatives, one meroterpenoid 12-epi-aspertetranone D, one aspyrone-related polyketide aspilactonol G, four novel compounds, and five known compounds were isolated from the Vietnamese maritime sediment-derived fungus *Aspergillus flocculosus*. The human prostate cancer cell line 22Rv1, the human breast cancer cell line MCF-7, and the mouse neuroblastoma Neuro-2-A cell line were used to test these compounds for anticancer activities. However, only insulicolideA demonstrated cytotoxic action against the 22Rv1 and Neuro-2-A cell lines with IC50 values of 3.0 and 4.9 μM (Yurchenko et al. 2019).

Penicillium allii-sativi a deep-sea fungus produced two new andrastane A compounds, 16-epi-citreohybriddione A, and one known meroterpene. Andrastane A is a rare meroterpene that has the cyclopentan-1, 3-dione group but lacks the lactone moiety. The antiproliferative activity of the three compounds was tested against Panc-1, HepG2, A549, BIU-87, BEL-7402, ECA-109, and HelaS3 cancer cells. However, only Andrastane A (IC50 = 7.8 μM) showed cytotoxic action against HepG2 together with a selective growth inhibition. On the other hand, research on the compound›s mode of action revealed that it directly activated caspase-3 through caspase-8 inducing death in HepG2 cells. Furthermore, the compound enhanced the reporter transcriptional activation of retinoid X receptor alpha (RXR) and decreased the trans-activity of RXR induced by 9-cis-RA, according to a dual luciferase reporter gene assay used to study the transcriptional activity of RXR (Xie et al. 2019a).

Aphidicolin is a DNA polymerase alpha inhibitor that has been studied in clinical trials for cancer therapy. Approximately 300 mutated aphidicolins have been found recently but none of them have demonstrated a detectable biological effect. However, seventy-one new aphidicolins A176-A246 as well as eight known aphidicolin derivatives were reported from the fungus *Botryotinia fuckeliana* MCCC 3A00494 in the western Pacific Ocean. Penicicindopene A, a novel indole diterpene, and seven known compounds were identified from the deep-sea fungus *Penicillium* sp. YPCMAC1. Penicindopene A, the first indole diterpene having a 3-hydroxyl-2-indolone moiety was found to be moderately cytotoxic to A549 and HeLa cell lines, with IC50 values of 15.2 and 20.5 μM, respectively (Liu et al. 2019).

Aspergillus flocculosus, a marine fungus derived from the Vietnamese seaweed Padina sp. was used to isolate five new sesterterpenes: 14,15-dehydro-6-epi-ophiobolin K, 14,15-dehydroophiobolin K, 14,15-dehydro-6-epi-ophiobolin G, 14,15-dehydro-ophiobolin G, and 14,15-dehydro-(Z)-14-ophiobolin G, as well as four known ophiobolins. The cytotoxicity of all the compounds was tested against six cancer cell lines: MDA-MB-231, ACHN, PC-3, NCI-H23, HCT-15, and NUGC-3. The results showed that all the compounds were potently cytotoxic to the cancer cell lines, with an IC50 value range of 0.14 to 2.01 μM (Choi et al. 2019b).

Two novel polyketides, xerucitrinin A and coniochaetone M, one new steroid, 16-_-methylpregna-17, 19-dihydroxy-(9,11)-epoxy-4-ene-3,18-dione-20-acetoxy, and eleven known analogues were obtained from the isolation of the fungus *Penicillium citrinum* SCSIO 41017, which has been associated with the sponge *Callyspongia* sp. 19-dihydroxy-(9, 11)-epoxy-4-ene-3,18-dione-20-acetoxy showed modest activity with an IC50 range of 13.5–18.0 μM against all cell lines (Salendra et al. 2021).

The deep-sea derived fungus *Penicillium chrysogenum* MCCC 3A00292 yielded three new versiol-type derivatives (peniciversiols A–C), two unique lactone derivatives (penicilactones A and B), and eleven known polyketides upon chemical treatment of the solid cultures. These substances were tested for their ability to inhibit the growth of five human cancer cell lines: BIU-87, ECA109, BEL-7402, Panc-1, and HeLa-S3. Moreover, the proliferations of the cell lines ECA109, BIU-87, and BEL-7402 was suppressed by penicilactones A and B (Niu et al. 2019).

Five new anthraquinone derivatives, auxarthrols D, E, F, G, and H, as well as two recognized compounds were isolated from the marine fungus *Sporendonema casei*. The seven compounds were tested for anticancer efficacy against ten human cancer cell lines, including HL-60 Hela, HCT-116, MGC-803, HO8910, MDA-MB-231, SH-SY5Y, PC-3, BEL-7402, and K562, as well as a non-cancer cell line, L-02. Auxarthrols D and F shown cytotoxic activity with IC50 values ranging from 4.5 μM to 22.9 μM (Ge et al. 2019). The eight new nitrogenated azaphilones, as well as two known compounds, chaetoviridin A and chaetoviridin E, were isolated using the deep-sea fungus *Chaetomium globosum* MP4-S01-7 culture. The compounds were tested for cytotoxic activity in vitro against the gastric cancer cell lines MGC803 and AGS. Nitrogenated azaphilones had the most potent cytotoxic action on cancer cells, with IC50 values less than 1.0 μM (Wang et al. 2020).

DC1149Ba, a new epidithiodiketopiperazine isolated from *Trichoderma lixii*, an unidentified sponge from Mentawai, Indonesia, reduced Panc-1 growth with an IC50 of 0.02 μM. This cell line was cultivated in a glucose-deficient medium, and the selectivity index was 35,500 times greater for glucose-starved cells than for ordinary culture conditions. Similarly, this drug decreased endoplasmic reticulum stress signaling, which might be explained by blocking complex II in the mitochondrial electron transport chain (Tang et al. 2020).

Five new ergostanes, penicisteroids D–H, and twenty-seven known compounds were identified from a liquid culture of *Penicillium granulatum* MCCC 3A00475, a deep-sea fungus. Some substances had mild cytotoxic effects on twelve cancer cell lines, with IC50 values ranging from 5 to 16.6 μM. The compounds may trigger apoptosis by retinoid X receptor (RXR)-dependent modulation of RXR transcriptional expression, which promotes poly-ADP-ribose polymerase (PARP) cleavage. Furthermore, they may be cytotoxic by causing cell cycle arrest in the G0/G1 phase (Xie et al. 2019b).

Twenty-four compounds (five of which are novel) were obtained from deep-sea sediment fungus *Penicillumcitreonigrum* XT20-134 (MCCC 3A00956). These compounds include adeninylpyrenocine, 2-hydroxyl-3-pyrenocine-thio propanoic acid, ozazino-cyclo-(2,3-dihydroxyl-trp-tyr), 5,5-dichloro-1-(3,5-dimethoxyphenyl)-1,4-dihydroxypentan-2-one, and 2,3,4-trihydroxybutyl cinnamate. The rest are known compounds. All the compounds were evaluated for anticancer activity against Bel7402 and HT1080 human cancer cells. Tang et al. (2019) found that the only compounds that inhibited the proliferation of those cancer cells were 2,3,4-trihydroxybutyl cinnamate containing a heteroatom and 2-hydroxyl-3-pyrenocine-thio propanoic acid, with IC50 values of 7.63, 13.14 μM and 10.22, 16.53 μM, respectively.

Seven culture media of *Clonostachys rosea* MMS1090 had been modified to improve the production of eburicol by using the one-strain-many-compounds strategy. NMR structural analysis and purification were followed by testing eburicol against four human cancer cell lines: MCF-7, MDA-MB-231, NSCLC-N6-L16 and A549. On the other hand, MCF-7 and MDA-MB-231 cells showed growth inhibition against eburicol with 2.0 and 15.7 μM IC50 values, respectively. The marine fungus strain *C. rosea* has been found to accumulate eburicol for the first time, which supports the strain's ability to synthesize bioactive lipids (Dos Santos Dias et al. 2019).

The marine-derived fungus *Penicillium* sp. L129 produced thirteen known substances, including two new dimeric 1, 4-benzoquinone derivatives, peniquinone A and peniquinone B, a new dibenzofuran, penizofuran A, and a new pyrazinoquinazoline derivative, quinadoline D. Peniquinone A suppressed the proliferation of MCF-7, U87, and PC3 human neoplasms with the corresponding IC50 (μM) values of 12.39, 9.01, and 14.59; however, dibenzofuran penizofuran A suppressed their growth with higher IC50 values of 25.32, 13.45, and 19.93 μM, respectively (Zhang et al. 2019b).

Phomopsis lithocarpus FS508, a marine-derived fungus, produced the novel lithocarols A–F with a highly oxygenated isobenzofuran core and a related known isoprenylisobenzofuran A. Four human malignant tumor cell lines were found to be moderately cytotoxic by lithocarols A–E, with an IC50 range of 10.5–38.7 μM (Xu et al. 2019 a). Through a search for novel anticancer agents in marine samples, Fan et al. (2020) identified several novel anticancer agents through their research in marine samples. They isolated three new derivatives of decalinoylspirotetramic acid, called pyrenosetins A, B, and C, from the C18-SPE fractions of the trichloromethane sub-extract of the endophyte *Pyrenochaetopsis* sp. FVE-001, which is a fungus derived from the brown alga *Fucus vesiculosus*. Additionally, they identified one known decalin tetramic acid, phomasetin. Pyrenosetins A and B demonstrated anticancer efficacy against the melanoma cell line A-375, with IC50 values of 2.8 and 6.3 μM, respectively (Fan et al. 2020).

The effect of cultivating the marine-derived fungus *Trichoderma erinaceum* F1-1 in glucose-peptone-yeast (GPY) medium enriched with L-phenylalanine, a single amino acid, or not was examined by Guo et al. (2019). Eighteen aromatic compounds, including six new compounds named trichoderolides A–F from the GPY medium with L-phenylalanine, and seven terpenoids and one polyketide, of which four were new compounds, namely, harziandione A, cyclonerodiols A and B, and trichodermaerin A, were obtained from the GPY medium without L-phenylalanine. The compounds were found to be cytotoxic to the MDA-MB-435 human melanocyte cancer cell and the A549 human alveolar basal epithelial cell adenocarcinoma (Guo et al. 2019).

The fungus *Emericella dentata* Nq45 yielded meleagrin as well as other well-known substances, including haenamindole, isorugulosuvine, secalonic acid D, ergosterol, and cerebroside A. Maleagrin were found to be cytotoxic to human cervix cancer cell line KB-3-1 and its multi-drug-resistant sub-clone KB-V1, with IC50 values of 3.07 and 6.07 μM, respectively (Hamed et al. 2021). From the marine shellfish-associated fungus *Monascusruber* BB5, the following three new monarubins—A, B, and C as well as ten known compounds—four alkaloids, two isocoumarins, and four polyketides were isolated. They were then tested for antiproliferative activity against human nasopharyngeal (CNE1, CNE2, CNE1, and HONE1) and hepatocellular (QGY7701 and HepG2) carcinoma cell lines. Lunatinin was moderately cytotoxic to HepG2, QGY770, and SUNE1, with IC50 values of 9.60, 7.12, and 28.12 μM, respectively; monarubin B was potently cytotoxic to HepG2 and QGY7701, with IC50 values of 1.72 and 0.71 μM (Ran and associates 2020).

Nine novel prenylated p-terphenyls, namely prenylterphenyllins F–J and prenylcandidusin D–G were isolated from *Aspergillus candidus* LDJ-5, an endophytic fungus. Data from NMR and MS were used to determine their structures. Unlike a study on p-terphenyls that was previously published, prenylcandidusin D features an uncommon 6, 5, 6, 6-fused ring structure. Prenylterphenyllins J and prenylcandidusin G were cytotoxic to a panel of nine human cancer cell lines with IC50 values within this range (0.40–16.9 μM) (Zhou et al. 2020b).

Aspergillus sp. LS34, a sponge-derived fungus, was cultivated in two distinct media (a solid-rice medium and potato dextrose broth (PDB)). Two new compounds, asperspin A and asperther A, as well as seven recognized compounds, were extracted by employing the one-strain-many-compound (OSMAC) technique. The compounds were generated by the fungal extracts cultivated in both the PDA medium and the rice medium. The compounds demonstrated significant antiproliferative activity against CCRF-CEM and K562 cells, with IC50 values of 1.22 and 10.58 μM, respectively (Lin et al. 2021). Three recognized and one novel p-terphenyl derivatives were extracted from the *Aspergillus candidus*, a fungus found in deep seawater. p-terphenyl derivatives showed strong anticancer activity against four human cancer cell lines: HeLa, Eca-109, Bel-7402, and Panc-1, with IC50 values ranging from 5.5 to 9.4 1 μM (Silchenko et al. 2019).

The fungal strain *Periconiabyssoides* OUPS-N133, which was isolated from the GI tract of the sea hare *Aplysia kurodai*, produces the cytotoxic metabolites known as second pericosines A, B, and D. These metabolites are part of a small class of cyclohexenoids. Pericosine A also showed selective and strong cytotoxicity against HBC-5 and SNB-75 human cancer cell lines, whereas the three fungal metabolites demonstrated strong *in vitro* cytotoxic action against murine P388 cells (Yamada et al. 2007).

Spiroxin A is a cytotoxic antibiotic that was isolated from fungus derived from a soft coral sample taken from Dixon Bay, Canada. When Spiroxin A was tested on a panel of 25 different cell lines, some of which were purportedly partially produced from single-stranded DNA cleavage, it showed high cytotoxicity (IC50 = 0.09 μg/mL). Spiroxin A can act chemically like a quinone epoxide and cause DNA cleavage through an oxidative stress mechanisms involving thiol conjugates because of its unique bisnaphthospiroketaloctacyclic ring structure with the spiroketal carbon oxidation state (McDonald et al. 1999).

Micro algae

Marine cyanobacteria are thought to be a significant source of naturally occurring compounds with a variety of structural and biological properties. Of all marine determined compounds, the potential of marine cyanobacteria as anticancer agents has been studied the most. The cyanobacteria population comprises 150 species and about 2000 types of decent variety. A few mixtures have emerged as models for the development of novel anticancer drugs, aside from their cytotoxicity in tumor cell lines. For example, cyanobacteria species include *Nostoc*, *Calothrix, Lyngbya, Symposia*, and more (Table 4) (Gerwick et al. 2001).

Table 4. Classification and sources of anticancer compounds from marine Cyanobacteria.

Type	Compound	Source	Reference
Alkaloids	Calothrixin	*Calothrix*	Rickards et al. 1999
Peptide	Scytonemin	*Symploca* sp.	Williams et al. 2003b
	Dolastatin 10	*Symploca* sp.	Kalemkerian et al. 1999
	Symplostatin 1	*Symploca* sp.	Mooberry et al. 2003
	Odoamide	*Okeania* sp.	Sueyoshi et al. 2016
	Galaxamide	*Galaxaura filamentosa*	Xu et al. 2008
	Microcolins	Moorea producens	Yu et al. 2019
	Largazole	*Symploca* sp.	Taori et al. 2008
	Apratoxins	*Lyngbya boulloni*	Luesch et al. 2001
	Coibamide A	*Leptolyngbya*	Medina et al. 2008
	Lyngbyabellin B	*Lyngbya majuscula*	Marquez et al. 2002
	Cryptophycin 1	*Nostoc* sp. GSV 224	Moore 1996
	Wewakazole B	Moorea-developed cyanobacteria	Lopez et al. 2016
	Ulithiacyclamide	*Prochloron* spp.	Williams et al. 1989
	Patellamide	*Lissoclinum patella*	Williams et al. 1989
Protein	Scytonemin	*Stigonema* sp.	Stevenson et al. 2002
Polyketide	Borophycin	*Nostoc linckia* and *N. Spongiaeforme* var. *Tenue*	Banker and Carmeli 1998
	Curacin-A	*Lyngbya majuscula*	Carte 1996
	Dragocins	Dark-red wooly textured Cyanobacterium	Choi et al. 2018
Oxadiazine	Nocuolin A	*Nodularia* sp. LEGE 06071	Zhang et al. 2016a

Over 50% of marine cyanobacteria have the potential to be used for the extraction of bioactive compounds that either induce apoptosis in cancer cells or alter cell signaling by activating members of the protein kinase-c family of signaling enzymes. A chloroquine-resistant *Plasmodium falciparum* and human HeLa cancer cells are inhibited *in vitro* in a dose-dependent manner by the cell extracts of *Calothrix* isolates (Rickards et al. 1999). Calothrixin A (I) and B (II), pentacyclic metabolites with an indole (3, 2 - j) phenanthridine alkaloids that inhibit growth at nanomolar doses were isolated and structurally characterized using bioassay-directed fractions of the extracts. Calothrixins A and B demonstrated strong anticancer activity towards human HeLa tumor cells *in vitro* exhibiting IC50 values of 40 and 350 nM, respectively (Rickards et al. 1999).

Tasiamide B is a peptide molecule that contains 4-amino-3-hydroxy-5-phenylpentanoic acid. It was obtained from the marine Cyanobacterium *Symploca* sp. Williams et al. (2003a). Tasipeptins A and B, two novel cytotoxic depsipeptides were also discovered in the marine cyanobacterium *Symploca* sp. Williams et al. (2003b). Scytonemin is a yellow-green ultraviolet sunscreen pigment that was isolated from the cyanobacterium *Stigonema* sp. (Stevenson et al. 2000a). It has been found to be present in the extracellular sheaths of various genera of aquatic and terrestrial blue-green algal cells. Scytonemin suppresses the proliferation of human fibroblasts and endothelial cells and controls the development of mitotic spindles and enzyme kinases involved in cell cycle regulation. As a result, scytonemin may be an effective treatment for protein kinase inhibitors with anti-inflammatory and antiproliferative effects (Stevenson et al. 2002b).

Furthermore, Curacin-A was extracted from the organic extracts of *Lyngbyamajuscule*. Curacao, is a very strong antiproliferative agent because it selectively inhibits the polymerization of tubulin and shows inhibitory activity on cell lines derived from colon, renal, and breast cancer (Carte 1996). Largazole, a novel chemical scaffold produced by Symploca sp. exhibited excellent antiproliferative action (Taori et al. 2008). Another class of cyanobacterial chemicals that inhibit different cancer cell lines at nanomolar doses are the apratoxins. An adenocarcinoma exhibits cytotoxicity against the parental chemical, apratoxin A, which was extracted from a strain of *Lyngbyaboulloni* (Luesch et al. 2001).

Coibamide A is a substance that is obtained from a strain of *Leptolyngbya* (Medina et al. 2008). It is highly cytotoxic to mouse neuro-2a and NCIH460 lung cells. Many cyanobacterial chemicals have cytotoxicity as a common mode of action (Gerwick et al. 2001). Human Burkitt lymphoma cells have demonstrated cytotoxic effects from Lyngbyabellin B, which was isolated from *Lyngbyamajuscula* (Marquez et al. 2002).

Borophycin is a boron-containing metabolite identified from marine cyanobacterial strains *Nostoc linckia* and *N. spongiaeforme* var. tenue (Banker and Carmeli 1998). The substance is highly cytotoxic towards human epidermoid carcinoma (LoVo) and human colorectal adenocarcinoma (KB) cell lines (Davidson 1995). Cryptophycin 1 is an effective fungicide derived from *Nostoc* sp. ATCC 53789. It was not considered a natural product lead because of its severe toxicity. Subsequently, the same chemical derived from *Nostoc* sp. GSV 224 demonstrated strong cytotoxicity against human tumor cell lines as well as action against a wide range of drug-sensitive and drug-resistant murine and human solid tumors (Moore 1996). However, cryptophycin 1 appeared to be too toxic to be a therapeutic candidate. This lead to a thorough structure-function analysis, which resulted in the discovery of cryptophycin 8, a semisynthetic counterpart with higher therapeutic efficacy and lower toxicity than cryptophycin 14 *in vivo* (Carmichael 1992).

Dolastatin 10 and Symplostatin 1 are linear pentapeptides isolated from *Symploca* sp. and have been shown to have cytotoxic effects on human lung cancer and breast cancer cell lines via Bcl-2 phosphorylation and Caspase-3 protein activation (Kalemkerian et al. 1999, Mooberry et al. 2003). Furthermore, anticancer peptides from *Lyngbya* sp. and *Nostoc* sp. have demonstrated action against cancer on numerous cell lines via microfilament breakdown, secretory pathway inhibition, and other mechanisms (Costa et al. 2012). Cyanobactin wewakazole B produced from Moorea-developed cyanobacteria has shown to be cytotoxic to human cancer cell lines (Lopez et al. 2016). Odoamide belongs to a cytotoxic cyclodepsipeptide. The cytotoxin is derived from the marine cyanobacterium *Okeania* sp., which demonstrated strong cytotoxicity against HeLa S3 human cervical cancer cells with an IC50 value of 26.3 nM (Sueyoshi et al. 2016).

Galaxamide, a rare cyclic pentapeptide, has been discovered from the sea algae *Galaxaurafilamentosa*. Galaxamide's preliminary bioassay demonstrated remarkable *in vitro* antiproliferative activity against the GRC-1 and HepG2 cell lines (Xu et al. 2008). Dolastatin 10 and curacin A, two notable anti-microtubule medicines produced from cyanobacteria, have been clinically evaluated for cancer treatment and served as lead structures for synthetic derivatives (Tan 2007). Cyanobacteria-derived substances such as ulithiacyclamide and patellamide, derived from *Prochloron* spp. and *Lissoclinum patella*, were found to have substantial cytotoxic action against a human nasopharyngeal cancer cell line, with IC50 concentrations of 17 and 3000 ng/mL, respectively (Williams et al. 1989).

Moorea producens, a marine cyanobacterium, was used to isolate nine new linear lipopeptides: microcolins E–M, as well as four known related compounds, microcolins A, B, C, and D. Catalytic hydrogenation of microcolins A, B, C, and D yielded two known compounds, 3,4-dihydromicrocolins A and B, and two novel derivatives, 3,4-dihydromicrocolins C and D, respectively. The structures of the novel compounds were determined using a mix of spectroscopy and sophisticated Marfey analysis. Surprisingly, the compounds with structurally unique amino acid units, 4-methyl-2-(methylamino)-pent-3-enoic (Mpe) acid were identified; these are rare residues in naturally occurring peptides. The analogues demonstrated cytotoxic action against H-460 human lung cancer cells with IC50 values ranging from 0.037–5.0 μM (Yu et al. 2019).

A dark-red, wooly textured Cyanobacterium was found in the aquatic environment of Boca del Drago, Panama. It yielded four novel dragocins A–D following extraction with a 2:1 dichloromethane: methanol mixture and further chemical treatments. Dragocins A–C include 2, 3-dihydroxypyrrolidine, 1-hydroxy-5-O-Me-benzoyl, and 40-substituted-ribofuranose moieties, which have been united to form a nine-membered macrocyclic ring. Dragocins A, B, and C have been identified by substituting a ribofuranose unit at the C-40 position. Dragocin A was found to be cytotoxic to human H-460 lung cancer cells at an IC50 of 7.6 μM (Choi et al. 2018).

The previously obtained oxadiazine, nocuolin A (NocA), from the cyanobacterial strain *Nodularia* sp. LEGE 06071 was tested for anticancer efficacy against a colon cancer cell line (HCT116) and an immortalized epithelial cell line (hTERT RPE-1). NocA had a strong cytotoxic effect on cancer cells that were growing exponentially, but not on non-proliferating immortalized cells (Zhang et al. 2016a). It caused apoptosis in HCT116 multicellular spheroids by inhibiting the overexpression of the bcl genes. However, the transcriptome analysis of the HCT116 cells did not correspond to any molecule in the CMap. Subsequently, it pointed to stress reactions and cell starvation, which could be attributed to a decrease in adenosine triphosphate (ATP) generation. Within one hour of therapy, autophagy and a decrease in the mitochondrial respiration parameter were observed. A similar study found that nutritionally starved spheroid cells are sensitive to reduced mitochondrial energy generation because to limited metabolic flexibility, indicating NocA as a mitochondrial toxin to HCT-116 (Sousa et al. 2019).

Lyngbya sp., a marine cyanobacterium isolated from Okinawa, Japan, produces bisebromoamide, a linear non-ribosomal peptide (Teruya et al. 2009). At the Japanese Foundation for Cancer Research, bisebromoamide was found to be cytotoxic to HeLa cells as well as 39 human cancer cells (known as JFCR39). Interestingly, bisebromoamide reduces the phosphorylation of ERK (extracellular signal-regulated protein kinase) in NRK (normal rat kidney cells) cells stimulated with PDGF (platelet-derived growth factor) (Teruya et al. 2009). Furthermore, bisebromoamide has been seen to promote apoptosis in RCC (renal cell cancer) cells by inhibiting the Raf/MEK/ERK and PI3K/Akt/mTOR pathways (Suzuki et al. 2013).

Conclusion

The incidence of cancer is still rising due to a number of environmental factors, malnutrition, and increased global warming. Natural derivatives play a crucial role in mitigating cancer fatalities, while synthetic medicines often come with a side effects. The diverse biochemical profiles of marine microbial flora including bacteria, fungi, actinomycetes and cyanobacteria, hold immense potential for the development of efficient, cost-effective, and safer anticancer drugs. Numerous marine-derived natural products such as peptides, polyketides, alkaloids, and quinones have exhibited anticancer activity against a broad spectrum of tumor cell lines in vitro, including those derived from lymphoid cancer, renal, lung, prostate, bladder, melanoma, osteosarcoma, and breast cancer. These marine-sourced chemicals demonstrate extraordinary potency in inhibiting tumor cell culture, often within the nanogram range. Furthermore, the majority of studies on the mechanism of action of marine products in reducing tumor growth both *in vitro* and *in vivo*, suggest that it is mediated by apoptosis and necrosis. In particular for patients with solid tumors, the combination of novel molecules with unique mechanisms of action turns into new cancer treatment approaches and eventually greater efficacy. Furthermore, the manipulation of microbial metabolic processes through genetic engineering has resulted in the production of interesting novel compounds. Therefore, marine organisms represent a rich source of genetic and metabolic variation that is primed to be applied and will certainly make anti-cancer drug development much easier.

References

Abdel-Lateff, A., König, G.M., Fisch, K.M., Höller, U., Jones, P.G. and Wright, A.D. (2002). New antioxidant hydroquinone derivatives from the algicolous marine fungus Acremonium sp. Journal of Natural Products 65(11): 1605–1611.

Abdel-Lateff, A., Klemke, C., König, G.M. and Wright, A.D. (2003). Two new xanthone derivatives from the algicolous marine fungus Wardomycesanomalus. Journal of Natural Products 66(5): 706–708.

Abdel-Wahab, N.M., Scharf, S., Özkaya, F.C., Kurtán, T., Mándi, A., Fouad, M.A. et al. (2019). Induction of secondary metabolites from the marine-derived fungus Aspergillus versicolor through co-cultivation with *Bacillus subtilis*. Planta Medica 85(06): 503–512.

Alves, C., Silva, J., Pinteus, S., Gaspar, H., Alpoim, M.C., Botana, L.M. et al. (2018). From marine origin to therapeutics: The antitumor potential of marine algae-derived compounds. Frontiers in Pharmacology 9: 777.

Asolkar, R.N., Freel, K.C., Jensen, P.R., Fenical, W., Kondratyuk, T.P., Park, E.J. et al. (2009). Arenamides A–C, cytotoxic NFκB inhibitors from the marine actinomycete Salinisporaarenicola. Journal of Natural Products 72(3): 396–402.

Baindara, P., Gautam, A., Raghava, G.P.S. and Korpole, S. (2017). Anticancer properties of a defensin like class IId bacteriocin Laterosporulin10. Scientific Reports 7(1): 46541.

Banker, R. and Carmeli, S. (1998). Tenuecyclamides A–D, Cyclic Hexapeptides from the Cyanobacterium *Nostocs pongiaeforme* var. tenue. Journal of Natural Products 61(10): 1248–1251.

Baz, J.P., Ca nedo, L.M., Fernández Puentes, J.L. and Silva Elipe, M.V. (1997). Thiocoraline, a novel depsipeptide with antitumor activity produced by a marine Micromonospora. II. Physico-chemical properties and structure determination. J. Antibiot. 50(9): 738–741.

Bhattacharya, B. and Mukherjee, S. (2015). Cancer therapy using antibiotics. Journal of Cancer Therapy 6(10): 849.

Bitzer, J., Große, T., Wang, L., Lang, S., Beil, W. and Zeeck, A. (2006). New aminophenoxazinones from a marine Halomonas sp.: fermentation, structure elucidation, and biological activity. The Journal of Antibiotics 59(2): 86–92.

Bringmann, G., Gulder, T.A., Lang, G., Schmitt, S., Stöhr, R., Wiese, J. et al. (2007). Large-scale biotechnological production of the antileukemic marine natural product sorbicillactone A. Marine Drugs 5(2): 23–30.

Brucoli, F., Natoli, A., Marimuthu, P., Borrello, M.T., Stapleton, P., Gibbons, S. et al. (2012). Efficient synthesis and biological evaluation of proximicins A, B and C. Bioorganic & Medicinal Chemistry 20(6): 2019–2024.

Butler, M.S., Robertson, A.A. and Cooper, M.A. (2014). Natural product and natural product derived drugs in clinical trials. Natural Product Reports 31(11): 1612–1661.

Canedo, L.M., Puentes, J.L.F., Baz, J.P., Huang, X.H. and Rinehart, K.L. (2000). IB-96212, a novel cytotoxic macrolide produced by a marine Micromonospora II. Physico-chemical properties and structure determination. The Journal of Antibiotics 53(5): 479–483.

Carmichael, W.W. (1992). Cyanobacteria secondary metabolites—the cyanotoxins. Journal of Applied Bacteriology 72(6): 445–459.

Carte, B.K. (1996). Biomedical potential of marine natural products. Bioscience 46(4): 271–286.

Chauhan, D., Catley, L., Li, G., Podar, K., Hideshima, T., Velankar, M. et al. (2005). A novel orally active proteasome inhibitor induces apoptosis in multiple myeloma cells with mechanisms distinct from Bortezomib. Cancer Cell 8(5): 407–419.

Chen, Y., Liu, Z., Huang, Y., Liu, L., He, J., Wang, L. et al. (2019). Ascomylactams A–C, cytotoxic 12- or 13-membered-ring macrocyclic alkaloids isolated from the mangrove endophytic fungus Didymella sp. CYSK-4, and structure revisions of phomapyrrolidones A and C. Journal of Natural Products 82(7): 1752–1758.

Chi, L.P., Li, X.M., Li, L., Li, X. and Wang, B.G. (2020). Cytotoxic thiodiketopiperazine derivatives from the deep sea-derived fungus Epicoccum nigrum SD-388. Marine Drugs 18(3): 160.

Cho, J.Y., Williams, P.G., Kwon, H.C., Jensen, P.R. and Fenical, W. (2007). Lucentamycins A–D, cytotoxic peptides from the marine-derived actinomycete Nocardiopsislucentensis. Journal of Natural Products 70(8): 1321–1328.

Choi, B.K., Lee, H.S., Kang, J.S. and Shin, H.J. (2019a). Dokdolipids A–C, hydroxylated rhamnolipids from the marine-derived Actinomycete Actinoalloteichushymeniacidonis. Marine Drugs 17(4): 237.

Choi, B.K., Trinh, P.T.H., Lee, H.S., Choi, B.W., Kang, J.S., Ngoc, N.T.D. et al. (2019b). New ophiobolin derivatives from the marine fungus Aspergillus flocculosus and their cytotoxicities against cancer cells. Marine Drugs 17(6): 346.

Choi, E.J., Park, J.S., Kim, Y.J., Jung, J.H., Lee, J.K., Kwon, H.C. et al. (2011). Apoptosis-inducing effect of diketopiperazine disulfides produced by *Aspergillus* sp. KMD 901 isolated from marine sediment on HCT116 colon cancer cell lines. Journal of Applied Microbiology 110(1): 304–313.

Choi, H., Engene, N., Byrum, T., Hwang, S., Oh, D.C. and Gerwick, W.H. (2018). Dragocins A–D, structurally intriguing cytotoxic metabolites from a panamanian marine cyanobacterium. Organic Letters 21(1): 266–270.

Costa, M., Costa-Rodrigues, J., Fernandes, M.H., Barros, P., Vasconcelos, V. and Martins, R. (2012). Marine cyanobacteria compounds with anticancer properties: a review on the implication of apoptosis. Marine Drugs 10(10): 2181–2207.

Davidson, B.S. (1995). New dimensions in natural products research: cultured marine microorganisms. Current Opinion in Biotechnology 6(3): 284–291.

Davies-Bolorunduro, O.F., Adeleye, I.A., Akinleye, M.O. and Wang, P.G. (2019). Anticancer potential of metabolic compounds from marine actinomycetes isolated from Lagos Lagoon sediment. Journal of Pharmaceutical Analysis 9(3): 201–208.

Dos Santos Dias, A.C., Couzinet-Mossion, A., Ruiz, N., Lakhdar, F., Etahiri, S., Bertrand, S. et al. (2019). Steroids from marine-derived fungi: Evaluation of antiproliferative and antimicrobial activities of eburicol. Marine Drugs 17(6): 372.

Du, L., Zhu, T., Fang, Y., Liu, H., Gu, Q. and Zhu, W. (2007). Aspergiolide A, a novel anthraquinone derivative with naphtho (1, 2, 3-de) chromene-2, 7-dione skeleton isolated from a marine-derived fungus *Aspergillus glaucus*. Tetrahedron 63(5): 1085–1088.

Du, L., Feng, T., Zhao, B., Li, D., Cai, S., Zhu, T. et al. (2010). Alkaloids from a deep ocean sediment-derived fungus Penicillium sp. and their antitumor activities. The Journal of Antibiotics 63(4): 165–170.

Dudhagara, D.R., Javia, B.M. and Vala, A.K. (2023). Exploiting marine fungi in the removal of hazardous pollutants and biomass valorisation. pp. 117–146. *In*: Marine Organisms: A Solution to Environmental Pollution? Uses in Bioremediation and in Biorefinery. Cham: Springer International Publishing.

Elborai, A., Sayed, R., Farag, A. and Elassar, S. (2023). A highly purified l-glutaminase from immobilized Pseudomonas SP. RAS123 cultures with antitumor and antibacterial activities. Journal of Microbiology, Biotechnology and Food Sciences e5637–e5637.

Erba, E., Mascellani, E., Pifferi, A. and D'Incalci, M. (1995). Comparison of cell-cycle phase perturbations induced by the DNA-minor-groove alkylator tallimustine and by melphalan in the SW626 cell line. International Journal of Cancer 62(2): 170–175.

Erba, E., Bergamaschi, D., Ronzoni, S., Faretta, M., Taverna, S., Bonfanti, M. et al. (1999). Mode of action of thiocoraline, a natural marine compound with anti-tumour activity. British Journal of Cancer 80(7): 971–980.

Fan, B., Dewapriya, P., Li, F., Blümel, M. and Tasdemir, D. (2020). Pyrenosetins A–C, new decalinoylspirotetramic acid derivatives isolated by bioactivity-based molecular networking from the seaweed-derived fungus Pyrenochaetopsis sp. FVE-001. Marine Drugs 18(1): 47.

Fan, Y., Zhou, Y., Du, Y., Wang, Y., Fu, P. and Zhu, W. (2019). Circumdatin-aspyrone conjugates from the coral-associated *Aspergillus ochraceus* LCJ11-102. Marine Drugs 17(7): 400.

Fenical, W. and Jensen, P.R. (1993). Marine microorganisms: a new biomedical resource. pp. 419–457. *In*: Pharmaceutical and Bioactive Natural Products. Boston, MA: Springer US.

Fenical, W., Jensen, P.R., Palladino, M.A., Lam, K.S., Lloyd, G.K. and Potts, B.C. (2009). Discovery and development of the anticancer agent salinosporamide A (NPI-0052). Bioorganic & Medicinal Chemistry 17(6): 2175–2180.

Fiedler, H.P., Bruntner, C., Riedlinger, J., Bull, A.T., Knutsen, G., Goodfellow, M. et al. (2008). Proximicin A, B and C, novel aminofuran antibiotic and anticancer compounds isolated from marine strains of the actinomycete Verrucosispora. The Journal of Antibiotics 61(3): 158–163.

Fu, P., Legako, A., La, S. and MacMillan, J.B. (2016). Discovery, characterization, and analogue synthesis of bohemamine dimers generated by non-enzymatic biosynthesis. Chemistry–A European Journal 22(10): 3491–3495.

Furumai, T., Takagi, K., Igarashi, Y., Saito, N. and Oki, T. (2000). Arisostatins A and B, new members of tetrocarcin class of antibiotics from *Micromonospora* sp. TP-A0316 I. Taxonomy, fermentation, isolation and biological properties. The Journal of Antibiotics 53(3): 227–232.

García-Caballero, M., Cañedo, L., Fernández-Medarde, A., Medina, M.Á. and Quesada, A.R. (2014). The marine fungal metabolite, AD0157, inhibits angiogenesis by targeting the Akt signaling pathway. Marine Drugs 12(1): 279–299.

Ge, X., Sun, C., Feng, Y., Wang, L., Peng, J., Che, Q. et al. (2019). Anthraquinone derivatives from a marine-derived fungus Sporendonema casei HDN16-802. Marine Drugs 17(6): 334.

Gerwick, W.H., Tan, L.T. and Sitachitta, N. (2001). Nitrogen-containing metabolites from marine cyanobacteria.

Goldin, B.R. and Gorbach, S.L. (1992). Probiotics for humans. Probiotics: The Scientific Basis 355: 376.

Goldin, B. and Gorbach, S.L. (1977). Alterations in fecal microflora enzymes related to diet, age, lactobacillus supplements, and dimethylhydrazine. Cancer 40(S5): 2421–2426.

Gomes, N.G., Lefranc, F., Kijjoa, A. and Kiss, R. (2015). Can some marine-derived fungal metabolites become actual anticancer agents?. Marine Drugs 13(6): 3950–3991.

Gomes, N.G., Dasari, R., Chandra, S., Kiss, R. and Kornienko, A. (2016). Marine invertebrate metabolites with anticancer activities: Solutions to the "supply problem". Marine Drugs 14(5): 98.

Guerriero, A., D'Ambrosio, M., Cuomo, V. and Pietra, F. (1991). A novel, degraded polyketidic lactone, leptosphaerolide, and its likely diketone precursor, leptosphaerodione. Isolation from cultures of the marine ascomycete Leptosphaeria oraemaris (Linder). Helvetica Chimica Acta 74(7): 1445–1450.

Guo, Y.W., Gong, B.Q., Yuan, J., Li, H.J., Mahmud, T., Huang, Y. et al. (2019). L-phenylalanine alters the privileged secondary metabolite production in the marine-derived fungus Trichoderma erinaceum F1-1. Journal of Natural Products 83(1): 79–87.

Hamed, A., Abdel-Razek, A.S., Araby, M., Abu-Elghait, M., El-Hosari, D.G., Frese, M. et al. (2021). Meleagrin from marine fungus Emericella dentata Nq45: crystal structure and diverse biological activity studies. Natural Product Research 35(21): 3830–3838.

Han, X., Liu, Z., Zhang, Z., Zhang, X., Zhu, T., Gu, Q. et al. (2017). Geranylpyrrol A and Piericidin F from *Streptomyces* sp. CHQ-64 Δ rdmF. Journal of Natural Products 80(5): 1684–1687.

Han, Y., Wang, Y., Yang, Y. and Chen, H. (2020). Shellmycin A–D, novel bioactive tetrahydroanthra-γ-pyrone antibiotics from marine *Streptomyces* sp. Shell-016. Marine Drugs 18(1): 58.

Hawas, U.W., Shaaban, M., Shaaban, K.A., Speitling, M., Maier, A., Kelter, G. et al. (2009). Mansouramycins A–D, cytotoxic isoquinolinequinones from a marine Streptomycete. Journal of Natural Products 72(12): 2120–2124.

Hawksworth, D.L. (2012). Global species numbers of fungi: are tropical studies and molecular approaches contributing to a more robust estimate?. Biodiversity and Conservation 21: 2425–2433.

Heimann, D.M. and Rosenberg, S.A. (2003). Continuous intravenous administration of live genetically modified salmonella typhimurium in patients with metastatic melanoma. Journal of Immunotherapy 26(2): 179–180.

Igarashi, Y., Takagi, K., Kan, Y., Fujii, K., Harada, K.I., Furumai, T. et al. (2000). Arisostatins A and B, new members of tetrocarcin class of antibiotics from Micromonospora sp. TP-A0316 II. Structure determination. The Journal of Antibiotics 53(3): 233–240.

Javia, B., Gadhvi, M., Vyas, S., Dudhagara, P., Shyu, D.J., Chen, Y.Y. et al. (2023). Bioprospecting of a thermostable L-methioninase from Alcaligenes aquatilis BJ-1 in agro-industrial waste. Microbiology Research 14(3): 959–976.

Javia, B.M., Gadhvi, M.S., Vyas, S.J., Ghelani, A., Wirajana, N. and Dudhagara, D.R. (2024). A review on L-methioninase in cancer therapy: Precision targeting, advancements and diverse applications for a promising future. International Journal of Biological Macromolecules 130997.

Jensen, P.R., Kauffman, C.A. and Fenical, W. (1996). High recovery of culturable bacteria from the surfaces of marine algae. Marine Biology 126: 1–7.

Kalemkerian, G.P., Ou, X., Adil, M.R., Rosati, R., Khoulani, M.M., Madan, S.K. et al. (1999). Activity of dolastatin 10 against small-cell lung cancer *in vitro* and *in vivo*: Induction of apoptosis and bcl-2 modification. Cancer Chemotherapy and Pharmacology 43: 507–515.

Kamaliya, D.B., Javia, B.M., Gadhvi, M.S. and Dudhagara, D.R. (2023). Isolation and characterization of probiotic bacteria from fruit pulp, screening of probiotic properties and production of bacteriocin and probiotic curd. Open Biol Sci J. 8(1): 033–042.

Kamata, K., Okamoto, S., Oka, S.I., Kamata, H., Yagisawa, H. and Hirata, H. (2001). Cycloprodigiosinhydrocloride suppresses tumor necrosis factor (TNF) α-induced transcriptional activation by NF-κB. Febs Letters 507(1): 74–80.

Kanoh, K., Kohno, S., Asari, T., Harada, T., Katada, J., Muramatsu, M. et al. (1997). (−)-Phenylahistin: a new mammalian cell cycle inhibitor produced by *Aspergillus ustus*. Bioorganic & Medicinal Chemistry Letters 7(22): 2847–2852.

Karim, M.R.U., Harunari, E., Oku, N., Akasaka, K. and Igarashi, Y. (2020). Bulbimidazoles A–C, antimicrobial and cytotoxic alkanoylimidazoles from a marine gammaproteobacterium Microbulbifer species. Journal of Natural Products 83(4): 1295–1299.

Kathiresan, K. and Duraisamy, A. (2005). Current issue of microbiology. ENVIS Centre Newsletters 4: 3–5.

Kathiresan, K. and Thiruneelakandan, G. (2008). Prospects of lactic acid bacteria of marine origin.

Kawahara, T., Izumikawa, M., Kozone, I., Hashimoto, J., Kagaya, N., Koiwai, H. et al. (2018). Neothioviridamide, a polythioamide compound produced by heterologous expression of a Streptomyces sp. cryptic RiPP biosynthetic gene cluster. Journal of Natural Products 81(2): 264–269.

Kim, M.C., Machado, H., Jang, K.H., Trzoss, L., Jensen, P.R. and Fenical, W. (2018). Integration of genomic data with NMR analysis enables assignment of the full stereostructure of neaumycin B, a potent inhibitor of glioblastoma from a marine-derived Micromonospora. Journal of the American Chemical Society 140(34): 10775–10784.

Kim, S.K. and Kalimuthu, S. (2015). Introduction to anticancer drugs from marine origin. Handbook of Anticancer Drugs From Marine Origin 1–13.

Kim, Y.H., Shin, H.C., Lee, S.H., Furumai, T., Park, J.W. and Kwon, T.K. (2003). Arisostatins A induces apoptosis through the activation of caspase-3 and reactive oxygen species generation in AMC-HN-4 cells. Biochemical and Biophysical Research Communications 309(2): 449–456.

Kiuru, P., D'Auria, M., Muller, C., Tammela, P., Vuorela, H. and Yli-Kauhaluoma, J. (2014). Exploring marine resources for bioactive compounds. Planta Medica 80(14): 1234–1246.

Kodama, M., Ogata, T., Sato, S. and Sakamoto, S. (1990). Possible association of marine bacteria with paralytic shellfish toxicity of bivalves. Marine Ecology Progress Series 61: 203–206.

Kokoulin, M.S., Kuzmich, A.S., Romanenko, L.A., Menchinskaya, E.S., Mikhailov, V.V. and Chernikov, O.V. (2018). Sulfated O-polysaccharide with anticancer activity from the marine bacterium Poseidonocellasedimentorum KMM 9023T. Carbohydrate Polymers 202: 157–163.

Kramer, M.G., Masner, M., Ferreira, F.A. and Hoffman, R.M. (2018). Bacterial therapy of cancer: promises, limitations, and insights for future directions. Frontiers in Microbiology 9: 16.

Krishnan, K., Mani, A. and Jasmine, S. (2014). Cytotoxic activity of bioactive compound 1, 2-benzene dicarboxylic acid, mono 2-ethylhexyl ester extracted from a marine derived Streptomyces sp. VITSJK8. International Journal of Molecular and Cellular Medicine 3(4): 246.

Lange, L. (1996). Microbial metabolites-an infinite source of novel chemistry. Pure and Applied Chemistry 68(3): 745–748.

Li, W., Gan, J. and Ma, D. (2009). Total synthesis of piperazimycin A: a cytotoxic cyclic hexadepsipeptide. AngewandteChemie 121(47): 9053–9057.

Liang, X., Zhang, X.Y., Nong, X.H., Wang, J., Huang, Z.H. and Qi, S.H. (2016). Eight linear peptides from the deep-sea-derived fungus Simplicilliumobclavatum EIODSF 020. Tetrahedron 72(22): 3092–3097.

Lin, C.C., Tantisantisom, W. and McAlpine, S.R. (2013). Total synthesis and biological activity of natural product urukthapelstatin A. Organic Letters 15(14): 3574–3577.

Lin, Y.K., Xie, C.L., Xing, C.P., Wang, B.Q., Tian, X.X., Xia, J.M. et al. (2021). Cytotoxic p-terphenyls from the deep-sea-derived *Aspergillus candidus*. Natural Product Research 35(10): 1627–1631.

Lin, Z., Koch, M., Pond, C.D., Mabeza, G., Seronay, R.A., Concepcion, G.P. et al. (2014). Structure and activity of lobophorins from a turrid mollusk-associated *Streptomyces* sp. The Journal of Antibiotics 67(1): 121–126.

Liu, H., Zhang, L., Chen, Y., Li, S., Tan, G., Sun, Z. et al. (2017). Cytotoxic pimarane-type diterpenes from the marine sediment-derived fungus Eutypella sp. FS46. Natural Product Research 31(4): 404–410.

Liu, L., Xu, W., Li, S., Chen, M., Cheng, Y., Yuan, W. et al. (2019). Penicindopene A, a new indole diterpene from the deep-sea fungus Penicillium sp. YPCMAC1. Natural Product Research 33(20): 2988–2994.

Liu, S.W., Jin, J., Chen, C., Liu, J.M., Li, J.Y., Wang, F.F. et al. (2013). PJS, a novel isocoumarin with hexahydropyrimidine ring from *Bacillus subtilis* PJS. The Journal of Antibiotics 66(5): 281–284.

Lopez, J.A.V., Al-Lihaibi, S.S., Alarif, W.M., Abdel-Lateff, A., Nogata, Y., Washio, K. et al. (2016). Wewakazole B, a cytotoxic cyanobactin from the cyanobacterium Moorea producens collected in the Red Sea. Journal of Natural Products 79(4): 1213–1218.

Luesch, H., Moore, R.E., Paul, V.J., Mooberry, S.L. and Corbett, T.H. (2001). Isolation of dolastatin 10 from the marine cyanobacterium Symploca species VP642 and total stereochemistry and biological evaluation of its analogue symplostatin 1. Journal of Natural Products 64(7): 907–910.

Luo, X.W., Gao, C.H., Lu, H.M., Wang, J.M., Su, Z.Q., Tao, H.M. et al. (2020). HPLC-DAD-guided isolation of diversified chaetoglobosins from the coral-associated fungus Chaetomium globosum C2F17. Molecules 25(5): 1237.

Lv, Q., Fan, Y., Tao, G., Fu, P., Zhai, J., Ye, B. et al. (2019). Sekgranaticin, a SEK34b-granaticin hybrid polyketide from *Streptomyces* sp. 166#. The Journal of Organic Chemistry 84(14): 9087–9092.

Ma, Z., Wang, N., Hu, J. and Wang, S. (2012). Isolation and characterization of a new iturinic lipopeptide, mojavensin A produced by a marine-derived bacterium Bacillus mojavensis B0621A. The Journal of Antibiotics 65(6): 317–322.

Marquez, B.L., Watts, K.S., Yokochi, A., Roberts, M.A., Verdier-Pinard, P., Jimenez, J.I. et al. (2002). Structure and absolute stereochemistry of hectochlorin, a potent stimulator of actin assembly. Journal of Natural Products 65(6): 866–871.

Maskey, R.P., Helmke, E., Kayser, O., Fiebig, H.H., Maier, A., Busche, A. et al. (2004a). Anti-cancer and antibacterial trioxacarcins with high anti-malaria activity from a marine Streptomycete and their absolute stereochemistry. The Journal of Antibiotics 57(12): 771–779.

Maskey, R.P., Sevvana, M., Usón, I., Helmke, E. and Laatsch, H. (2004b). Gutingimycin: a highly complex metabolite from a marine streptomycete. AngewandteChemie International Edition 43(10): 1281–1283.

Mathew, D.E., Vala, A.K., Dineshkumar, R., Niharika, J., Singh, R.P., Shinde, P.B. et al. (2023). Performance evaluation and yield optimization of L-glutaminase free L-asparaginase from seaweed-associated bacteria. Bioresource Technology Reports 23: 101534.

Matsuo, Y., Kanoh, K., Imagawa, H., Adachi, K., Nishizawa, M. et al. (2007). Urukthapelstatin A, a novel cytotoxic substance from marine-derived Mechercharimycesasporophorigenens YM11-542. The Journal of Antibiotics 60(4): 256–260.

McDonald, L.A., Abbanat, D.R., Barbieri, L.R., Bernan, V.S., Discafani, C.M., Greenstein, M. et al. (1999). Spiroxins, DNA cleaving antitumor antibiotics from a marine-derived fungus. Tetrahedron Letters 40(13): 2489–2492.

Medina, R.A., Goeger, D.E., Hills, P., Mooberry, S.L., Huang, N., Romero, L.I. et al. (2008). Coibamide A, a potent antiproliferative cyclic depsipeptide from the Panamanian marine cyanobacterium Leptolyngbya sp. Journal of the American Chemical Society 130(20): 6324–6325.

Mengesha, A., Dubois, L., Chiu, R.K., Paesmans, K., Wouters, B.G., Lambin, P. et al. (2007). Potential and limitations of bacterial-mediated cancer therapy. Frontiers in Bioscience-Landmark 12(10): 3880–3891.

Miller, E.D., Kauffman, C.A., Jensen, P.R. and Fenical, W. (2007). Piperazimycins: cytotoxic hexadepsipeptides from a marine-derived bacterium of the genus Streptomyces. The Journal of Organic Chemistry 72(2): 323–330.

Mitall, B.K. and Garg, S.K. (1995). Anticarcinogenic, hypocholesterolemic, and antagonistic activities of *Lactobacillus acidophilus*. Critical Reviews in Microbiology 21(3): 175–214.

Molinski, T.F., Dalisay, D.S., Lievens, S.L. and Saludes, J.P. (2009). Drug development from marine natural products. Nature Reviews Drug Discovery 8(1): 69–85.

Mooberry, S.L., Leal, R.M., Tinley, T.L., Luesch, H., Moore, R.E. and Corbett, T.H. (2003). The molecular pharmacology of symplostatin 1: A new antimitotic dolastatin 10 analog. International Journal of Cancer 104(4): 512–521.

Moore, R.E. (1996). Cyclic peptides and depsipeptides from cyanobacteria: a review. Journal of Industrial Microbiology 16: 134–143.

Nallar, S.C., Xu, D.Q. and Kalvakolanu, D.V. (2017). Bacteria and genetically modified bacteria as cancer therapeutics: Current advances and challenges. Cytokine 89: 160–172.

Nalli, Y., Gupta, S., Khajuria, V., Singh, V.P., Sajgotra, M., Ahmed, Z. et al. (2017). TNF-α and IL-6 inhibitory effects of cyclic dipeptides isolated from marine bacteria Streptomyces sp. Medicinal Chemistry Research 26: 93–100.

Nie, Y.L., Wu, Y.D., Wang, C.X., Lin, R., Xie, Y., Fang, D.S. et al. (2018). Structure elucidation and antitumour activity of a new macrolactam produced by marine-derived actinomycete Micromonospora sp. FIM05328. Natural Product Research 32(18): 2133–2138.

Niu, S., Xia, M., Chen, M., Liu, X., Li, Z., Xie, Y. et al. (2019). Cytotoxic polyketides isolated from the deep-sea-derived fungus Penicillium chrysogenum MCCC 3A00292. Marine Drugs 17(12): 686.

Oku, N., Adachi, K., Matsuda, S., Kasai, H., Takatsuki, A. and Shizuri, Y. (2008). Ariakemicins A and B, novel polyketide-peptide antibiotics from a marine gliding bacterium of the genus Rapidithrix. Organic Letters 10(12): 2481–2484.

Patyar, S., Joshi, R., Byrav, D.P., Prakash, A., Medhi, B. and Das, B.K. (2010). Bacteria in cancer therapy: a novel experimental strategy. Journal of Biomedical Science 17: 1–9.

Pérez, M., Crespo, C., Schleissner, C., Rodríguez, P., Zúñiga, P. and Reyes, F. (2009). Tartrolon D, a cytotoxic macrodiolide from the marine-derived actinomycete Streptomyces sp. MDG-04-17-069. Journal of Natural Products 72(12): 2192–2194.

Pérez, M., Schleissner, C., Fernández, R., Rodríguez, P., Reyes, F., Zuñiga, P. et al. (2016). PM100117 and PM100118, new antitumor macrolides produced by a marine Streptomyces caniferus GUA-06-05-006A. The Journal of Antibiotics 69(5): 388–394.

Ran, Y.Q., Lan, W.J., Qiu, Y., Guo, Q., Feng, G.K., Deng, R. et al. (2020). Monarubins A–C from the marine shellfish-associated fungus Monascusruber BB5. Marine Drugs 18(2): 100.

Ravi, L., Ragunathan, A. and Krishnan, K. (2017). Marine Streptomyces paradoxus VITALK03 derived gancidin W mediated cytotoxicity through Ras-Raf-MEK-ERK signalling pathway.

Ravikumar, S., Fredimoses, M. and Gnanadesigan, M. (2012). Anticancer property of sediment actinomycetes against MCF–7 and MDA–MB–231 cell lines. Asian Pacific Journal of Tropical Biomedicine 2(2): 92–96.

Rickards, R.W., Rothschild, J.M., Willis, A.C., de Chazal, N.M., Kirk, J., Kirk, K. et al. (1999). Calothrixins A and B, novel pentacyclic metabolites from Calothrix cyanobacteria with potent activity against malaria parasites and human cancer cells. Tetrahedron 55(47): 13513–13520.

Rodrigues, A.M., Rohée, C., Fabre, T., Batailler, N., Sautel, F., Carletti, I. et al. (2017). Cytotoxic indole alkaloids from Pseudovibriodenitrificans BBCC725. Tetrahedron Letters 58(32): 3172–3173.

Ruiz-Ruiz, C., Srivastava, G.K., Carranza, D., Mata, J.A., Llamas, I., Santamaría, M. et al. (2011). An exopolysaccharide produced by the novel halophilic bacterium Halomonasstenophila strain B100 selectively induces apoptosis in human T leukaemia cells. Applied Microbiology and Biotechnology 89: 345–355.

Ruiz-Torres, V., Encinar, J.A., Herranz-Lopez, M., Pérez-Sánchez, A., Galiano, V., Barrajón-Catalán, E. et al. (2017). An updated review on marine anticancer compounds: The use of virtual screening for the discovery of small-molecule cancer drugs. Molecules 22(7): 1037.

Sagar, S., Esau, L., Holtermann, K., Hikmawan, T., Zhang, G., Stingl, U. et al. (2013). Induction of apoptosis in cancer cell lines by the Red Sea brine pool bacterial extracts. BMC Complementary and Alternative Medicine 13(1): 1–12.

Saha, S., Zhang, W., Zhang, G., Zhu, Y., Chen, Y., Liu, W. et al. (2017). Activation and characterization of a cryptic gene cluster reveals a cyclization cascade for polycyclic tetramate macrolactams. Chemical Science 8(2): 1607–1612.

Salendra, L., Lin, X., Chen, W., Pang, X., Luo, X., Long, J. et al. (2021). Cytotoxicity of polyketides and steroids isolated from the sponge-associated fungus Penicillium citrinum SCSIO 41017. Natural Product Research 35(6): 900–908.

Schleissner, C., Canedo, L.M., Rodríguez, P., Crespo, C., Zuniga, P., Penalver, A. et al. (2017). Bacterial production of a pederin analogue by a free-living marine Alphaproteobacterium. Journal of Natural Products 80(7): 2170–2173.

Schneider, K., Keller, S., Wolter, F.E., Röglin, L., Beil, W., Seitz, O. et al. (2008). Proximicins A, B, and C—antitumor furan analogues of netropsin from the marine actinomycete Verrucosispora induce upregulation of p53 and the cyclin kinase inhibitor p21. AngewandteChemie International Edition 47(17): 3258–3261.

Sedighi, M., Zahedi Bialvaei, A., Hamblin, M.R., Ohadi, E., Asadi, A., Halajzadeh, M. et al. (2019). Therapeutic bacteria to combat cancer; current advances, challenges, and opportunities. Cancer Medicine 8(6): 3167–3181.

Selim, M.S., Mounier, M.M., Abdelhamid, S.A., Hamed, A.A., Abo Elsoud, M.M. and Mohamed, S.S. (2024). Characterization, modeling, and anticancer activity of L. arginase production from marine Bacillus licheniformis OF2. BMC Biotechnology 24(1): 1–16.

Shaaban, K.A., Shaaban, M., Meiners, M., Schüffler, A., Kelter, G., Fiebig, H.H. et al. (2021). Boshramycinones AC: New anthracyclinones produced by a marine-derived Streptomyces sp.: isolation, structure elucidation and biological activities. Natural Product Research 35(8): 1281–1291.

Shen, T., Li, L.M., Xu, Z.Y., Wang, Y.D. and Xie, W.D. (2021). Julichrome derivatives and gliotoxin from a soil derived Streptomyces sp. Natural Product Research 35(1): 34–40.

Shin, D., Byun, W.S., Moon, K., Kwon, Y., Bae, M., Um, S. et al. (2018). Coculture of marine *Streptomyces* sp. with *Bacillus* sp. produces a new piperazic acid-bearing cyclic peptide. Frontiers in Chemistry 6: 498.

Silchenko, A.S., Kalinovsky, A.I., Avilov, S.A., Kalinin, V.I., Andrijaschenko, P.V., Dmitrenok, P.S. et al. (2019). Structures and bioactivities of psolusosides B1, B2, J, K, L, M, N, O, P, and Q from the sea cucumber Psolusfabricii. The first finding of tetrasulfated marine low molecular weight metabolites. Marine Drugs 17(11): 631.

Song, S., Vuai, M.S. and Zhong, M. (2018). The role of bacteria in cancer therapy–enemies in the past, but allies at present. Infectious Agents and Cancer 13(1): 1–7.

Sousa, M.L., Preto, M., Vasconcelos, V., Linder, S. and Urbatzka, R. (2019). Antiproliferative effects of the natural oxadiazine nocuolin A are associated with impairment of mitochondrial oxidative phosphorylation. Frontiers in Oncology 9: 224.

Sproule, A., Correa, H., Decken, A., Haltli, B., Berrué, F., Overy, D.P. et al. (2019). Terrosamycins A and B, bioactive polyether ionophores from Streptomyces sp. RKND004 from Prince Edward Island sediment. Marine Drugs 17(6): 347.

Stevenson, C.S., Capper, E.A., Roshak, A.K., Marquez, B., Eichman, C., Jackson, J.R. et al. (2002a). The identification and characterization of the marine natural product scytonemin as a novel antiproliferative pharmacophore. Journal of Pharmacology and Experimental Therapeutics 303(2): 858–866.

Stevenson, C.S., Capper, E.A., Roshak, A.K., Marquez, B., Grace, K., Gerwick, W.H. et al. (2002b). Scytonemin-a marine natural product inhibitor of kinases key in hyperproliferative inflammatory diseases. Inflammation Research 51(2): 112.

Sueyoshi, K., Kaneda, M., Sumimoto, S., Oishi, S., Fujii, N., Suenaga, K. et al. (2016). Odoamide, a cytotoxic cyclodepsipeptide from the marine cyanobacterium Okeania sp. Tetrahedron 72(35): 5472–5478.

Suja, M., Vasuki, S. and Sajitha, N. (2014). Anticancer activity of compounds isolated from marine endophytic fungus Aspergillus terreus. World J. Pharm. Pharm. Sci, 3(6): 661–72.

Sun, Y., Tomura, T., Sato, J., Iizuka, T., Fudou, R. and Ojika, M. (2016). Isolation and biosynthetic analysis of haliamide, a new PKS-NRPS hybrid metabolite from the marine myxobacterium *Haliangium ochraceum*. Molecules 21(1): 59.

Sun, C., Ge, X., Mudassir, S., Zhou, L., Yu, G., Che, Q. et al. (2019). New glutamine-containing azaphilone alkaloids from deep-sea-derived fungus Chaetomium globosum HDN151398. Marine Drugs 17(5): 253.

Suzuki, K., Mizuno, R., Suenaga, K., Teruya, T., Tanaka, N., Kosaka, T. et al. (2013). B isebromoamide, an extract from L yngbya species, induces apoptosis through ERK and mTOR inhibitions in renal cancer cells. Cancer Medicine 2(1): 32–39.

Tan, L.T. (2007). Bioactive natural products from marine cyanobacteria for drug discovery. Phytochemistry 68(7): 954–979.

Tang, R., Kimishima, A., Ishida, R., Setiawan, A. and Arai, M. (2020). Selective cytotoxicity of epidithiodiketopiperazine DC1149B, produced by marine-derived Trichoderma lixii on the cancer cells adapted to glucose starvation. Journal of Natural Medicines 74: 153–158.

Tang, X.X., Liu, S.Z., Yan, X., Tang, B.W., Fang, M.J., Wang, X.M. et al. (2019). Two new cytotoxic compounds from a deep-sea Penicillumcitreonigrum XT20-134. Marine Drugs 17(9): 509.

Tao, L.Y., Zhang, J.Y., Liang, Y.J., Chen, L.M., Zhen, L.S., Wang, F. et al. (2010). Anticancer effect and structure-activity analysis of marine products isolated from metabolites of mangrove fungi in the South China Sea. Marine Drugs 8(4): 1094–1105.

Taori, K., Paul, V.J. and Luesch, H. (2008). Structure and activity of largazole, a potent antiproliferative agent from the Floridian marine cyanobacterium Symploca sp. Journal of the American Chemical Society 130(6): 1806–1807.

Tareq, F.S., Kim, J.H., Lee, M.A., Lee, H.S., Lee, Y.J., Lee, J.S. et al. (2012). Ieodoglucomides A and B from a marine-derived bacterium *Bacillus licheniformis*. Organic Letters 14(6): 1464–1467.

Teruya, T., Sasaki, H., Fukazawa, H. and Suenaga, K. (2009). Bisebromoamide, a potent cytotoxic peptide from the marine cyanobacterium Lyngbya sp.: isolation, stereostructure, and biological activity. Organic Letters 11(21): 5062–5065.

Trischman, J.A., Jensen, P.R. and Fenical, W. (1994). Halobacillin: a cytotoxic cyclic acylpeptide of the iturin class produced by a marine Bacillus. Tetrahedron Letters 35(31): 5571–5574.

Vala, A.K., Sachaniya, B., Dudhagara, D., Panseriya, H.Z., Gosai, H., Rawal, R. et al. (2018). Characterization of L-asparaginase from marine-derived *Aspergillus niger* AKV-MKBU, its antiproliferative activity and bench scale production using industrial waste. International Journal of Biological Macromolecules 108: 41–46.

Varghese, T.A., Jayasri, M.A. and Suthindhiran, K. (2015). Marine Actinomycetes as potential source for histone deacetylase inhibitors and epigenetic modulation. Letters in Applied Microbiology 61(1): 69–76.

Vasko, R.C., Rodriguez, R.A., Cunningham, C.N., Ardi, V.C., Agard, D.A. and McAlpine, S.R. (2010). Mechanistic studies of Sansalvamide A-amide: an allosteric modulator of Hsp90. ACS Medicinal Chemistry Letters 1(1): 4–8.

Wang, D., Wang, Y., Ouyang, Y., Fu, P. and Zhu, W. (2019). Cytotoxic p-terphenyls from a marine-derived Nocardiopsis species. Journal of Natural Products 82(12): 3504–3508.

Wang, L., Groβe, T., Stevens, H., Brinkhoff, T., Simon, M., Liang, L. et al. (2006). Bioactive hydroxyphenylpyrrole-dicarboxylic acids from a new marine Halomonas sp.: Production and structure elucidation. Applied Microbiology and Biotechnology 72: 816–822.

Wang, L., Wang, Q., Tian, X. and Shi, X. (2018). Learning from Clostridium novyi-NT: how to defeat cancer. Journal of Cancer Research and Therapeutics 14(Suppl 1): S1–S6.

Wang, Q., Zhang, Y., Wang, M., Tan, Y., Hu, X., He, H. et al. (2017). Neo-actinomycins A and B, natural actinomycins bearing the 5 H-oxazolo (4, 5-b) phenoxazine chromophore, from the marine-derived Streptomyces sp. IMB094. Scientific Reports 7(1): 3591.

Wang, W., Yang, J., Liao, Y.Y., Cheng, G., Chen, J., Cheng, X.D. et al. (2020). Cytotoxic nitrogenated azaphilones from the deep-sea-derived fungus *Chaetomium globosum* MP4-S01-7. Journal of Natural Products 83(4): 1157–1166.

Weerakkody, L.R. and Witharana, C. (2019). The role of bacterial toxins and spores in cancer therapy. Life Sciences 235: 116839.

Williams, D.E., Moore, R.E. and Paul, V.J. (1989). The structure of ulithiacyclamide B. Antitumor evaluation of cyclic peptides and macrolides from *Lissoclinum patella*. Journal of Natural Products 52(4): 732–739.

Williams, P.G., Yoshida, W.Y., Moore, R.E. and Paul, V.J. (2003a). The isolation and structure elucidation of Tasiamide B, a 4-amino-3-hydroxy-5-phenylpentanoic acid containing peptide from the marine Cyanobacterium *Symploca* sp. Journal of Natural Products 66(7): 1006–1009.

Williams, P.G., Yoshida, W.Y., Moore, R.E. and Paul, V.J. (2003b). Tasipeptins A and B: new cytotoxic depsipeptides from the marine cyanobacterium *Symploca* sp. Journal of Natural Products 66(5): 620–624.

Williams, D.E., Dalisay, D.S., Patrick, B.O., Matainaho, T., Andrusiak, K., Deshpande, R. et al. (2011). Padanamides A and B, highly modified linear tetrapeptides produced in culture by a *Streptomyces* sp. isolated from a marine sediment. Organic Letters 13(15): 3936–3939.

Wollowski, I., Rechkemmer, G. and Pool-Zobel, B.L. (2001). Protective role of probiotics and prebiotics in colon cancer. The American Journal of Clinical Nutrition 73(2): 451s–455s.

World Health Organization. (2020). WHO report on cancer: setting priorities, investing wisely and providing care for all.

Xie, C.L., Xia, J.M., Lin, T., Lin, Y.J., Lin, Y.K., Xia, M.L. et al. (2019a). Andrastone A from the deep-sea-derived fungus Penicillium allii-sativi acts as an inducer of Caspase and RXRα-dependent apoptosis. Frontiers in Chemistry 7: 692.

Xie, C.L., Zhang, D., Xia, J.M., Hu, C.C., Lin, T., Lin, Y.K. et al. (2019b). Steroids from the deep-sea-derived fungus *Penicillium granulatum* MCCC 3A00475 induced apoptosis via retinoid X receptor (RXR)-α pathway. Marine Drugs 17(3): 178.

Xu, J., Liu, Z., Chen, Y., Tan, H., Li, H., Li, S. et al. (2019a). Lithocarols AF, six tenellone derivatives from the deep-sea derived fungus *Phomopsis lithocarpus* FS508. Bioorganic Chemistry 87: 728–735.

Xu, T., Shi, L., Zhang, Y., Wang, K., Yang, Z. and Ke, S. (2019b). Synthesis and biological evaluation of marine alkaloid-oriented β-carboline analogues. European Journal of Medicinal Chemistry 168: 293–300.

Xu, W.J., Liao, X.J., Xu, S.H., Diao, J.Z., Du, B., Zhou, X.L. et al. (2008). Isolation, structure determination, and synthesis of galaxamide, a rare cytotoxic cyclic pentapeptide from a marine algae *Galaxaura filamentosa*. Organic Letters 10(20): 4569–4572.

Yamada, T., Iritani, M., Ohishi, H., Tanaka, K., Minoura, K., Doi, M. et al. (2007). Pericosines, antitumour metabolites from the sea hare-derived fungus *Periconia byssoides*. Structures and biological activities. Organic & Biomolecular Chemistry 5(24): 3979–3986.

Yang, L., Tan, R.X., Wang, Q., Huang, W.Y. and Yin, Y.X. (2002). Antifungal cyclopeptides from *Halobacillus litoralis* YS3106 of marine origin. Tetrahedron Letters 43(37): 6545–6548.

Yang, T., Yamada, K., Zhou, T., Harunari, E., Igarashi, Y., Terahara, T. et al. (2019). Akazamicin, a cytotoxic aromatic polyketide from marine-derived *Nonomuraea* sp. The Journal of Antibiotics 72(4): 202–209.

Ye, X., Anjum, K., Song, T., Wang, W., Liang, Y., Chen, M. et al. (2017). Antiproliferative cyclodepsipeptides from the marine actinomycete *Streptomyces* sp. P11-23B downregulating the tumor metabolic enzymes of glycolysis, glutaminolysis, and lipogenesis. Phytochemistry 135: 151–159.

Yu, H.B., Glukhov, E., Li, Y., Iwasaki, A., Gerwick, L., Dorrestein, P.C. et al. (2019). Cytotoxic microcolin lipopeptides from the marine cyanobacterium *Moorea producens*. Journal of Natural Products 82(9): 2608–2619.

Yu, Z., Lang, G., Kajahn, I., Schmaljohann, R. and Imhoff, J.F. (2008). Scopularides A and B, cyclodepsipeptides from a marine sponge-derived fungus, *Scopulariopsis brevicaulis*. Journal of Natural Products 71(6): 1052–1054.

Yurchenko, A.N., Trinh, P.T.H., Girich, E.V., Smetanina, O.F., Rasin, A.B., Popov, R.S. et al. (2019). Biologically active metabolites from the marine sediment-derived fungus *Aspergillus flocculosus*. Marine Drugs 17(10): 579.

Zhang, D., Yi, W., Ge, H., Zhang, Z. and Wu, B. (2019a). Bioactive streptoglutarimides A–J from the marine-derived *Streptomyces* sp. ZZ741. Journal of Natural Products 82(10): 2800–2808.

Zhang, D., Yi, W., Ge, H., Zhang, Z. and Wu, B. (2021). A new antimicrobial indoloditerpene from a marine-sourced fungus *Aspergillus versicolor* ZZ761. Natural Product Research 35(18): 3114–3119.

Zhang, G.F., Li, C.X., Liu, Z.Q., Ma, S. and Chen, H.B. (2016a). Cancer stem cell targets—a review. European Review for Medical & Pharmacological Sciences 20(10).

Zhang, H.L., Hua, H.M., Pei, Y.H. and Yao, X.S. (2004). Three new cytotoxic cyclic acylpeptides from marine *Bacillus* sp. Chemical and Pharmaceutical Bulletin 52(8): 1029–1030.

Zhang, H.M., Ju, C.X., Li, G., Sun, Y., Peng, Y., Li, Y.X. et al. (2019b). Dimeric 1, 4-benzoquinone derivatives with cytotoxic activities from the marine-derived fungus *Penicillium* sp. L129. Marine Drugs 17(7): 383.

Zhang, J.Y., Tao, L.Y., Liang, Y.J., Yan, Y.Y., Dai, C.L., Xia, X.K. et al. (2009). Secalonic acid D induced leukemia cell apoptosis and cell cycle arrest of G1 with involvement of GSK-3β/β-catenin/c-Myc pathway. Cell Cycle 8(15): 2444–2450.

Zhang, J.Y., Tao, L.Y., Liang, Y.J., Chen, L.M., Mi, Y.J., Zheng, L.S. et al. (2010). Anthracenedione derivatives as anticancer agents isolated from secondary metabolites of the mangrove endophytic fungi. Marine Drugs 8(4): 1469–1481.

Zhang, P., Li, X.M., Mao, X.X., Mándi, A., Kurtán, T. and Wang, B.G. (2016b). Varioloid A, a new indolyl-6, 10b-dihydro-5aH-(1) benzofuro (2, 3-b) indole derivative from the marine alga-derived endophytic fungus *Paecilomyces variotii* EN-291. Beilstein Journal of Organic Chemistry 12(1): 2012–2018.

Zhang, W., Che, Q., Tan, H., Qi, X., Li, J., Li, D. et al. (2017). Marine *Streptomyces* sp. derived antimycin analogues suppress HeLa cells via depletion HPV E6/E7 mediated by ROS-dependent ubiquitin–proteasome system. Scientific Reports 7(1): 42180.

Zhang, Y.X., Yu, H.B., Xu, W.H., Hu, B., Guild, A., Zhang, J.P. et al. (2019c). Eutypellacytosporins A–D, meroterpenoids from the arctic fungus *Eutipella* sp. D-1. Journal of Natural Products 82(11): 3089–3095.

Zhou, B., Jiang, Y.J., Ji, Y.Y., Zhang, H.J. and Shen, L. (2020a). Lactoquinomycin C and D, two new medermycin derivatives from the marine-derived *Streptomyces* sp. SS17A. Natural Product Research 34(9): 1213–1218.

Zhou, G., Chen, X., Zhang, X., Che, Q., Zhang, G., Zhu, T. et al. (2020b). Prenylated p-terphenyls from a mangrove endophytic fungus, *Aspergillus candidus* LDJ-5. Journal of Natural Products 83(1): 8–13.

Zhou, L.M., Kong, F.D., Xie, Q.Y., Ma, Q.Y., Hu, Z., Zhao, Y.X. et al. (2019). Divergolides T–W with apoptosis-inducing activity from the mangrove-derived Actinomycete *Streptomyces* sp. KFD18. Marine Drugs 17(4): 219.

Zhou, S.L., Wang, M., Zhao, H.G., Huang, Y.H., Lin, Y.Y., Tan, G.H. and Chen, S.L. (2016). Penicilazaphilone C, a new antineoplastic and antibacterial azaphilone from the Marine Fungus *Penicillium sclerotiorum*. Archives of Pharmacal Research 39: 1621–1627.

Chapter 9

Biotechnological Application of Probiotic and Biocontrol Properties of Marine *Bacillus* Species in Developing the Blue Economy

Shubhra Singh[1,2] and *Douglas J. H. Shyu*[2,*]

Introduction

Background

The world's oceans, covering over 70% of the Earth's surface, hold immense potential for economic growth, environmental sustainability, and improved livelihoods. The concept of the blue economy, which emphasizes the responsible use of ocean resources, has gained significant attention in recent years. Central to this paradigm shift is the integration of cutting-edge biotechnology, unlocking the latent benefits of marine organisms for a variety of applications (FAO 2022).

In this context, one of the most exciting developments is the utilization of marine *Bacillus* species, which possess remarkable probiotic and biocontrol properties. These microorganisms, found abundantly in marine environments, have demonstrated their

[1] Department of Tropical Agriculture and International Cooperation, National Pingtung University of Science and Technology, Pingtung 912301, Taiwan.
[2] Department of Biological Science and Technology, National Pingtung University of Science and Technology, Pingtung 912301, Taiwan.
Email: shubhrasingh6@gmail.com
* Corresponding author: dshyu@mail.npust.edu.tw

ability to positively impact aquaculture systems, sustainable seafood production, and the overall health of marine ecosystems. Probiotics have long been recognized for their beneficial effects on human health, but their application has now expanded to the realm of aquatic organisms. Marine *Bacillus* species exhibit exceptional probiotic properties that are being increasingly harnessed for the enhancement of aquaculture practices. By optimizing water quality, boosting the immune responses of aquatic organisms, and reducing the need for antibiotics, these probiotics play a pivotal role in ensuring the health and productivity of farmed marine species (Buruiana et al. 2014).

Biocontrol, a sustainable alternative to chemical pest and disease management, finds a perfect ally in marine *Bacillus* species. These microorganisms have evolved potent antimicrobial compounds that can suppress harmful pathogens in aquaculture systems. By mitigating the risks of disease outbreaks, biocontrol *Bacillus* strains not only protect the farmed organisms but also contribute to the health of wild marine populations, preserving biodiversity and ecosystem integrity (Jayakumar et al. 2019). The synergistic application of probiotic and biocontrol properties of marine *Bacillus* species presents a transformative opportunity for the blue economy. From bolstering the sustainability of aquaculture operations, which are a vital component of the global seafood supply, to safeguarding the balance of marine ecosystems, these biotechnological advancements offer a path to economic growth, environmental stewardship, and food security (Ye et al. 2022). In this chapter, we will delve into the multifaceted benefits of marine *Bacillus* species in the context of the blue economy. We will examine their role in sustainable aquaculture, ecosystem health, and economic prosperity. By understanding the diverse applications and innovative approaches that leverage the remarkable capabilities of these microorganisms, we can chart a course towards a more resilient, prosperous and sustainable blue economy.

Objectives

The objectives of this chapter include (i) conducting a comprehensive analysis of various marine *Bacillus* species to identify those with robust probiotic and biocontrol properties; (ii) investigating the reports that mention the probiotic potential of selected marine *Bacillus* strains by assessing their impact on aquatic organism health that involves their ability to enhance immune responses, promote growth, and mitigate stressors, ultimately contributing to the development of sustainable aquaculture practices; (iii) evaluating the biocontrol efficacy of the chosen *Bacillus* strains against key aquatic pathogens including conducting *in vitro* and *in vivo* experiments to measure the inhibition of pathogen growth and the prevention of disease outbreaks in aquaculture systems (iv) developing efficient and environmentally friendly methods for applying marine *Bacillus* probiotics and biocontrol agents in aquaculture and related sectors of the blue economy; (v) implementing long-term monitoring of water quality parameters in aquaculture systems where marine *Bacillus* probiotics are applied (vi) assessing the impact of these probiotics on organic matter breakdown, pollutant reduction, and overall water quality enhancement (vii) providing practical guidelines and recommendations for incorporating marine *Bacillus* probiotics and biocontrol agents into aquaculture operations (viiii) emphasizing the reduction of antibiotic use, disease prevention, and improved overall sustainability in the production of aquatic organisms; (ix) highlighting the role of marine *Bacillus* species in the development of the blue economy by presenting the economic benefits and ecological advantages of their application (x) emphasizing how these biotechnological interventions can contribute to increased productivity, biodiversity conservation, and economic growth within the blue economy framework along with practical insights with stakeholders in the aquaculture sector, policymakers, environmental organizations, and the scientific community to facilitate the adoption of these biotechnological solutions for blue economy development.

Marine *Bacillus* species: diversity and adaptation

Overview of Marine Bacillus species

Marine *Bacillus* species are a diverse and ecologically significant group of bacteria that inhabit the vast and varied marine environments across the globe. These microorganisms have evolved remarkable adaptations to thrive in the challenging conditions of the ocean, making them essential players in the aquatic ecosystem. From the sunlit surface waters to the mysterious depths of the abyssal zone, Marine *Bacillus* species play pivotal roles in nutrient cycling, organic matter decomposition, and the overall ecological balance of aquatic ecosystems. Their ability to withstand a wide range of salt concentrations, temperature fluctuations, and nutrient availability underscores their resilience in the face of ever-changing oceanic conditions. These bacteria exhibit a plethora of metabolic capabilities, allowing them to utilize a diverse array of organic compounds present in the marine environment. This metabolic versatility contributes to the recycling of organic matter, which is crucial for sustaining aquatic life. Additionally, certain Marine *Bacillus* species produce bioactive compounds with potential applications beyond their ecological roles. These compounds have been studied for their antimicrobial, enzymatic, and biotechnological properties. The bioactive molecules synthesized by Marine *Bacillus* species have garnered attention for their potential in medical research, pharmaceutical development, and environmental bioremediation (Caulier et al. 2019, Chen et al. 2016).

In the context of the blue economy, the biotechnological applications of Marine *Bacillus* species hold significant promise. Their probiotic properties can revolutionize sustainable aquaculture practices by improving the health and growth of cultured aquatic organisms while reducing the need for antibiotics. Furthermore, their biocontrol capabilities offer environmentally friendly alternatives for managing disease outbreaks in aquaculture systems, mitigating the negative impacts of pathogens on both farmed and wild marine species. As the blue economy continues to grow, the multifaceted contributions of Marine *Bacillus* species become increasingly important. Their ecological significance, metabolic diversity, and biotechnological potential highlight their role in achieving the dual objectives of economic growth and environmental sustainability. By understanding and harnessing the capabilities of these remarkable bacteria, we can navigate the challenges and opportunities of the blue economy, ensuring a harmonious coexistence with the ocean and unlocking the potential of this vast and valuable resource for generations to come (Balqadi et al. 2015).

Studies have shown that the genus *Bacillus* has been considered as a probiotic in water treatment. *Bacillus* has the ability to convert organic matter into carbon dioxide and the isolate from *Bacillus* decreases nitrite, nitrate, and ammonium concentrations in ornamental fish water. A similar phenomenon has been observed with *Bacillus subtilis*, *B. cereus* and *B. licheniformis* that are responsible for bioaccumulation, bioassimilation and nitrification (Suman et al. 2020). *Bacillus subtilis* at the dose of 10^8 CFU/mL in the rearing of shrimp (*Litopenaeus vannamei*) for eight weeks have been seen beneficial effects on shrimp culture from water quality along with growth enhancement, digestive enzyme activity improvement and improvement in the immune response and in disease resistance (Table 1). Additionally, administration of *B. subtilis* strains increased the total protein, protease and amylase activity of treated shrimp after colonization in the gastrointestinal tract. According to research, the administration of *Bacillus* probiotic strains (*B. subtilis*, *B. licheniformis*, *B. polymyxa*, *B. laterosporus* and *B. circulans*) elevated the digestive enzymes activity, survival and growth of white shrimp (*Fenneropenaeus indicus*) (Buruiana et al. 2014).

Table 1. *Bacillus* as probiotics to improve aquatic animal growth and digestive processes.

Probiotic genus or species	Hosts	Activity	References
Bacillus	*Oncorhynchus mykiss*	Enhances the immune system, modulates the digestive microbiota, increases growth and anti-vibrio activity, increases weight, efficient in food and protein intake	Tuan et al. 2013, Seifzadeh et al. 2019
	Catla catla		
	Acipencer persicus		
	Cyprinus carpio		
	Carassius auratus		
	Xiphophorus helleri		
	Epinephelus coioides		
	Litopenaeus vannamei		
	Penaeus monodon		
	Penaeus vannamei		
	Macrobrachium rosenbergii		
	Babylonia areolate		
	Oreochromis niloticus		

Adaptation to marine environments

The adaptation of marine *Bacillus* species is a captivating story of biological evolution shaped by the demanding conditions of the world's oceans. These microorganisms have successfully colonized a wide range of marine habitats, each with its unique set of challenges. Their ability to thrive in such diverse environments (as shown in Figure 1) speaks of their remarkable genetic versatility and finely tuned responses to the complex interplay of factors in marine ecosystems. One of the most crucial

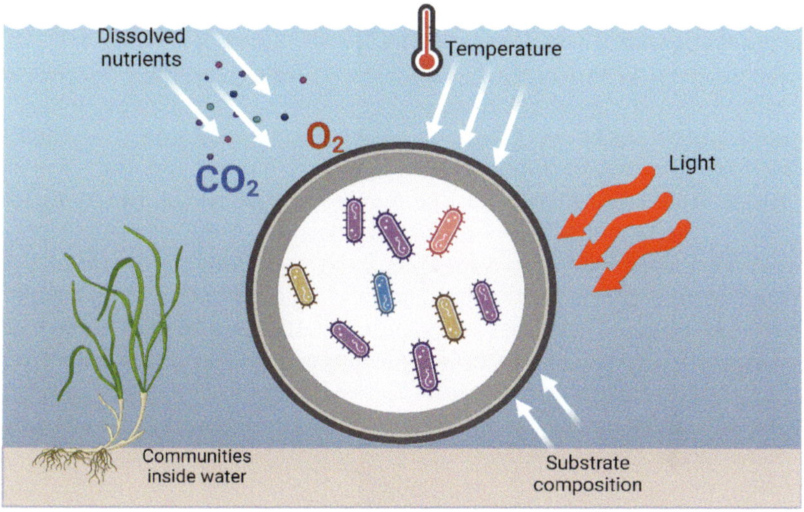

Figure 1. Environmental factors affect the genetic composition of *Bacillus* species (created with BioRender.com).

adaptations of marine *Bacillus* species is their halotolerance and osmoregulation mechanisms. The ocean's salinity can vary significantly, from the relatively low salt content of estuaries to the high salinity of open ocean waters. Marine *Bacillus* species have evolved specialized cellular processes that allow them to adjust their internal salt concentrations, ensuring osmotic balance and preventing dehydration or overhydration of their cells. This adaptability enables them to flourish in a wide range of salinity conditions, making them versatile inhabitants of the ever-changing marine environment (Bahaddad et al. 2023).

Temperature resilience is another noteworthy aspect of their adaptation. The ocean's temperature gradient extends from the frigid waters of polar regions to the warmth of tropical seas, and marine *Bacillus* species have evolved diverse strategies to cope with this variability. Some species have become psychrophiles, adapted to thrive in cold waters, while others are thermophiles, capable of withstanding the extreme heat near hydrothermal vents. These adaptations reflect their ability to function optimally within specific temperature ranges, enabling them to colonize various oceanic zones. The nutrient-poor nature of many marine environments has also driven the evolution of efficient nutrient utilization in marine *Bacillus* species. These microorganisms possess versatile metabolic pathways that allow them to extract essential elements from the limited organic and inorganic compounds available in the ocean (Prodhan et al. 2022). This adaptation not only contributes to their own survival but also plays a critical role in nutrient cycling, benefiting the broader marine ecosystem. Their capacity to adapt to variable oxygen levels further showcases their resilience. Some marine *Bacillus* species are aerobic, thriving in oxygen-rich waters, while others have evolved as facultative anaerobes, capable of surviving in low-oxygen or anaerobic conditions. This flexibility in oxygen usage enables them to inhabit various oceanic depths and adapt to changing oxygen gradients. Perhaps one of their most intriguing adaptations is their production of bioactive compounds. These compounds, including antibiotics and enzymes, serve as powerful weapons in the struggle for survival within the competitive and sometimes hostile marine environment. By inhibiting the growth of competing microorganisms and acting as defenses against pathogens, these compounds offer a distinct advantage to marine *Bacillus* species (Buruiana et al. 2014).

The intricate adaptations of marine *Bacillus* species reveal their evolutionary competence, honed through ages of natural selection. These adaptations position them as vital contributors to marine ecosystems, influencing nutrient cycling, participating in symbiotic relationships, and playing a critical role in the health and balance of the oceans. Their adaptability and unique traits also hold great potential for biotechnological applications, particularly in the context of the blue economy, where their probiotic and biocontrol properties can be harnessed to enhance aquaculture, promote sustainable practices, and contribute to the responsible development of marine resources (Thompson et al. 2013). Understanding and appreciating these adaptations not only deepen our appreciation for the wonders of marine life but also inspire us to unlock the potential of these microorganisms for the benefit of both humanity and the oceans.

Metabolic capabilities and biotechnological potential

The metabolic capabilities of marine *Bacillus* species are central to their biotechnological potential in the development of the blue economy, particularly in the application of probiotic and biocontrol properties. These capabilities encompass a range of biochemical processes that enable these microorganisms to interact with their environment, produce bioactive compounds, and play essential roles in nutrient cycling. Leveraging these metabolic traits offers promising opportunities for sustainable practices in aquaculture, environmental management, and the broader context of the blue economy (Sharma et al. 2021). *Bacillus* species have evolved diverse metabolic pathways that allow them to efficiently utilize various organic and inorganic compounds found in marine environments. Their ability to break down complex molecules contributes to nutrient cycling, influencing the availability of essential nutrients like nitrogen and phosphorus. This aspect is particularly relevant in aquaculture systems, where optimal nutrient levels are crucial for the growth of aquatic organisms (Kumar et al. 2017). One of the most significant biotechnological potentials of marine *Bacillus* species lies in their production of bioactive compounds. These compounds can have antimicrobial, antiviral, and enzymatic properties. Harnessing their ability to produce such compounds offers opportunities for natural disease management in aquaculture. These compounds can inhibit the growth of harmful pathogens, reducing the need for chemical treatments and promoting eco-friendly biocontrol strategies. *Bacillus* species possess probiotic potential, considered as a key aspect of their biotechnological application in aquaculture. Probiotics are a combination of beneficial microorganisms that, when administered to aquatic organisms, enhance their health, improve digestion, and boost immune responses. The probiotic capabilities of specific *Bacillus* strains can improve the overall well-being of cultured species, leading to increased survival rates, faster growth, and reduced stress. The metabolic versatility of marine *Bacillus* species extends to their role in environmental remediation. They can participate in the breakdown of organic matter, contribute to the degradation of pollutants and play a role in maintaining water quality. Applying selected strains in aquaculture systems can aid in the removal of excess nutrients, organic waste, and even certain contaminants, reducing the environmental impact of aquaculture operations. The integration of marine *Bacillus* probiotics and biocontrol agents in aquaculture systems aligns with the goals of sustainable practices in the blue economy. By improving water quality, reducing disease risks, and decreasing the use of antibiotics and chemical treatments, these microbial solutions contribute to environmentally responsible and economically viable aquaculture (Huan et al. 2020, Ringø et al. 2022). Therefore, the metabolic capabilities of marine *Bacillus* species underpin their biotechnological potential in the blue economy, particularly in the application of probiotic and biocontrol properties. By understanding and harnessing these capabilities, we can develop innovative and sustainable strategies that benefit both aquaculture production and the broader marine ecosystem, fostering economic growth while safeguarding the health of our oceans.

Probiotic properties of marine *Bacillus* species

Probiotics in aquaculture

Probiotics, hailed as a game-changer in aquaculture have emerged as a powerful ally in shaping a sustainable and efficient blue economy. In the intricate world of aquaculture, where disease outbreaks, water quality management, and stress resilience are paramount concerns, probiotics offer a natural and environmentally friendly solution. These beneficial microorganisms, primarily bacteria, act as guardians of aquatic health, bolstering the immune systems of farmed organisms and inhibiting the proliferation of harmful pathogens. By reducing disease risks, probiotics decrease the reliance on antibiotics, promoting responsible practices that align with the principles of the blue economy. Furthermore, probiotics play a pivotal role in maintaining water quality within aquaculture systems. Their ability to break down organic matter and reduce pollutants ensures a cleaner and healthier environment for aquatic organisms to thrive. The reduction of waste products, such as ammonia, not only enhances the well-being of cultured species but also contributes to the overall sustainability of aquaculture operations (Sayes et al. 2017).

Probiotics have demonstrated their potential to boost growth rates and survival in various aquaculture species. A healthier and stress-resistant population translates into a more efficient growth, leading to increased productivity and, consequently, economic growth within the blue economy sector. However, the presence of pathogenic microorganisms is often enhanced with probiotic bacteria proliferating in the same medium and this problem has been tackled by the application of probiotics at larval stage during pisciculture (Paul et al. 2021, Sayes et al. 2017). Nevertheless, the successful deployment of probiotics in aquaculture requires careful considerations. Strain selection is critical, ensuring that the chosen probiotics are tailored to the specific needs of the target species and the unique conditions of each aquaculture system. Bioactive compounds derived from marine microbe from different species of *Bacillus* such as Macrolactin from *Bacillus amyloliquefaciencs* possess antibacterial activity. Additionally, surfactin from *Bacillus subtilis* has antibacterial subtilis along with bacitracin from *Bacillus* spp. These species are important as marine microorganisms due to presence of bioactive natural products such as antibiotics, biocatalysts and other potential metabolites (Paul et al. 2021). It is also important to find the optimal dosage and application method that is essential to avoid unintended consequences. Additionally, regulatory and safety concerns must be addressed to guarantee the well-being of aquatic organisms and the consumers of aquaculture products. In embracing probiotics as a cornerstone of modern aquaculture, the blue economy takes a significant step toward a more balanced and harmonious coexistence with the ocean. By harnessing the preventive and ecological benefits of probiotics, aquaculture can navigate challenges, improve efficiency, and contribute to sustainable growth, ultimately ensuring that our precious marine resources thrive, not just for us but for generations to come.

Mechanisms of probiotic action

The mechanisms through which probiotics exert their beneficial effects encompass a sophisticated interplay of interactions between these friendly microorganisms and their host, whether in the gut of humans or in the aquatic environments of aquaculture. First and foremost, probiotics employ the principle of competitive exclusion, competing head-to-head with harmful pathogens for nutrients and binding sites, ultimately crowding out the invaders and reducing the risk of infections. This competition sets the stage for a healthier microbial balance. A key weapon in the probiotic arsenal is the production of antimicrobial compounds. These bioactive substances, including organic acids and bacteriocins, act as natural antibiotics, inhibiting the growth and virulence of pathogenic microorganisms. Such microbial control mechanisms contribute to the prevention of infections and support the maintenance of a more harmonious microbiome (Caulier et al. 2019).

Probiotics are also adept at enhancing the host's immune response, acting as modulators of the immune system. They stimulate the production of immune cells, promote balanced immune reactions and contribute to overall immune system resilience. By bolstering the body's defenses, probiotics create a less hospitable environment for pathogens to take hold of it, reducing the likelihood of illness and inflammation. Furthermore, the metabolism and nutrient-modifying capabilities of probiotics play another significant role. These microorganisms can break down complex carbohydrates that the host cannot digest, providing additional sources of energy and nourishment. This metabolic assistance not only benefits the host but also aids in the growth and establishment of beneficial bacteria, creating a positive feedback loop that supports a diverse and thriving microbiota (Balqadi et al. 2015).

Probiotics, through their diverse mechanisms, can disrupt the formation of harmful biofilms, which are protective matrices that shield pathogenic microorganisms from traditional treatments. By disrupting these biofilms, probiotics help prevent chronic infections and contribute to overall microbiome health. In addition, probiotics can influence the pH balance in their respective environments. By regulating pH levels, they create conditions that are less favorable for the growth of harmful microorganisms, further fortifying the host's defenses. Lastly, by supporting a healthy and diverse microbial community, probiotics contribute to the overall well-being of the host, whether it's a human digestive system or a cultured aquatic organism. This microbial harmony facilitates improved digestion, nutrient absorption, and overall resilience. In the realm of aquaculture, these multifaceted mechanisms are essential. Probiotics play a pivotal role in promoting the health and growth of aquatic species, improving water quality and reducing disease risks (Meroni et al. 2021).

Role of marine Bacillus species in fish health and disease resistance

Marine *Bacillus* species play a crucial role in fish health and disease resistance, particularly in aquaculture settings. These beneficial microorganisms, when applied as probiotics, contribute to the overall well-being of farmed fish by enhancing their immune responses, promoting a balanced gut microbiota, and providing protection against pathogenic invaders. Understanding the specific roles that marine *Bacillus* species play in fish health and disease resistance is essential for optimizing their application in aquaculture systems (Sayes et al. 2017).

Immune system enhancement: Marine *Bacillus* probiotics can stimulate and modulate the immune system of fish. They help increase the production of immune cells, such as white blood cells and antibodies, leading to improved defense mechanisms against pathogens. This immune enhancement is particularly valuable in reducing the susceptibility of fish to various infectious diseases (Khan et al. 2014).

Pathogen exclusion: Marine *Bacillus* species, when introduced into fish gut, compete for binding sites and nutrients and exclude harmful pathogens. This competitive exclusion mechanism prevents the establishment of pathogenic microorganisms in the digestive tract, reducing the risk of infections and diseases (Ahmad et al. 2019).

Production of antimicrobial compounds: Many marine *Bacillus* strains produce antimicrobial compounds, such as bacteriocins, which are effective in inhibiting the growth of pathogenic bacteria. These compounds act as natural antibiotics, helping to maintain a healthier microbial balance in the fish gut and the aquatic environment (Ahmad et al. 2019).

Gut microbiota modulation: A balanced gut microbiota is essential for fish health and digestion. Marine *Bacillus* probiotics contribute to this balance by promoting the growth of beneficial bacteria and suppressing the overgrowth of potential pathogens. This modulation of the gut microbiota supports efficient nutrient absorption and overall gut health in farmed fish.

Stress reduction: Fish in aquaculture settings often experience stress due to factors such as handling, transportation, and changes in water quality. Stress weakens the immune system and makes fish more susceptible to diseases. Marine *Bacillus* probiotics can help mitigate the negative effects of stress by bolstering the immune responses of fish (Figure 2), leading to improved stress tolerance and reduced disease susceptibility (Zhou et al. 2018).

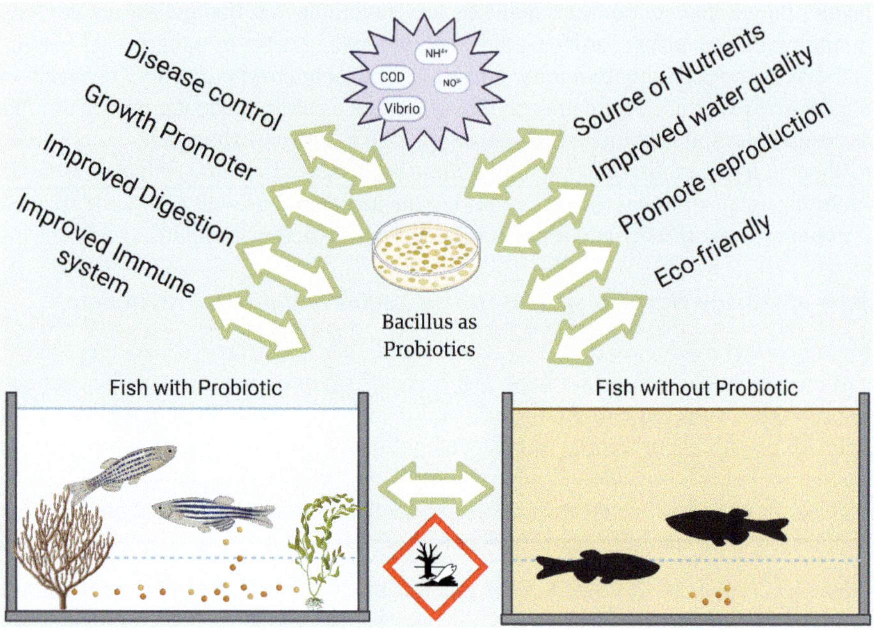

Figure 2. Enhancing aquaculture: The multifaceted benefits of *Bacillus* probiotics (Created with BioRender.com).

Natural disease management: By providing fish with a natural defense mechanism against pathogens, marine *Bacillus* probiotics reduce the need for antibiotics and chemical treatments in aquaculture operations. This aligns with the goals of sustainable aquaculture and the development of eco-friendly disease management strategies (Rahman et al. 2021).

In summary, marine *Bacillus* species, when utilized as probiotics in aquaculture, positively impact fish health and disease resistance through a combination of immune system enhancement, pathogen exclusion, antimicrobial compound production, gut microbiota modulation, stress reduction, and natural disease management. Their multifaceted role contributes to the overall sustainability and productivity of aquaculture systems while reducing the environmental impact associated with disease control measures.

Reduction of antibiotic dependence in aquaculture

The reduction of antibiotic dependence in aquaculture is a critical goal for the sustainable development of the industry, and this objective aligns with the principles of responsible aquaculture and the blue economy. The overuse of antibiotics in aquaculture can lead to antibiotic-resistant bacteria, environmental contamination, and potential health risks for consumers. As a result, finding alternative strategies to manage disease and promote fish health is essential. Probiotics, including marine

Bacillus species, represent a promising avenue for achieving this reduction in antibiotic dependence (Villela et al. 2019).

Probiotics act as a natural line of defense against pathogens, reducing the need for antibiotics by enhancing the immune system of the fish and creating an environment that is less hospitable to harmful microorganisms. For example, in shrimp farming, the application of specific marine *Bacillus* probiotics has shown to improve the survival rates and disease resistance in shrimp, reducing the reliance on antibiotics to combat diseases such as *Vibrio* infections. One key mechanism through which probiotics reduce antibiotic dependence is competitive exclusion. Marine *Bacillus* probiotics compete with pathogenic bacteria for nutrients and binding sites in the fish gut, effectively crowding out the harmful microbes. This competition reduces the opportunity for pathogenic bacteria to establish themselves, leading to a lower risk of infections. In tilapia aquaculture, the introduction of *Bacillus* probiotics has been associated with reduced mortality rates and a decrease in the use of antibiotics.

Moreover, the production of antimicrobial compounds by marine *Bacillus* probiotics provides a natural defense mechanism. These compounds, such as bacteriocins, inhibit the growth of pathogenic bacteria, offering an alternative to the use of antibiotics. For instance, in salmon farming, the supplementation of diets with specific *Bacillus* strains has been found to improve fish health and reduce the need for antibiotics to manage bacterial infections. The proactive use of probiotics, including marine *Bacillus* species, in aquaculture systems supports a more sustainable approach to disease management. By enhancing fish health, promoting a balanced microbiota, and reducing disease risks, probiotics contribute to healthier and more resilient fish populations. This shift away from antibiotic dependence not only safeguards the environment and preserves antibiotic effectiveness for human medicine but also aligns with the principles of the blue economy, fostering responsible resource management and the long-term viability of aquaculture practices. Through continued research, responsible application, and industry adoption, probiotics can play a pivotal role in achieving the reduction of antibiotic dependence while ensuring the health and productivity of aquaculture systems.

Biocontrol activities of marine *Bacillus* species

Biocontrol in marine environments

Biocontrol in marine environments refers to the use of natural predators, beneficial microorganisms, or other ecological mechanisms to manage the populations of harmful organisms, such as invasive species or aquatic pathogens. This approach aims to reduce the negative impacts of these organisms on native ecosystems, aquaculture, and other marine activities. Biocontrol is an important component of sustainable marine resource management and conservation efforts. Several biocontrol strategies are employed in marine environments, each tailored to specific challenges and ecological contexts. Some of the methods are introducing or enhancing the populations of natural predators that feed on harmful organisms and can be an effective biocontrol strategy. For example, in some locations, the introduction of specific predatory fish or invertebrates has been used to control the populations of invasive species that threaten native marine ecosystems. Another way is promoting the growth of specific biological control agents, such as parasites, pathogens, or competing species, to target harmful organisms. This method aims to disrupt the reproduction or survival of the target species without causing significant harm to non-target organisms. The use of beneficial microorganisms such as probiotics, including certain marine *Bacillus* species are employed in biocontrol to inhibit the growth of pathogenic bacteria or other harmful microorganisms. These beneficial microorganisms can help maintain a healthy microbial balance, preventing disease outbreaks in aquaculture systems and other marine environment. Some biocontrol strategies involve the use of natural compounds or biochemical agents that disrupt the life cycle or behavior of the target organisms. These compounds may affect the reproduction, feeding, or mobility of the harmful species, reducing their impact on marine ecosystems (Ren et al. 2021).

In certain cases, physical barriers or structures are used to control the movement of harmful organisms. For example, installing screens or barriers in aquaculture facilities can prevent the entry of invasive species or pathogens, reducing the need for chemical treatments. Creating or restoring natural habitats that support the growth of beneficial species can indirectly control the populations of harmful organisms. This approach enhances the overall biodiversity and ecological balance of marine environments. Successful biocontrol in marine environments requires careful consideration of the target species, the potential impact on non-target species, and the ecological dynamics of the marine ecosystem. Balancing the effectiveness of biocontrol measures with potential ecological and regulatory concerns is essential to ensure the responsible and sustainable management of marine resources. When properly implemented, biocontrol strategies can contribute to the conservation of native species, the prevention of economic losses, and the promotion of a healthy and balanced marine ecosystem (Noatch and Suski 2012).

Antagonistic properties of marine Bacillus species

The Marine *Bacillus* species exhibit a formidable array of antagonistic properties that position them as crucial players in maintaining the equilibrium and health of marine ecosystems (as shown in Figure 3). These microorganisms, often found in diverse marine environments, have evolved strategies to counter the threats posed by competing microorganisms and pathogens. Their antagonistic arsenal includes the production of potent antimicrobial compounds, acting as natural antibiotics to inhibit the growth of harmful bacteria, fungi, and other microorganisms. This antagonistic prowess extends to competitive exclusion, where marine *Bacillus* species outcompete rival microbes for vital resources and binding sites, preventing the establishment of potential pathogens. Additionally, they excel in disrupting the formation of biofilms, complex communities that can harbor pathogenic organisms. By disrupting biofilms, these *Bacillus* species prevent the adhesion of pathogens to marine surfaces, reducing the risk of biofilm-associated infections. This biocontrol potential positions marine *Bacillus* species as promising candidates for biotechnological applications, including the enhancement of aquaculture practices, where their antagonistic properties can bolster disease resistance in farmed aquatic organisms. In the broader context of

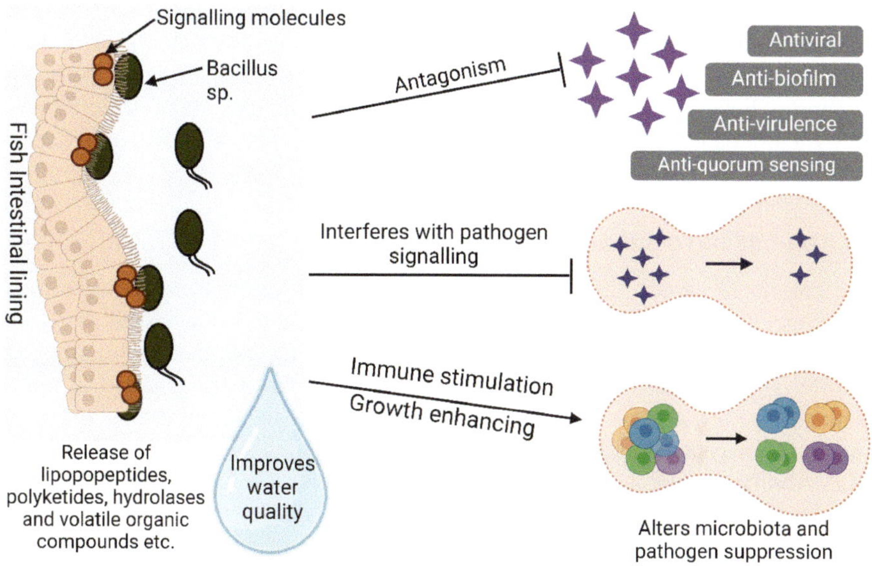

Figure 3. Mechanisms of *Bacillus* sp. in enhancing fish health and water quality in aquaculture (created with Biorender.com).

the blue economy, understanding and harnessing these antagonistic properties pave the way for sustainable resource management, ecosystem preservation, and the responsible utilization of marine resources (Gao et al. 2022).

Few examples illustrating the antagonistic properties of marine *Bacillus* species are as follows:

Aquaculture pathogen suppression: In shrimp farming, certain marine *Bacillus* strains have demonstrated antagonistic properties against pathogenic *Vibrio* species, which are a common cause of disease outbreaks in shrimp aquaculture. By producing antimicrobial compounds and competing with *Vibrio* for resources, these *Bacillus* strains help reduce the prevalence of *Vibrio* infections, leading to healthier and more resilient shrimp populations (Dabadé et al. 2015).

Biofilm prevention: Marine *Bacillus* species have been seen to disrupt the formation of biofilms on surfaces in marine environments, including equipment used in aquaculture facilities. By inhibiting the attachment and growth of biofilm-forming microorganisms, these *Bacillus* strains help prevent biofilm-related issues such as fouling and the potential harboring of pathogens (Álvarez-Mena et al. 2020).

Water quality improvement: Some marine *Bacillus* species play a role in improving water quality by competing with nutrient-consuming or potentially harmful microorganisms. For example, in recirculating aquaculture systems, specific *Bacillus* strains can outcompete nitrifying bacteria for ammonia and nitrite, helping maintain optimal water conditions for aquatic organisms (Ren et al. 2021).

Disease resistance enhancement: Marine *Bacillus* probiotics, when administered to fish in aquaculture settings, can enhance disease resistance through their immune-modulating effects. By stimulating the production of immune cells and promoting balanced immune responses, these probiotics make it possible for fish to ward off pathogens and thus reduce the need for antibiotics (Kumar Maurya et al. 2018).

Control of harmful algal blooms

The control of harmful algal blooms (HABs) is a critical concern in marine and freshwater ecosystems. HABs, often driven by excessive nutrient levels and favorable environmental conditions can lead to the rapid and uncontrolled proliferation of harmful algae species. These blooms pose significant threats to aquatic life, water quality, fisheries, and even human health due to the production of toxins. Effective strategies for controlling HABs are essential to mitigate their detrimental impacts. One promising approach for HAB control involves the utilization of natural biological agents, such as marine *Bacillus* species. These microorganisms, with their antagonistic properties, can play a pivotal role in reducing the occurrence and severity of HABs. Marine *Bacillus* strains have been seen to exert inhibitory effects on the growth of certain harmful algae through allelopathic interactions, production of antimicrobial compounds, and competitive exclusion. By disrupting the conditions that favor the growth of harmful algae, marine *Bacillus* species contribute to the prevention of HABs. For example, HAB are caused by massive and exceptional overgrowth of microalgae and cyanobacteria. Twenty-four strains of *Bacillus amyloliquefaciencs* FZB42 showed strongest bactericidal activity against Microcystis aeruginosa with a kill rate of 98.78%. At the genetic level, the synthesis of anticyanobacterial synthesis was dependent upon the sfp-independent antibiotic bacilysin and acted as a biological control agent to mitigate the effects of HAB (Wu et al. 2014). Additionally, some *Bacillus* strains possess the capability to interfere with the signaling and communication mechanisms of the target algae, further hampering their ability to form dense blooms. In aquaculture settings and coastal environments, the application of marine *Bacillus* probiotics offers a proactive and eco-friendly approach to HAB management, such as a novel *Bacillus* strain B1 initially showed 96.4% algicidal activity against *Phaeocystic globosa*, and showed 91.4, 90.7, and 90.6% algicidal activities against other harmful algal species (*Skeletonema costatum*, *Heterosigma akashiwo* and *Prorocentrum donghaeiense*), respectively. Different algae have different responses including change of algicidal activity and MDA contents with decrease in enzymes activities. However practical applications are required to track the effectiveness in the algicidal processes (Hou et al. 2015). These probiotic strains can be introduced to aquatic systems to help maintain a balanced microbial community, enhance water quality, and reduce the availability of nutrients that fuel HABs. *Bacillus* has also been presented as a valuable tool for the development of novel chemical-ecological methods to control harmful algae. The complete genome sequence of *B. subtilis* strain JA has demonstrated strong algicidal activities on the algae *Alexandrium minutum* with the inhibition rate exceeding 80% within forty-eight hours. Strain JA significantly reduced the photosynthetic ability of

A. minutum and caused extensive morphological damage to the algae potential regulated by the quorum sensing mechanism (Zhang et al. 2020). While marine *Bacillus* species provide a promising tool in the fight against HABs it is essential to integrate their use with other sustainable practices, such as nutrient management, watershed protection and monitoring programs. A comprehensive approach that combines biological control with responsible resource management can effectively reduce the frequency and impact of harmful algal blooms, safeguarding the health of aquatic ecosystems and supporting the goals of the blue economy.

Shellfish disease management

It is a critical aspect of ensuring the health and sustainability of shellfish populations in aquaculture and natural environments. Shellfish, including oysters, clams, mussels, and scallops, are valuable resources for both food production and ecosystem services. However, they can be susceptible to various diseases caused by pathogens, environmental stressors, and changing water quality. Effective disease management strategies are essential to prevent disease outbreaks, minimize economic losses, and protect the marine environment (Miget 2010). Several key approaches are employed for shellfish disease management:

Biosecurity measures: Implementing strict biosecurity protocols in shellfish aquaculture is crucial. This includes monitoring and controlling the movement of shellfish, equipment, and personnel to prevent the introduction and spread of diseases from one location to another. Proper site selection, quarantine procedures for new shellfish stock, and regular health assessments are integral components of biosecurity (Baker 2016).

Disease surveillance: Continuous monitoring of shellfish populations helps detect disease outbreaks early, allowing for rapid response and containment. Regular sampling, health assessments, and the use of diagnostic tools are essential for timely disease detection. This surveillance provides valuable data for informed decision-making in disease management (Soltani et al. 2019).

Pathogen management: Understanding the pathogens that affect shellfish is essential for effective disease management. Some strategies include deploying disease-resistant shellfish strains, optimizing water quality to reduce stress and vulnerability, and implementing treatments or biocontrol agents targeting specific pathogens (Dabadé et al. 2015).

Environmental stewardship: Promoting a healthy marine environment through responsible resource management is vital. Reducing pollution, mitigating the impact of climate change, and minimizing stressors such as poor water quality help bolster the natural resilience of shellfish populations to diseases (Shumway et al. 2003).

Probiotics and immune enhancement: The use of probiotics, including beneficial microorganisms like marine *Bacillus* species, can improve the overall health and disease resistance of shellfish. Probiotics can be administered to enhance the immune system, maintain a balanced microbiota, and prevent opportunistic pathogens from taking hold (Nayak 2010).

Research and innovation: Continuous research on shellfish diseases, including pathogen biology, transmission mechanisms, and host-pathogen interactions, is essential for developing new disease management strategies. Innovation in disease diagnosis, treatment, and prevention techniques is crucial for staying ahead of emerging diseases.

Public awareness and education: Educating shellfish farmers, the aquaculture industry stakeholders, and the public about disease risks, prevention measures, and responsible practices contribute to a comprehensive disease management approach.

By integrating these approaches, shellfish disease management aims to strike a balance between sustaining shellfish production, protecting the marine environment, and ensuring a safe supply of shellfish products for consumers. This holistic approach is essential for the continued success of shellfish aquaculture and the preservation of valuable coastal ecosystems.

Aquaculture-associated infection control

Aquaculture-associated infection control is a vital component of responsible and sustainable aquaculture operations. As the demand for seafood continues to rise, the aquaculture industry plays a crucial role in meeting this demand. However, the intensification of aquaculture systems can create conditions conducive to disease outbreaks among cultured species, posing significant risks to both economic viability and environmental integrity. Effective infection control measures encompass a comprehensive range of strategies aimed at preventing, detecting, and managing diseases within aquaculture settings. These strategies include robust biosecurity protocols, vigilant pathogen monitoring, optimization of water quality parameters, proper nutrition and feeding practices, consideration of environmental impacts, continuous education and training for industry professionals, and active collaboration between stakeholders and researchers to advance our understanding of disease dynamics and innovative control methods. By prioritizing infection control, the aquaculture industry can ensure the health and welfare of cultured organisms, minimize disease-related economic losses, reduce the environmental impact of aquaculture operations, and align with the sustainable principles of the blue economy, ultimately contributing to the responsible and long-term growth of the aquaculture sector (Robinson et al. 2016).

Biotechnological applications of marine *Bacillus* species

Production of enzymes

The production of enzymes by marine *Bacillus* species is a prominent aspect of their biotechnological applications, offering a wide range of valuable enzymatic activities that have significant industrial, environmental, and bioprocessing relevance. Marine *Bacillus* strains are renowned for their ability to secrete a diverse array of

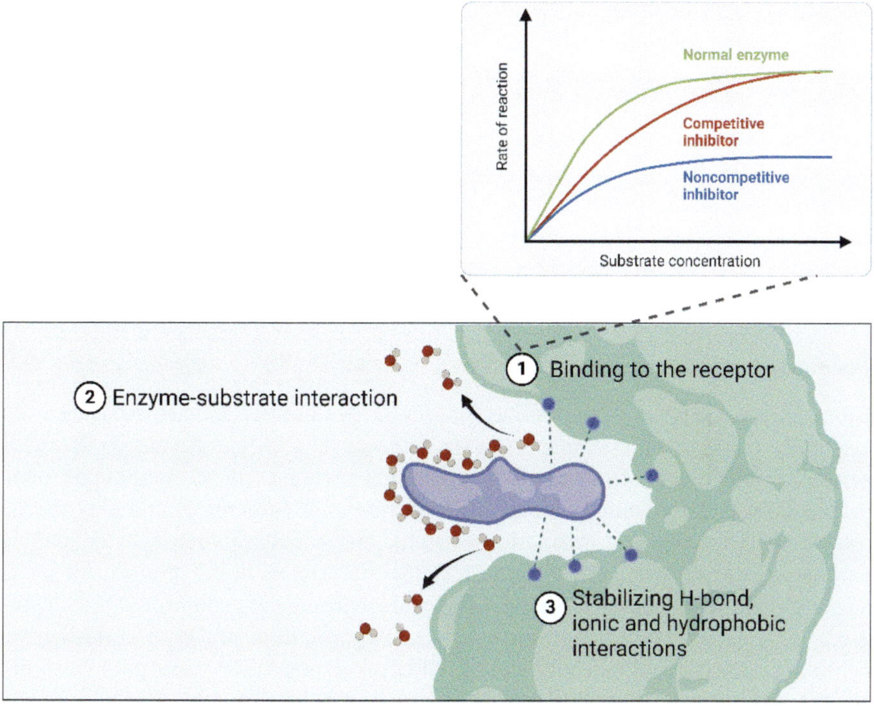

Figure 4. Mechanism of enzyme-substrate interaction in *Bacillus* species for marine pollution in bioremediation (created with BioRender.com).

extracellular enzymes with unique properties and stabilities, making them valuable candidates for enzyme production in various biotechnological processes. Marine *Bacillus* species are prolific producers of proteases, enzymes that catalyze the breakdown of proteins (as shown in Figure 4). These proteases have applications in various industries, including detergent formulation (for stain removal), food processing (meat tenderization), leather industry, and as a tool in bioremediation (to degrade proteinaceous waste). Amylases produced by marine *Bacillus* strains are utilized in starch hydrolysis, a critical process in the production of sugars, sweeteners, and biofuels. These enzymes also find applications in the food industry, such as in baking, brewing, and the production of high-fructose corn syrup. Lipases, enzymes that hydrolyze lipids (fats), are produced by marine *Bacillus* species and have applications in the production of biodiesel, as well as in food processing (e.g., cheese ripening), detergent industry (to remove lipid-based stains), and cosmetic formulations. Cellulases, which break down cellulose, are of interest for biomass conversion, aiding in the bioconversion of plant materials into biofuels and other value-added products. They are also used in textile processing to achieve a desired fabric finish. Chitinases, enzymes that degrade chitin, are significant in the bioremediation of chitin-rich waste, such as from seafood processing, and in the control of chitin-containing pests. They are also used in the production of chitosan, a biopolymer with diverse applications. Xylanases, enzymes that break down xylan, are employed in the paper and pulp industry to improve pulp bleaching and to reduce the use of harsh chemicals. They also find applications in animal feed, food processing, and bioethanol production from lignocellulosic materials. The biotechnological applications of enzymes produced by marine *Bacillus* species contribute to the efficient and sustainable production of various products, waste management, and the reduction of the environmental impact of several industries. Harnessing the enzymatic potential of these microorganisms aligns with the principles of the blue economy, emphasizing responsible and innovative utilization of marine resources for economic growth and ecological sustainability (Krueger et al. 2015, Wu et al. 2014).

Bioactive peptides and metabolites

Bioactive peptides and metabolites produced by marine *Bacillus* species offer a treasure trove of diverse and valuable compounds with significant implications across various fields, from medicine to biotechnology. These compounds often originate from the secondary metabolism of these microorganisms and play crucial roles in their survival and interactions with the marine environment. Understanding and harnessing the potential of these bioactive molecules holds promise for novel drug development, industrial applications, and environmental solutions. One notable group of bioactive peptides produced by marine *Bacillus* strains is antimicrobial peptides (AMPs) (as shown in Figure 5). These small peptides possess potent activity against a wide range of pathogenic microorganisms, including bacteria, fungi, and viruses. The appeal of AMPs stems from their unique mode of action, targeting microbial membranes or intracellular components, which reduces the likelihood of developing microbial resistance. Researchers are investigating the potential of marine *Bacillus*-derived AMPs for the development of new antimicrobial agents to combat the growing problem of antibiotic-resistant infections (Boottanun et al. 2017).

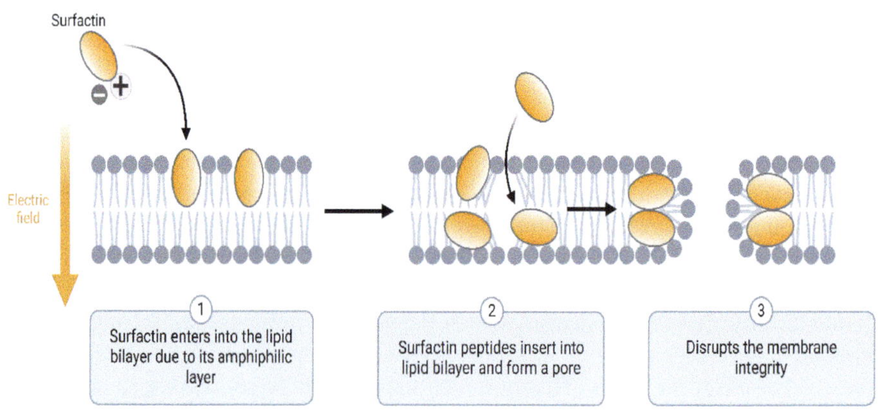

Figure 5. Marine pollution bioremediation involving Surfactin-mediated cell wall penetration (created with BioRender.com).

Another intriguing aspect of marine *Bacillus* metabolites is their antioxidative potential. Antioxidant compounds play a crucial role in neutralizing harmful free radicals in biological systems, thereby mitigating oxidative stress and its associated health risks. Marine *Bacillus* species have been found to produce antioxidant peptides and metabolites, which could have applications in functional foods and dietary supplements designed to promote human health and combat oxidative-related diseases, such as cardiovascular conditions and neurodegenerative disorders. Furthermore, the immunomodulatory properties of certain bioactive metabolites produced by marine *Bacillus* strains are garnering attention. These compounds have the ability to modulate the immune response, making them potential candidates for immunotherapies or treatments aimed at boosting the defense mechanisms of the immune system. Such compounds hold promise for addressing immune-related disorders and improving overall immune health (Alagappan et al. 2021).

Biosurfactants

The biotechnological applications of biosurfactants from marine *Bacillus* species play a pivotal role in advancing the principles of the blue economy by promoting sustainable resource utilization, environmental stewardship, and economic growth. These natural surfactants, derived from marine microorganisms, offer a multitude of environmentally friendly and economically viable applications that align with the goals of the blue economy (Zhen et al. 2023).

Biosurfactants from marine *Bacillus* species enhance the bioremediation of oil spills and contaminated sites, thereby reducing the environmental impact of such incidents. By improving the solubility and biodegradation of hydrophobic pollutants, these surfactants contribute to efficient cleanup processes thereby preserving marine ecosystems and maintaining the integrity of the oceans—a fundamental aspect of the blue economy. The use of biosurfactants in the production of green chemicals and sustainable manufacturing processes reduces reliance on synthetic, petrochemical-derived surfactants that can harm aquatic life and disrupt ecosystems. Marine *Bacillus*-derived biosurfactants serve as eco-friendly alternatives, supporting cleaner and more environmentally responsible industrial practices. The exploitation of marine *Bacillus*-derived biosurfactants in various biotechnological applications fosters the sustainable development of marine resources. Whether in pharmaceuticals, cosmetics, or food processing, these surfactants provide a renewable and environmentally compatible foundation for innovative products, contributing to the blue economy's emphasis on value-added marine-based industries. Biosurfactants from marine *Bacillus* species can be integrated into the development of marine bioproducts, adding value to marine resources while minimizing waste. The utilization of these surfactants in aquaculture, agriculture, and other sectors enhances the efficiency and sustainability of production practices, supporting the responsible exploitation of marine resources. By incorporating biosurfactants into blue growth strategies, economies can diversify their marine sectors while reducing environmental impacts. The innovative use of marine *Bacillus*-derived biosurfactants fosters new opportunities for economic growth in sectors that align with sustainable marine resource management. The responsible use of biosurfactants contributes to maintaining the natural beauty and biodiversity of coastal regions, promoting eco-tourism and the preservation of marine habitats. A thriving blue economy depends on the conservation and sustainable utilization of marine resources, and the application of biosurfactants from marine *Bacillus* species plays a crucial role in this effort.

The biotechnological applications of biosurfactants from marine *Bacillus* species offer a compelling pathway to create a blue economy. These natural surfactants address environmental challenges, support innovative industries, and pave the way for sustainable economic growth that is intrinsically linked to the health and vitality of marine ecosystems. As the blue economy continues to evolve, the responsible utilization of marine *Bacillus*-derived biosurfactants serves as a model for harmonizing economic progress with ecological sustainability.

Industrial applications in pharmaceuticals, cosmetics, and wastewater treatment

Marine *Bacillus* species play a significant and multifaceted role in various industrial applications, contributing to the creation of a blue economy by harnessing marine resources in a sustainable and responsible manner. Their unique capabilities have transformative implications in the fields of pharmaceuticals, cosmetics, and wastewater treatment, aligning with the blue economy's focus on utilizing marine assets for economic growth while preserving marine ecosystems. Marine *Bacillus* species have gained attention as valuable sources of bioactive compounds with pharmaceutical potential. These microorganisms produce bioactive metabolites, including antibiotics, anticancer agents, immunomodulators, and enzyme inhibitors, which have applications in drug discovery and development. By exploring the vast diversity of marine *Bacillus* strains, scientists can identify novel compounds that hold promise in treating diseases, including cancer, infections, and inflammatory disorders. The sustainable bioprospecting of marine *Bacillus*-derived compounds contributes to pharmaceutical innovation while adhering to the principles of the blue economy, which emphasize the responsible use of marine resources to drive economic growth. The cosmetic industry benefits from the unique properties of marine *Bacillus*-derived compounds. Biosurfactants, antimicrobial peptides, and antioxidant molecules produced by these microorganisms are used in skincare products, cosmetics, and personal care items. Marine *Bacillus*-based ingredients contribute to the development of environmentally friendly and biodegradable cosmetic formulations, reducing the ecological impact associated with synthetic additives. The emphasis of the blue economy on sustainable practices is embodied in the responsible use of these marine-derived compounds, aligning with the growing demand for eco-friendly and effective cosmetic products.

They also offer a natural and eco-friendly solution for wastewater treatment. Their ability to produce biosurfactants, enzymes, and other bioactive molecules aids in the degradation of organic pollutants and enhances the efficiency of wastewater treatment processes. These microorganisms can play a vital role in bioremediation projects, especially in industries that generate wastewater with high levels of organic contaminants. The application of marine *Bacillus*-based technologies in wastewater treatment supports the blue economy's commitment to sustainable resource management by promoting clean water practices and minimizing pollution in aquatic environments. In each of these industrial sectors, pharmaceuticals, cosmetics, and wastewater treatment marine *Bacillus* species contribute to the creation of a blue economy by providing innovative and sustainable solutions. Their role underscores the potential of marine biotechnology to drive economic growth while preserving the health and vitality of marine ecosystems, an essential principle for the responsible utilization of marine resources in a rapidly changing world.

Implications for the blue economy

Sustainable aquaculture development

Sustainable aquaculture development is a critical component of studying and advancing the implications of the blue economy. The blue economy encompasses the sustainable use of marine resources while promoting economic growth, social well-being, and the preservation of the marine environment. As aquaculture plays a significant role in seafood production and has the potential to contribute to the blue economy, understanding its implications within this framework is essential. It aligns with the emphasis of the blue economy on responsible resource management. By adopting practices that minimize environmental impact, such as proper site selection, reduced use of antibiotics and chemicals, and minimized effluent discharge, aquaculture operations contribute importance to the conservation of marine ecosystems. The implementation of best practices ensures that aquatic resources are used efficiently, supporting long-term resource availability and the resilience of coastal and marine environments (Pace et al. 2023).

Aquaculture, when conducted sustainably, contributes significantly to the blue economy by generating economic growth and livelihoods (as shown in Figure 6). Responsible aquaculture practices create jobs in coastal communities, enhances local economies, and provides a stable source of high-quality protein for consumption. As the demand for seafood continues to rise, well-managed aquaculture operations contribute to meeting this demand in a manner that is economically viable and socially inclusive. Sustainable aquaculture development recognizes the value of ecosystem services provided by healthy marine environments. Aquaculture practices that minimize habitat disruption, avoid over-extraction of wild fish for feed, and prioritize water quality contributes to the preservation of ecosystem services such as nutrient cycling, habitat provision, and biodiversity. By recognizing the intrinsic value of these services, the blue economy underscores the importance of sustainable aquaculture practices. Sustainable aquaculture development plays a role in enhancing climate resilience, another key aspect of the blue economy. By implementing climate-smart practices, such as designing farms to withstand extreme weather events and utilizing species that are resilient to changing environmental conditions, aquaculture can adapt to the challenges posed by climate change. This resilience contributes to the overall sustainability of the blue economy by ensuring the continued

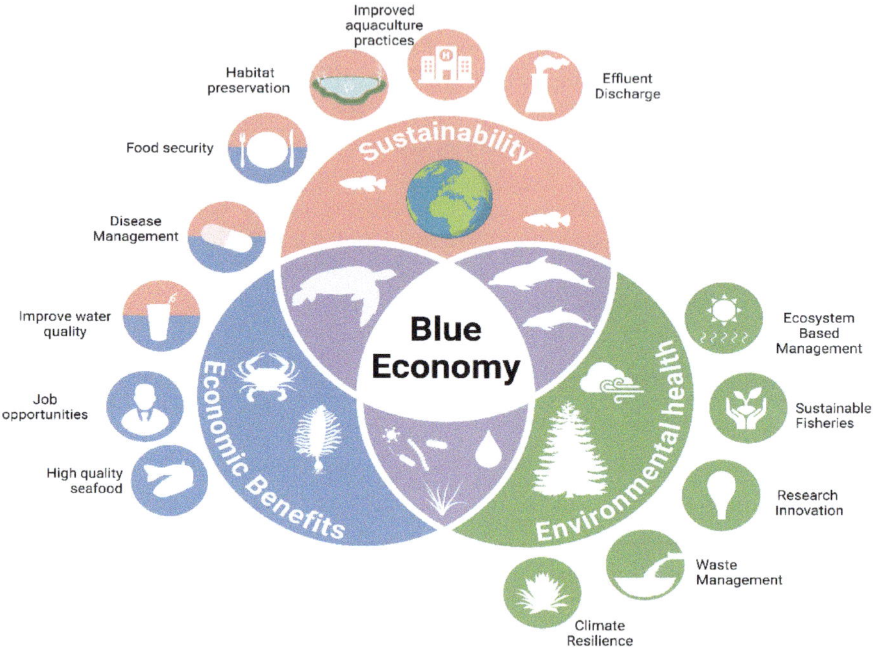

Figure 6. Incorporation of sustainable aquaculture practices with *Bacillus* that can have improved economic benefits and contribute to environmental health (created with BioRender.com).

availability of seafood resources. The blue economy encourages innovation and research to improve the efficiency and sustainability of marine-related industries. In aquaculture, this involves developing new technologies for efficient feed utilization, disease management, water quality monitoring, and reducing environmental impacts. These innovations not only enhance the productivity of aquaculture but also contribute to the overall goals of the blue economy. In conclusion, sustainable aquaculture development is a vital component for studying and realizing the implications of the blue economy. It embodies responsible resource management, economic growth, and environmental stewardship, all of which are central to the blue economy's vision of a sustainable and prosperous future that values and preserves the health of our oceans and coastal regions (Veríssimo et al. 2021).

Reduction of environmental impacts

Reducing environmental impacts is a core principle of practicing the blue economy, which seeks to harness marine resources while preserving the health and integrity of marine ecosystems. By adopting sustainable practices and prioritizing responsible resource management, the blue economy aims to minimize its ecological footprint.

Here are some key strategies and initiatives to achieve this reduction in environmental impacts:

Sustainable fisheries and aquaculture: Implementing sustainable fishing practices, such as science-based quotas, minimum size limits, and seasonal closures, ensures that fish stocks are not overexploited. Responsible aquaculture practices, including proper site selection, reduced use of antibiotics and chemicals, and minimizing effluent discharge, mitigate the negative impacts of intensive fish farming on local water bodies and coastal ecosystems.

Ecosystem-based management: Embracing an ecosystem-based approach in resource management considers the broader ecological context. It involves understanding the interdependencies within marine ecosystems and considering the impact of human activities on these systems. By incorporating ecosystem-based management into decision-making, the blue economy minimizes unintended consequences and promotes the long-term health of marine environments.

Marine spatial planning: Properly planned and managed marine areas can help avoid conflicts between various uses, such as fishing, shipping, tourism, and conservation. Marine spatial planning ensures that human activities are coordinated to minimize overlapping interests and reduce stress on vulnerable marine ecosystems.

Renewable energy: Harnessing renewable energy sources from the ocean, such as offshore wind, tidal, and wave energy, reduces dependence on fossil fuels and mitigates the environmental impact of energy generation. The blue economy encourages the development and deployment of sustainable ocean-based energy technologies.

Waste management and pollution prevention: Implementing effective waste management practices, reducing plastic pollution, and preventing harmful chemical discharges into the oceans are essential steps. By minimizing pollution, the blue economy helps maintain water quality and protects marine life.

Climate resilience: Addressing climate change impacts is a critical component of reducing environmental impacts. The blue economy promotes climate-resilient practices, including the protection of coastal ecosystems (e.g., mangroves, seagrasses, and coral reefs) that act as natural buffers against storm surges and rising sea levels.

Environmental education and public awareness: Raising awareness about the importance of marine conservation and sustainable practices is essential. Education empowers communities to make informed choices that support the blue economy's objectives, fostering a sense of stewardship for marine resources.

Research and innovation: Continued research on marine ecosystems, sustainable technologies, and best practices is essential. Innovations that enhance resource efficiency, minimize environmental impacts, and develop new sustainable solutions contribute to the success of the blue economy.

By integrating these strategies, the blue economy seeks to create a balance between economic growth and environmental protection, ensuring that future generations can benefit from the valuable resources the ocean provides while maintaining the health and vitality of marine ecosystems.

Economic benefits and opportunities

Marine *Bacillus* species offer a wide array of economic benefits and opportunities that are integral to the development of the blue economy. These microorganisms, found in diverse marine environments, have become valuable assets in multiple sectors, capitalizing on their unique properties and functions. Their significance in building a sustainable blue economy lies in their potential contributions to innovation, aquaculture, bioremediation, biosurfactant production, the creation of marine bioproducts, and aligning with blue growth initiatives. In the realm of biotechnological innovation, marine *Bacillus* species hold promise as sources of bioactive compounds with applications across industries. Their ability to produce enzymes, antimicrobial peptides, and other biotechnologically relevant molecules opens up opportunities in pharmaceuticals, cosmetics, and food processing. By harnessing the biotechnological potential of marine *Bacillus*, the blue economy can drive research and development, leading to the creation of novel products that improve human health and enhance industrial processes (Majed et al. 2016).

Aquaculture, a central component of the blue economy, benefits from marine *Bacillus*-based solutions. Probiotics and biocontrol agents derived from these microorganisms play a pivotal role in promoting fish health, disease resistance, and water quality in aquaculture operations. By integrating marine *Bacillus*-based technologies, the aquaculture sector can improve productivity, reduce environmental impact, and contribute to the sustainable production of high-quality seafood, a crucial aspect of the blue economy. In the realm of environmental sustainability, marine *Bacillus* species contribute significantly to bioremediation efforts. Their potential to clean up oil spills, contaminated sites, and wastewater treatment aligns with the blue economy's emphasis on responsible resource management and environmental conservation. The economic opportunities in providing efficient and environmentally friendly bioremediation solutions are evident, creating a specialized niche within the environmental services sector. Biosurfactant production from marine *Bacillus* strains is another avenue with economic and environmental advantages. These natural surfactants find applications in oil spill cleanup, enhanced oil recovery, and the formulation of eco-friendly detergents. By replacing traditional synthetic surfactants, marine *Bacillus*-derived biosurfactants not only reduce environmental impacts but also create market opportunities in industries seeking sustainable alternatives. The development of marine bioproducts offers a forward-looking dimension to the blue economy. Marine *Bacillus*-derived compounds, whether in functional foods, nutraceuticals, specialty chemicals, or bio-based materials, represent a frontier of innovation. As consumers increasingly seek sustainable and environmentally friendly products, the blue economy can capitalize on the unique properties of marine *Bacillus* species to create value-added bioproducts that cater to this demand. Marine *Bacillus* species also align with blue growth initiatives, as governments and organizations worldwide invest in the sustainable development of marine industries.

The scientific exploration of marine *Bacillus* for novel compounds and applications fits seamlessly into the blue economy's objectives, fostering research collaborations, intellectual property development, and economic growth based on the responsible utilization of marine biodiversity.

Future directions and challenges

Research and development needs

The future of research and development in the biotechnology application of probiotic and biocontrol properties of marine *Bacillus* species holds immense potential for the advancement of the blue economy. Key directions include the continuous exploration of diverse marine environments to discover new *Bacillus* strains with valuable traits, the deepening of our understanding of the functional mechanisms driving these traits, and the optimization of application methods to ensure efficacy and environmental safety. Understanding the ecological interactions of introduced *Bacillus* strains and assessing their environmental impact is crucial to strike a balance between innovation and ecosystem preservation. As these applications mature, integrating them into aquaculture, agriculture, and other industries, while ensuring market acceptance and regulatory compliance, becomes essential. Sustainable production methods and a commitment to environmental responsibility will be vital to avoid overexploitation of marine resources. By addressing these challenges and aligning research efforts with the principles of the blue economy, we can unlock the economic benefits of marine *Bacillus* species while safeguarding marine ecosystems and contributing to a sustainable and prosperous future.

Regulatory considerations

Regulatory considerations play a pivotal role in the successful integration of biotechnology applications involving marine *Bacillus* species within the blue economy. As these applications advance, it is essential to establish clear and robust regulatory frameworks that ensure product safety, efficacy, environmental compatibility, and ethical use. Several key aspects of regulatory considerations are vital in navigating the intersection of biotechnology, marine resources, and the blue economy. Regulatory agencies must conduct thorough safety assessments for products derived from marine *Bacillus* species. This includes evaluating potential risks to human health, the environment, and native species. Proper risk assessment protocols should be in place to identify any adverse effects that may arise from the introduction of these products. Understanding the potential environmental impact of marine *Bacillus*-based applications is critical. Regulatory bodies should assess the consequences of releasing these microorganisms into aquatic ecosystems, taking into account their potential interactions with native microbial communities, unintended ecological disruptions, and long-term consequences on marine biodiversity (Veríssimo et al. 2021).

Regulatory approval is necessary for the commercialization and market acceptance of marine *Bacillus*-derived products. Clear guidelines and procedures should be established to obtain approvals for specific applications, ensuring that products meet safety and efficacy standards before entering the market. This approval process builds trust and confidence among industries, consumers, and stakeholders. Regulatory frameworks should address intellectual property rights for novel marine *Bacillus* strains, bioactive compounds, and biotechnological innovations. Protecting intellectual property encourages innovation and provides incentives for researchers and industries to invest in the development of new products and technologies. Regulatory considerations should encompass ethical dimensions, including the responsible use of marine resources, respect for indigenous knowledge, and social equity. Regulations should promote sustainable practices and ensure that the benefits of marine *Bacillus*-based applications are shared equitably among communities, industries, and nations. In the context of the blue economy, harmonizing regulatory approaches at the international level is essential. Collaboration among regulatory agencies across nations promotes consistency in safety assessments, facilitates trade, and prevents regulatory discrepancies that could hinder the global adoption of marine *Bacillus*-based solutions. Regulatory oversight should extend beyond initial approvals to encompass ongoing monitoring and compliance with established guidelines. Continuous evaluation of the environmental, health, and safety aspects of marine *Bacillus*-based applications ensures that any emerging issues are addressed promptly. Transparent communication and public engagement are vital components of regulatory considerations. Involving the public, stakeholders, and affected communities in the decision-making process fosters a better understanding of the benefits and risks associated with marine *Bacillus*-based technologies, ensuring that regulatory decisions align with societal values and interests.

Integration of biotechnology in the blue economy

The integration of biotechnology into the blue economy is a transformative synergy that holds the key to unlocking sustainable growth, innovation, and responsible resource management. This integration capitalizes on the unique attributes of biotechnological advancements, offering novel solutions to enhance various sectors within the blue economy. From aquaculture to renewable energy, environmental monitoring to bio-based product development, biotechnology brings precision, efficiency, and eco-friendly practices to the forefront. By leveraging biotechnology, we can revolutionize aquaculture practices, boost productivity, and reduce environmental impacts. Moreover, the development of bio-based products derived from marine organisms addresses resource scarcity concerns while fostering economic growth. This integration promotes cross-sector collaboration, research innovation, and the responsible use of marine resources, aligning perfectly with the blue economy's principles. As we navigate the complexities of global challenges, climate resilience, and sustainable development, the integration of biotechnology emerges as a cornerstone of the blue economy, offering a pathway to a more prosperous, ecologically balanced, and future-oriented maritime industry (Pace et al. 2023).

Conclusions

In conclusion, the biotechnology application of probiotic and biocontrol properties derived from marine *Bacillus* species presents a promising avenue for driving the development of the blue economy while ensuring environmental sustainability. The unique characteristics of marine *Bacillus*, such as their adaptability to challenging marine environments and their bioactive capabilities, make them valuable assets in multiple sectors within the blue economy. By harnessing the potential of these microorganisms, we can enhance aquaculture practices, mitigate disease outbreaks, and improve water quality, contributing to the growth of the aquaculture industry. Additionally, the development of bio-based products and the integration of innovative biotechnological approaches align with the principles of resource efficiency and circular economy, essential components of the blue economy's long-term viability.

However, this journey is not without its challenges. Regulatory considerations, environmental impact assessment, and the careful management of introduced *Bacillus* strains are crucial to preventing unintended ecological disruptions. Striking a balance between economic benefits and environmental preservation is paramount, and clear guidelines must be established to ensure responsible use. Collaboration among researchers, industries, policymakers, and local communities is essential to navigate these challenges and to build a solid foundation for the sustainable integration of marine *Bacillus*-derived solutions into the blue economy. The potential benefits are significant—from increased aquaculture productivity to reduced dependence on harmful chemicals in agriculture and pollution clean-up. Marine *Bacillus* species hold the key to innovation, economic growth, and a more resilient and ecologically conscious blue economy. As we move forward, it is imperative to continue advancing our understanding of these microorganisms, optimizing application methods, and ensuring that biotechnological advancements align with the broader goals of the blue economy. By seizing this opportunity and addressing the challenges with diligence and foresight, we can forge a path toward a prosperous, sustainable, and harmonious blue economy that respects the delicate balance of marine ecosystems and benefits both present and future generations.

References

Ahmad, B., Ali, A., Raziq, S., Khan, A., Azi, A., Ullah, F. et al. (2019). Biochemical composition of fish and changes during processing and storage: A review. Bioscience Research 16(1): 367–374.

Alagappan, S., Chaliha, M., Sultanbawa, Y., Fuller, S., Hoffman, L.C.C., Netzel, G. et al. (2021). Nutritional analysis, volatile composition, antimicrobial and antioxidant properties of Australian green ants (Oecophylla smaragdina). Future Foods 3(July 2020): 100007. https://doi.org/10.1016/j.fufo.2020.100007.

Álvarez-Mena, A., Cámara-Almirón, J., de Vicente, A. and Romero, D. (2020). Multifunctional amyloids in the biology of gram-positive bacteria. Microorganisms 8(12): 1–20. https://doi.org/10.3390/microorganisms8122020.

Bahaddad, S.A., Almalki, M.H.K., Alghamdi, O.A., Sohrab, S.S. and Yasir, M. (2023). *Bacillus* species as direct-fed microbial antibiotic alternatives for monogastric production. Probiotics and Antimicrobial Proteins 15(5): 1–16. https://doi.org/10.1007/s12602-022-09909-5.

Baker, G.L. (2016). Food safety impacts from post-harvest processing procedures of molluscan shellfish. Foods 5(2): 1–12. https://doi.org/10.3390/foods5020029.

Balqadi, A.A., Salama, A.J., Mudarris, M.S.A. and Satheesh, S. (2015). Effect of a probiotic product, promarine on growth responses of Indian White Prawn, Penaeus (Fenneropenaeus) Indicus. 2015(1): 33–43.

Boottanun, P., Potisap, C., Hurdle, J.G. and Sermswan, R.W. (2017). Secondary metabolites from Bacillus amyloliquefaciens isolated from soil can kill Burkholderia pseudomallei. AMB Express 7(1). https://doi.org/10.1186/s13568-016-0302-0.

Buruiana, C., Profir, A.G. and Vizireanu, C. (2014). Effects of probiotic *Bacillus* species in aquaculture – an overview. The Annals of the University Dunarea de Jos of Galati 38(2): 9–17.

Caulier, S., Nannan, C., Gillis, A., Licciardi, F., Bragard, C. and Mahillon, J. (2019). Overview of the antimicrobial compounds produced by members of the *Bacillus subtilis* group. Frontiers in Microbiology 10(FEB): 1–19. https://doi.org/10.3389/fmicb.2019.00302.

Chen, Y., Li, J., Xiao, P., Zhu, W. and Mo, Z. (2016). The ability of marine *Bacillus* spp. isolated from fish gastrointestinal tract and culture pond sediment to inhibit growth of aquatic pathogenic bacteria. Iranian Journal of Fisheries Sciences 15(2): 701–714.

Dabadé, D.S., Zwietering, M.H., Hounhouigan, D.J. and Den Besten, H.M.W. (2015). Shrimp quality and safety management along the supply chain in Benin. In PQDT - Global (Issue 28233364). https://www.proquest.com/dissertations-theses/b-shrimp-quality-safety-management-along-supply/docview/2564855424/se-2?accountid=26642.

FAO. (2022). The State of World Fisheries and Aquaculture: Towards Blue Transformation. In In Brief to The State of World Fisheries and Aquaculture 2022. https://doi.org/10.4060/cc0463en.

Gao, Z., Zhang, Y., Wang, Q., Ren, H. and Gao, G. (2022). Preliminary screening, identification and biological characteristic analysis of *Bacillus* probiotics isolated from Cynoglossus semilaevis. The Israeli Journal of Aquaculture 4: 1–12.

Hou, S., Shu, W., Tan, S., Zhao, L. and Yin, P. (2015). Exploration of the antioxidant system and photosynthetic system of a marine algicidal *Bacillus* and its effect on four harmful algal bloom species. Canadian Journal of Microbiology 62(1): 49–59. https://doi.org/10.1139/cjm-2015-0425.

Huan, Y., Kong, Q., Mou, H. and Yi, H. (2020). Antimicrobial peptides: classification, design, application and research progress in multiple fields. Frontiers in Microbiology 11(October): 1–21. https://doi.org/10.3389/fmicb.2020.582779.

Jayakumar, A., Krishna, A., Mohan, M., Nair, I.C. and Radhakrishnan, E.K. (2019). Plant growth enhancement, disease resistance, and elemental modulatory effects of plant probiotic endophytic *Bacillus* sp. Fcl1. Probiotics and Antimicrobial Proteins 11(2): 526–534. https://doi.org/10.1007/s12602-018-9417-8.

Khan, M.N., Lin, H., Wang, J., Li, M. and Mirani, Z.A. (2014). recA-based identification and antagonistic potential against fish pathogens by marine isolate DK1-SA11. Indian Journal of Geo-Marine Sciences 43(May): 731–736.

Krueger, M.C., Harms, H. and Schlosser, D. (2015). Prospects for microbiological solutions to environmental pollution with plastics. Applied Microbiology and Biotechnology 99(21): 8857–8874. https://doi.org/10.1007/s00253-015-6879-4.

Kumar Maurya, A., Faizabad Uttar Pradesh, K., Hari Om Verma, I., Pandey, G., Pal, J., Shukla, B. et al. (2018). A review on role of fish in human nutrition with special emphasis to essential fatty acid. International Journal of Fisheries and Aquatic Studies 6(2): 427–430. https://www.delamaris.si/healthy-diet/that-.

Kumar, V., Kumar, M., Sharma, S. and Prasad, R. (2017). Probiotics and plant health. In Probiotics and Plant Health. https://doi.org/10.1007/978-981-10-3473-2.

Majed, R., Faille, C., Kallassy, M. and Gohar, M. (2016). Bacillus cereus Biofilms-same, only different. Frontiers in Microbiology 7(JUL): 1–16. https://doi.org/10.3389/fmicb.2016.01054.

Meroni, G., Panelli, S., Zuccotti, G., Bandi, C., Drago, L. and Pistone, D. (2021). Probiotics as therapeutic tools against pathogenic biofilms: have we found the perfect weapon? Microbiology Research 12(4): 916–937. https://doi.org/10.3390/microbiolres12040068.

Miget, R. (2010). Shellfish Handling Practices - Shrimp and Molluscs. SRAC Publication, 4902, 6. https://srac.tamu.edu/index.cfm/event/getFactSheet/whichfactsheet/222/%5Cnhttp://www.aces.edu/anr/fish/aquaculture/download/species/marine_shrimp/SRAC 4902.pdf.

Nayak, S.K. (2010). Probiotics and immunity: A fish perspective. Fish and Shellfish Immunology 29(1): 2–14. https://doi.org/10.1016/j.fsi.2010.02.017.

Noatch, M.R. and Suski, C.D. (2012). Non-physical barriers to deter fish movements. Environmental Reviews 20(1): 71–82. https://doi.org/10.1139/a2012-001.

Pace, L.A., Saritas, O. and Deidun, A. (2023). Exploring future research and innovation directions for a sustainable blue economy. Marine Policy 148(November 2022): 105433. https://doi.org/10.1016/j.marpol.2022.105433.

Paul, S.I., Majumdar, B.C., Ehsan, R., Hasan, M. and Baidya, A. (2021). Bioprospecting potential of marine microbial natural bioactive compounds. Journal of Applied Biotechnology Reports 8(2): 96–108. https://doi.org/10.30491/JABR.2020.233148.1232.

Prodhan, M.M.H., Khan, M.A., Palash, M.S. and Rahman, M.T. (2022). Nature, extent, and causes of post-harvest losses at fisher and farmer level: An in-depth study. Aquaculture 550(June 2021): 737856. https://doi.org/10.1016/j.aquaculture.2021.737856.

Rahman, A., Hossain, S., Shefat, T., Chowdhury, M.A. and Khan, S.U. (2021). Effects of probiotic *Bacillus* on growth performance, immune response and disease resistance in aquaculture. Journal of Aquaculture Research & Development 12(634): 1–10.

Ren, W., Wu, H., Guo, C., Xue, B. and Long, H. (2021). Multi-strain tropical *Bacillus* spp. as a potential probiotic biocontrol agent for large-scale enhancement of mariculture water quality. Frontiers in Microbiology 12(August): 1–14. https://doi.org/10.3389/fmicb.2021.699378.

Ringø, E., Li, X., Doan, H. Van and Ghosh, K. (2022). Interesting probiotic bacteria other than the more widely used lactic acid bacteria and bacilli in finfish. Frontiers in Marine Science 9(June): 1–27. https://doi.org/10.3389/fmars.2022.848037.

Robinson, G., Caldwell, G.S., Wade, M.J., Free, A., Jones, C.L.W. and Stead, S.M. (2016). Profiling bacterial communities associated with sediment-based aquaculture bioremediation systems under contrasting redox regimes. Scientific Reports 6(December): 1–13. https://doi.org/10.1038/srep38850.

Sayes, C., Leyton, Y., Sayes, C., Leyton, Y. and Riquelme, C. (2017). Probiotic Bacteria as an Healthy Alternative for Fish Aquaculture. https://doi.org/10.5772/intechopen.71206.

Seifzadeh, M., Khorasgani, M.R. and Shafiei, R. (2019). Study on probiotic potential of *Bacillus* species isolated from the intestine of farmed rainbow trout *Oncorhynchus mykiss*. International Journal of Aquatic Biology 7(3): 166–174.

Sharma, R., Shrivas, V.L. and Sharma, S. (2021). Effect of substitution of chemical fertilizer by bioinoculants on plant performance and rhizospheric bacterial community: case study with Cajanus cajan. Brazilian Journal of Microbiology 52(1): 373–386. https://doi.org/10.1007/s42770-020-00418-7.

Shumway, S.E., Davis, C., Downey, R., Karney, R., Kraeuter, J., Parsons, J. et al. (2003). Shellfish aquaculture—in praise of sustainable economies and environments. World Aquaculture 34(4): 15–17.

Soltani, M., Ghosh, K., Hoseinifar, S.H., Kumar, V., Lymbery, A.J., Roy, S. et al. (2019). Genus *Bacillus*, promising probiotics in aquaculture: Aquatic animal origin, bio-active components, bioremediation and efficacy in fish and shellfish. Reviews in Fisheries Science and Aquaculture 27(3): 331–379. https://doi.org/10.1080/23308249.2019.1597010.

Suman, T.Y., Jia, P.P., Li, W.G., Junaid, M., Xin, G.Y., Wang, Y. et al. (2020). Acute and chronic effects of polystyrene microplastics on brine shrimp: First evidence highlighting the molecular mechanism through transcriptome analysis. Journal of Hazardous Materials 400: 123220. https://doi.org/10.1016/j.jhazmat.2020.123220.

Thompson, C.C., Chimetto, L., Edwards, R.A., Swings, J., Stackebrandt, E. and Thompson, F.L. (2013). Microbial genomic taxonomy. BMC Genomics 14(1). https://doi.org/10.1186/1471-2164-14-913.

Tuan, T.N., Duc, P.M. and Hatai, K. (2013). Overview of the use of probiotics in aquaculture. International Journal of Research in Fisheries and Aquaculture 3(3): 89–97.

Veríssimo, N.V., Mussagy, C.U., Oshiro, A.A., Mendonça, C.M.N., Santos-Ebinuma, V.D.C., Pessoa, A. et al. (2021). From green to blue economy: Marine biorefineries for a sustainable ocean-based economy. Green Chemistry 23(23): 9377–9400. https://doi.org/10.1039/d1gc03191k.

Villela, H.D.M., Vilela, C.L.S., Assis, J.M., Varona, N., Burke, C., Coil, D.A. et al. (2019). Prospecting Microbial Strains for Bioremediation and Probiotics Development for Metaorganism Research and Preservation. October, 1–12. https://doi.org/10.3791/60238.

Wu, L., Wu, H., Chen, L., Xie, S., Zang, H., Borriss, R. et al. (2014). Bacilysin from *Bacillus* amyloliquefaciens FZB42 has specific bactericidal activity against harmful algal bloom species. Applied and Environmental Microbiology 80(24): 7512–7520. https://doi.org/10.1128/AEM.02605-14.

Ye, B., Chen, J., Fu, L. and Wang, Y. (2022). Application of nondestructive evaluation (NDE) technologies throughout cold chain logistics of seafood: Classification, innovations and research trends. Lwt 158: 113127. https://doi.org/10.1016/j.lwt.2022.113127.

Zhang, S.J., Du, X.P., Zhu, J.M., Meng, C.X., Zhou, J. and Zuo, P. (2020). The complete genome sequence of the algicidal bacterium *Bacillus subtilis* strain JA and the use of quorum sensing to evaluate its antialgal ability. Biotechnology Reports 25: e00421. https://doi.org/10.1016/j.btre.2020.e00421.

Zhen, C., Ge, X., Lu, Y. and Liu, W. (2023). Chemical structure, properties and potential applications of surfactin, as well as advanced strategies for improving its microbial production. AIMS Microbiology 9(March): 195–217. https://doi.org/10.3934/microbiol.2023012.

Zhou, S., Xia, Y., Zhu, C. and Chu, W. (2018). Isolation of marine *Bacillus* sp. with antagonistic and organic-substances-degrading activities and its potential application as a fish probiotic. Marine Drugs. https://doi.org/10.3390/md16060196.

Index